Methods of Astrodynamics

METHODS OF ASTRODYNAMICS

PEDRO RAMON ESCOBAL

Assistant Manager

Mission Design Department

TRW Systems

John Wiley & Sons, Inc.

New York · London · Sydney

Library of Congress Catalog Card Number: 68-24795
SBN 471 24528 3
Printed in the United States of America

To

Ramon Castroviejo, M.D.

If you would be a real seeker after truth, it is necessary that at least once in your life you doubt, as far as possible, all things.

RENE DESCARTES

Preface

Astrodynamics is the science of applying the techniques of celestial mechanics to the solution of space engineering problems. Today, as man prepares to make the first voyage between celestial bodies, the importance of this field continues to expand in scope. Perhaps this interesting field has such deep-rooted importance because the design and construction of all extraterrestrial spacecraft is initially dependent on the satisfaction of the laws of celestial mechanics. Since the historic creation of man's first artificial satellite late in 1957 the knowledge of astrodynamics has continued to increase. The classical methods of celestial mechanics have been exploited, and a considerable amount of new technology has been developed to cope with present-day problems. *Methods of Astrodynamics*, which follows *Methods of Orbit Determination*, attempts to relate in detailed fashion many of the techniques currently in use in the area of space mechanics. In essence, the purpose of this book is to continue the introduction provided by *Methods of Orbit Determination* and to discuss more advanced astrodynamical techniques from an engineering point of view.

As in *Methods of Orbit Determination*, it is hoped that the presentation of each chapter of this work will require minimum back reading in order to utilize the desired astrodynamical technique. The notation used approximately follows the traditional astronomical notation, with deviations and changes kept to a minimum.

Chapter 1, which could be classified as empirical in nature, describes the sizes and locations of the primary planets of our solar system. To give it engineering value great care was utilized to obtain accurate time-dependent functions that can be used to predict explicitly the positions and velocities of the members of the solar system. These analytic expressions should be of considerable value to space engineers in the execution of interplanetary missions. The second half of this chapter also relates the necessary fundamentals and atmospheric models that will be required in space flights that approach the neighboring members of our solar system. It may seem a little incongruous to treat both planetary ephemerides and atmospheres in one chapter; however, both subjects are empirical in nature and are related to the process of interplanetary flight from one body to another. Therefore, on second thought, after examining this combination and obtaining the opinion of several people, I integrated both topics into a single chapter.

Commencing with Chapter 2, the next three chapters deal with transfer mechanics, and nearly all the emphasis is aimed at trajectory optimization. Chapter 2 begins with an elementary treatment of the maximum and minimum of functions of a single variable and quickly proceeds to much more advanced concepts of optimization. I purposely started this chapter at an elementary level in order to show the reader the step-by-step expansion of the theory to several variables. It discusses the optimization of functions subject to equality and inequality constraints in full detail and thus provides all the necessary mathematical background required in later chapters.

Optimum planetocentric transfer mechanics is discussed in Chapter 3, which collects a large number of optimum orbital transfer problems about a central planet. Both minimum fuel and minimum time transfer techniques are discussed in detail from an analytical point of view. Wherever possible the orbit determination techniques for one-, two-, or three-impulse transfers are reduced to algebraic polynomials, which, in some cases, are solvable in closed form. Emphasis is placed on obtaining accurate dependable solutions. The main highlight of this chapter is the reduction of the three-dimensional weighted two-impulse transfer trajectory between two general orbits to the solution of an algebraic polynomial of degree twelve.

Chapter 4 treats the optimum interplanetary transfer operation. In this chapter, as in Chapter 3, the one-, two-, and three-impulse transfer procedure is analyzed. Optimum midcourse correction of interplanetary vehicles is discussed in detail and should prove to be of considerable interest to the reader. Minimum time interplanetary transfer techniques are also discussed at length. The techniques developed in the beginning of this chapter can be used to obtain the time and fuel critical abort modes required in operational manned interplanetary missions. At the end of this chapter I have introduced the reader to interplanetary systems analysis trajectory methods, that is, the determination of minimum weight out-of-orbit interplanetary transfers.

Chapters 3 and 4 should give the reader a substantial insight into the subjects of planetocentric and interplanetary transfer maneuvers.

Lunar trajectories are discussed in Chapter 5. The reader may wonder why the chapter on lunar trajectories was developed after interplanetary analysis was introduced. Basically, my reason for this device was to maintain the parallel development of Chapters 3 and 4. Chapter 5 starts with a detailed discussion of the patched conic technique and its limitations. The restricted three-body problem is introduced in detail, and several new approaches to the determination of Earth to Moon trajectories are discussed. The subject matter of this chapter has been an area of much mathematical investigation and has attracted the research activity of the greatest mathematicians in history. I believe the student will find this chapter valuable in the pursuit of research activities. The methods and techniques developed are of considerable interest today, especially for the Apollo project which requires rapid analytic techniques for the computation of lunar trajectories.

A subject of continuing interest is introduced in Chapter 6, namely, communications analysis. In this chapter the earliest and latest communications times of satellites relative to one another, as well as interplanetary vehicles and satellites in orbit about distant planets, are investigated from an astrodynamical point of view. I have stressed the closed and semiclosed analytic solution to these problems whenever possible. It is my belief that this chapter provides the mission engineer with the next level of analysis that must be applied after the material discussed in Chapters 2, 3, 4, and 5 has been utilized to yield the basic mission. Furthermore, these chapters will give the practicing engineer suitable analytic methods for performing large mission studies and for obtaining suitable starting values for the precise integration methods discussed in Chapter 7.

Accurate trajectory integration techniques are introduced in Chapter 7, and the two most prominent methods currently in use are discussed. Modifications are incorporated into the method of Encke, and a new technique, the hybrid patched conic, is introduced. I have tried to develop the subject of special perturbations from the point of view that it is a refinement of the methods developed in Chapters 3, 4, and 5. This is especially true for the hybrid technique.

Chapter 8 enters into the description of the complicated selenographic coordinate transformation. This chapter relates in full detail the necessary transformations required to locate space vehicles relative to the lunar surface. The analysis presented was performed in support of trajectory computations for the Apollo mission by TRW Systems and in my belief is the most detailed analytical treatment of this topic available at present. As exploratory flights are made to the Moon, the constants utilized in these coordinate rotations will no doubt be improved; however, the functional form of the transformation should remain functionally unchanged.

As in *Methods of Orbit Determination*, the reader who uses this book should have a firm understanding of the differential and integral calculus. Elementary vector and matrix methods are used throughout the text in order to preserve clarity and conciseness of presentation. An attempt was made to introduce vector and matrix analysis techniques at logical points only when it was truly felt that their inclusion would be beneficial. The reader familiar with dot and cross products and with the definition of the gradient of a function should have little trouble with the analysis. The reader is assumed to have a working knowledge of matrix multiplication and inversion techniques.

The material in this book could be used to teach an advanced astrodynamics course that would emphasize orbital transfer and rendezvous procedures. The student would read Chapter 1 quickly to gain familiarity with the solar system, methods of computing the location of its major bodies, and the modeling of planetary atmospheres. Depending on the level of mathematical background in optimization techniques, a suitable entry could be made into Chapter 2. Chapters 3 through 7 would then form the basic core of the course. Chapter 8 would be studied as special material if time permitted.

Problems that stress the practical concepts and ideas of our space age are included at the end of each chapter to illustrate additional material. Some exercises are theoretical in nature, others are numerical, but some of the numerical analyses, in order to be meaningful, are by their very nature, complex and should be handled with modern calculating machine equipment. The availability of this equipment, both in industry and at the universities, prompted the incorporation of these problems or projects into this book.

Methods of Astrodynamics, by its very nature, could not have been possible without the assistance of many people actively working in this field. It is a distinct pleasure to thank Harvey L. Roth for his contributions to this work, especially to Chapter 8. Other people among the many who have helped *Methods of Astrodynamics* to reach fruition are Kurt Forster, Hans Lieske, Thomas Mucha, George Stern, Robert Chase, and many others. I wish to thank Jody Custer and Etta Monfore for their assistance and an excellent typing effort that took many hours. To TRW Systems for its overall assistance I shall always be grateful.

P. R. Escobal

Redondo Beach, California
July 1968

Contents

Methods of Astrodynamics

1 The Solar System

Direct your telescope to a point on the ecliptic in the constellation of Aquarius, in longitude 326 degrees and you will find within a degree of that place a new planet of about the ninth magnitude and having a perceptible disk.

LE VERRIER (*The Discovery of Neptune*) [25]

1.1 PHYSICAL ENVIRONMENT OF THE PLANETS

Mercury, a small scorched planet devoid of atmosphere, would present a blinding picture to a space explorer. The intensity of light would be enormous. This planet, because of the tremendous gravitational strength of the Sun, has become slightly more massive on one side and thus, with time, slowed in axial rotation until one side always greets the Sun. The days last forever on the light side, which is unbearably hot. Temperatures are so high that pools of liquid lead might be found in abundance along the craggy ground. Lack of an atmosphere would show the Sun bright and strong, possessed of a magnificent corona set against a jet black sky. Indeed, Mercury would be a cruel planet to an Earthman.

Venus, sometimes referred to as the Earth's sister planet owing to its physical dimensions, would be quite bizarre by terrestrial standards. The surface is red hot, yet the atmospheric pressure so high that water would not boil. On Earth, for example, bromine is a gas, but on Venus it would remain a liquid. The sky viewed from the supposedly inhospitable dry or sandy surface is probably yellow or orange green. Perhaps the Sun would be either obliterated or a deep red, filtered and diminished by the ever-present thick cloud layer. A Venusian day would last a great length of time. Approximately 120 Earth days would pass between successive sunrises. For comfort and survival, at least for a few eons until the Sun cools sufficiently, Earthmen must look elsewhere for more rewarding worlds worthy of conquest.

Earth is a strange place, even though it is our home. Approximately three-quarters of our planet is covered by water, and even more falls from the sky when a storm erupts. Electrical discharges are frequently seen overhead and high winds of tremendous energy originate with great frequency. Surface temperatures fluctuate readily and can be unbearably high or low. Earth and Moon form a unique system; in truth the Earth-Moon system could be classified as a binary system, for they actually rotate about each other. Indeed, Earth can be both cruel and kind to those who inhabit its boundless surface.

Mars, in contrast, is quite small and its atmosphere is very thin. Nights and days are nearly equal in duration to their terrestrial counterparts. The sky is a chilly royal blue with faint wisps of sparse clouds. In the summers the ice caps vaporize but no water flows to the land. Atmospheric pressure is very low and water, if any, is scarce. The seasonal change transforms the crater-riddled land from a red arid ground mass to a field of low-growing green vegetation. At night two small moons race across the very dark skies in opposite directions shedding almost no light on very cold ground. During a summer day at the equator the temperature rises rapidly from a cold night to reach a comfortable 60 or 70 degrees at high noon. Earthmen could feasibly live here, but only by exerting much caution and patience, and if they received a helpful hand from the third planet.

Jupiter is a giant planet whose diameter exceeds the Earth's by a factor of 11, even though its mean density is absurdly low. The Sun, as seen at arm's length from the surface of this great planet, would appear to be a disk only an eighth of an inch across. It is a cold world. The gravitational force would make an Earthman weigh two and a half times his terrestrial standard. Atmospheric bands at varying latitudes rotate at different angular speeds. Winds of more than 2 miles a second flow across the surface. The atmosphere is extensive and composed of many gases; methane and hydrogen exist in considerably amounts. A huge satellite system composed of twelve moons surrounds Jupiter; some are as large as the planet Mercury. The satellite orbits are not well-behaved paths, but undergo fluctuations due to the complex system of perturbations present. Man may, one day, land on the larger satellites, Io, Europa, Ganymede, or Callisto, to observe this giant planet, but future landings are not foreseen. It is not known whether the surface of this planet is solid. Indeed, the atmosphere may continuously increase in density in the direction of the planetary center.

Saturn, with its unique system of rings, makes a picture of great beauty. In dimensions it is a little smaller than Jupiter and is possessed of a complex satellite system of 11 known moons. Saturn has the largest moon of all the planets, Titan, which is larger than the planet Mercury. The Saturnian ring system, probably a decomposed 12th moon, is extremely thin and very spread out. Great shadows and various twilight belts are generated on the planet's surface. The rings are yellow in hue, illuminated by a Sun which, if seen from the surface, would appear to be the size of a small nail head. The superb view of the ring system from one of Saturn's nearest moons would be awe inspiring. The atmosphere of the parent planet is composed of various gases of which ammonia forms a good percentage. The mean density of the planet is so low that, as for Jupiter, it is easy to suspect that there is a fundamental core of hydrogen. Indeed, Saturn is beautiful, but only some of its moons will form adequate landing areas for visitors from the third planet.

Uranus, the 7th planet from the Sun, is roughly half the size of Saturn. It apparently possesses five moons. The planet rolls along like a drunkard with an inclination of its equator to orbit of over 90 degrees. It is such a cold world that ammonia has condensed out of the atmosphere. The clouds are probably composed of methane and nitrogen.

Neptune, romantically discovered without telescopes as a result of perturbed behavior of Uranus, is over three times the size of the Earth. This planet has two moons, one of them, Triton, is the same size as Mercury; the other is small. Light on Neptune is very feeble. The Sun, as viewed from the cold ground, appears like a yellowish pinhead in the sky. Methane and nitrogen clouds shroud the planetary surface.

Pluto is the outermost planet. Its dimensions are about the same as the Earth's. On this cold and black planet the Sun would appear as a faint star. The atmosphere lies frozen to Pluto's craggy surface. It is a lonely outpost at the farthest reaches of the solar system. This planet moves slowly in space at more than three and a half thousand million miles from the dynamical center of the solar system.

Beyond lie the stars and about some of them rotate other planetary systems, some of which, with almost certain probability, are reflections of our own image. Beyond the eternal stars lie the galaxies, a void without end.

In this chapter the dimensions, locations, and adoptable atmospheric models of the members of the solar system are discussed. Most of the information, by its very nature, is empirical.

1.2 BODE'S LAW

A clock is ticking. The gyrations and movement of its internal mechanisms proceed in an orderly fashion, attempting to keep pace with the precision of the solar system.

The solar system is basically composed of nine known planets and an extremely large number of smaller planetoids. In this and subsequent sections the members of the solar system are discussed from an astrodynamical point of view. Expressions presented will enable the mean orbital elements of the more important planets to be computed for any desired date. First, however, let the members of the solar system be formally introduced: the central body, the Sun, and the major planets Mercury, Venus, Earth, Mars, Jupiter, Saturn, Uranus, Neptune, and Pluto. Between the orbits of Mars and Jupiter exists a belt of small planetoids. The most prominent ones have such intriguing names as Ceres, Pallas, Juno, and Vesta.

A brief preview of scale in the solar system can be obtained as follows. Consider the accompanying scheme. The addition of the first row to the

Mercury	Venus	Earth	Mars	Planetoids	Jupiter	Saturn	Uranus
0.4	0.4	0.4	0.4	0.4	0.4	0.4	0.4
0.0	0.3	0.6	1.2	2.4	4.8	9.6	19.2
0.4	0.7	1.0	1.6	2.8	5.2	10.0	19.6

second, which is merely a doubling of each number from left to right, yields a close approximation of the distance of the planets from the Sun in astronomical units. Mathematically this scheme can be represented by

$$a = \tfrac{1}{10}[4 + 3 \cdot 2^n], \qquad n = -\infty, 0, 1, 2, \ldots, 8. \tag{1.1}$$

The empirical relationship (1.1) is known in the literature as Bode's law, and it holds quite well up to the planet Uranus. After this point it deviates markedly from the true orbital distances of the planets. These distances and other related data are collected in Table 1.1. The satellites of the planetary members are given in Table 1.2.

Bode's law provides a simple way to remember the approximate distances of the closer planets from the Sun. It appears, however, that a correction term is needed for the more distant outer planets. A distinction is usually made between the *inner planets*, Mercury, Venus, Earth, and Mars, and the *outer planets*, Jupiter, Saturn, Uranus, Neptune, and Pluto.

1.3 MEAN PLANETARY ORBITAL ELEMENTS AS A FUNCTION OF TIME

For purposes of position and velocity determination for use in space targeting, much more accurate empirical relationships must be utilized. In this section the mean orbital elements of the planets are tabulated as a function of elapsed time in Julian centuries since some particular epoch. For the major planets it is possible to obtain a mean model of planetary elements with accuracies to within about 1 minute in position.

Certain of the time-varying mean planetary elements are listed in [2], but some of the more important major planets are missing. This section presents a consistent set of mean planetary time varying elements to be used in analytic ephemeris generation. The analytic ephemeris of the Moon is recorded in

Table 1.1 Approximate Mean Planetary Data of the Solar System

Planet	Mean Distance from Sun (million miles)	Length of Year	Mean Daily Motion (deg)	Eccentricity	Inclination to Ecliptic
Mercury (☿)	36.0	88.0 days	4.092	0.2056	7°0′12″
Venus (♀)	67.2	224.7 days	1.602	0.0068	3°23′38″
Earth (⊕)	93.0	365.25 days	0.986	0.0167	0°0′0″
Mars (♂)	141.5	1.88 yr	0.524	0.0934	1°51′0″
Jupiter (♃)	483.3	11.86 yr	0.083	0.0485	2°29′29″
Saturn (♄)	886.1	29.46 yr	0.034	0.0516	1°18′26″
Uranus (⛢)	1,782.8	84.02 yr	0.012	0.0443	0°46′22″
Neptune (♆)	2,793.5	164.79 yr	0.006	0.0073	1°46′37″
Pluto (♇)	3,675.0	248.43 yr	0.004	0.2481	17°08′38″

Planet	Orbital Speed (miles/sec)	Equatorial Escape Speed (miles/sec)	Gravity at Surface (Earth = 1)	Period of Rotation	Inclination of Equator to Orbit
Mercury (☿)	29.7	2.2	0.29	88 days?	near zero
Venus (♀)	21.7	6.3	0.86	120 days	?
Earth (⊕)	18.5	7.0	1.00	1 day	23°27′
Mars (♂)	15.0	3.1	0.37	24 hr 37 min	25°10′
Jupiter (♃)	8.1	37.0	2.64	9 hr 55 min	3°7′
Saturn (♄)	6.0	22.0	1.17	10 hr 14 min	26°47′
Uranus (⛢)	4.2	13.0	0.91	10 hr 40 min	98°
Neptune (♆)	3.4	14.0	1.12	15 hr 40 min	151°
Pluto (♇)	2.7?	6.5?	0.79?	?	?

Planet	Mass (Earth = 1)	Volume (Earth = 1)	Density (water = 1)	Diameter (miles)	Diameter of Iron Sphere of Equal Weight (miles)
Mercury (☿)	0.04	0.055	2.86	3,000	2,410
Venus (♀)	0.8	0.876	4.86	7,500	6,540
Earth (⊕)	1.0	1.000	5.52	7,900	7,040
Mars (♂)	0.11	0.151	3.96	4,100	3,370
Jupiter (♃)	317.0	1,312.0	1.34	86,800	48,000
Saturn (♄)	95.0	763.0	0.71	71,500	32,120
Uranus (⛢)	14.7	59.0	1.27	30,000	17,250
Neptune (♆)	17.2	72.0	1.58	27,000	18,160
Pluto (♇)	0.7	0.9?	5.3?	8,100?	6,500

Table 1.2[a] Approximate Satellite Data of the Solar System

Planet	Satellite	Discoverer	Year of Discovery	Mean Distance from Primary (miles)	Sidereal Period of Revolution (days)	Diameter (miles)
Earth	Moon	Pithecan-thropus	Prehistoric	239,000	27.32	2,160
Mars	Phobos	Asaph Hall	1877	5,800	0.32	10
	Deimos	Asaph Hall	1877	14,600	1.26	5
Jupiter	V	Barnard	1892	112,600	0.50	100
	I, Io	Galilei	1610	261,800	1.77	2,300
	II, Europa	Galilei	1610	416,600	3.55	2,000
	III, Ganymede	Galilei	1610	664,200	7.15	3,200
	IV, Callisto	Galilei	1610	1,169,000	16.69	3,200
	VI	Perrine	1904	7,114,000	250.57	100
	VII	Perrine	1905	7,292,000	259.65	40
	X	Nicholson	1938	7,350,000	263.55	15?
	XI	Nicholson	1938	14,040,000	692.5	15?
	VIII	Melotte	1908	14,600,000	738.9	40
	IX	Nicholson	1914	14,880,000	758.0	20
	XII	—	—	13,180,000	631.1	—
Saturn	Janus	Dollfus	1966	97,800	0.75	?
	Mimas	Herschel	1789	115,000	0.94	370
	Enceladus	Herschel	1789	148,000	1.37	460
	Tethys	Cassini	1684	183,000	1.89	750
	Dione	Cassini	1684	234,000	2.74	900
	Rhea	Cassini	1672	327,000	4.52	1,150
	Titan	Huyghens	1655	759,000	15.97	3,550
	Hyperion	Bond	1848	920,000	21.32	300
	Japetus	Cassini	1671	2,210,000	79.92	1,000
	Phoebe	Pickering	1898	8,034,000	523.7	200
	Thermis	Pickering	1905	—	—	—
Uranus	Miranda	Kuiper	1948	80,800	1.41	150
	Ariel	Lassell	1851	119,100	2.52	600
	Umbriel	Lassell	1851	165,900	4.14	400
	Titania	Herschel	1787	272,000	8.71	1,000
	Oberon	Herschel	1787	364,000	13.46	900
Neptune	Triton	Lassell	1846	220,000	5.88	3,000
	Nereid	Kuiper	1948	3,465,000	362.0	180

[a] For more details on satellite data consult [2], p. 492.

Section 1.4. The expressions collected here can be found in [3], which offers a very clear French exposition of the somewhat mystical English version [2] of the motions of the major planets. According to Block [31], the best formulas to be utilized are Newcomb's values for the inner planets with the corrected elements of Ross [32] for the planet Mars. The outer planetary elements are to be calculated from the formulas of Le Verrier and Gaillot [3]. The osculating elements of Pluto, due to Bower [33], are substituted for the corresponding mean values. It is stressed that, with the exception of those of Pluto, these elements are mean and not osculating elements; hence, if velocities are to be determined rigorously, the appropriate time rate of change of these elements should be included in the representation process.

The argument of these expansions is the parameter

$$T_u \equiv \frac{(\text{J.D.}) - (\text{J.D.})_0}{36525.0}, \tag{1.2}$$

where the epoch Julian date for January 0.5, 1900 is $(\text{J.D.})_0 = 2415020.0$ and J.D. is the Julian date at instant. For consistency the masses of these planets should be taken as listed in Table 1.3.

Table 1.3 Masses of the Planets (Sun = 1)
$(k_\odot \equiv 0.017202\ 09895\ (\text{a.u.})^{3/2}/\text{day})$

Mercury	1/6,000,000	Jupiter	1/1047.355
Venus	1/408,000	Saturn	1/3501.6
Earth	1/333,432	Uranus	1/22,869
Earth + Moon	1/329,390	Neptune	1/19,314
Mars	1/3,093,500	Pluto	1/360,000

1.3.1 The Major Planets

The standard elements ([1], Chapter 3) tabulated here are defined relative to the ecliptic coordinate system ([1], Chapter 4) in the following fashion.

$\Omega \equiv$ ecliptic longitude of the ascending node,
$\omega \equiv$ ecliptic argument of perihelion,
$M \equiv$ mean anomaly,
$L \equiv$ the mean longitude $= \Omega + \omega + M$,
$\tilde{\omega} \equiv$ the longitude of perihelion $= \Omega + \omega$,
$e \equiv$ the orbital eccentricity,
$i \equiv$ the orbital inclination to the ecliptic plane,
$a \equiv$ the semimajor axis in astronomical units,
$T \equiv$ the time of perifocal passage,
$n \equiv$ the mean sidereal motion.

In terms of (1.2), mean planetary elements[1] of the major planets, according to the recommendations of Block, can be obtained from observational astronomy as follows.

MERCURY

$$\begin{aligned} L &= 178°10'44''.68 + 538106654''.80T_u + 1''.084T_u^2, \\ \tilde{\omega} &= 75°53'58''.91 + 5599''.76T_u + 1''.061T_u^2, \\ \Omega &= 47°8'45''.40 + 4266''.75T_u + 0''.626T_u^2, \\ e &= 0.20561421 + 0.00002046T_u - 0.000000030T_u^2, \\ i &= 7°0'10''.37 + 6''.699T_u - 0''.066T_u^2, \\ a &= 0.3870984. \end{aligned} \tag{1.3}$$

VENUS

$$\begin{aligned} L &= 342°46'1''.39 + 210669162''.88T_u + 1''.1148T_u^2, \\ \tilde{\omega} &= 130°9'49''.8 + 5068''.93T_u - 3''.515T_u^2, \\ \Omega &= 75°46'46''.73 + 3239''.46T_u + 1''.476T_u^2, \\ e &= 0.00682069 - 0.00004774T_u + 0.000000091T_u^2, \\ i &= 3°23'37''.07 + 3''.621T_u - 0''.0035T_u^2, \\ a &= 0.72333015. \end{aligned} \tag{1.4}$$

EARTH

$$\begin{aligned} L &= 99°41'48''.04 + 129602768''.13T_u + 1''.089T_u^2, \\ \tilde{\omega} &= 101°13'15''.0 + 6189''.03T_u + 1''.63T_u^2 + 0''.012T_u^3, \\ \Omega &= 0°, \\ e &= 0.01675104 - 0.00004180T_u - 0.000000126T_u^2, \\ i &= 0°, \\ a &= 1.00000013. \end{aligned} \tag{1.5}$$

MARS

$$\begin{aligned} L &= 293°44'51''.46 + 68910117''.33T_u + 1''.1184T_u^2, \\ \tilde{\omega} &= 334°13'5''.53 + 6626''.73T_u + 0''.4675T_u^2 - 0''.0043T_u^3, \\ \Omega &= 48°47'11''.19 + 2775''.57T_u - 0''.005T_u^2 - 0''.0192T_u^3, \\ e &= 0.09331290 + 0.000092064T_u - 0.000000077T_u^2, \\ i &= 1°51'1''.20 - 2''.430T_u + 0''.0454T_u^2, \\ a &= 1.52368839. \end{aligned} \tag{1.6}$$

JUPITER

$$\begin{aligned} L &= 238°2'57''.32 + 10930687''.148T_u + 1''.20486T_u^2 - 0''.005936T_u^3, \\ \tilde{\omega} &= 12°43'15''.34 + 5795''.862T_u + 3''.80258T_u^2 - 0''.01236T_u^3, \\ \Omega &= 99°26'36''.19 + 3637''.908T_u + 1''.2680T_u^2 - 0''.03064T_u^3, \\ e &= 0.04833475 + 0.000164180T_u - 0.0000004676T_u^2 - 0.0000000017T_u^3, \\ i &= 1°18'31''.45 - 20''.506T_u + 0''.014T_u^2, \\ a &= 5.202561. \end{aligned} \tag{1.7}$$

[1] Because these equations are empirically derived, effects such as relativity have been absorbed into the coefficients.

SATURN

$$
\begin{aligned}
L &= 266^\circ 33'51''.76 + 4404635''.5810T_u + 1''.16835T_u^2 - 0''.021T_u^3, \\
\tilde{\omega} &= 91^\circ 5'53''.38 + 7050''.297T_u + 2''.9749T_u^2 + 0''.0166T_u^3, \\
\Omega &= 112^\circ 47'25''.40 + 3143''.5025T_u - 0''.54785T_u^2 - 0''.0191T_u^3, \\
e &= 0.05589232 - 0.00034550T_u - 0.000000728T_u^2 + 0.00000000074T_u^3, \\
i &= 2^\circ 29'33''.07 - 14''.108T_u - 0''.05576T_u^2 + 0''.00016T_u^3, \\
a &= 9.554747.
\end{aligned} \tag{1.8}
$$

URANUS

$$
\begin{aligned}
L &= 244^\circ 11'50''.89 + 1547508''.765T_u + 1''.13774T_u^2 - 0''.002176T_u^3, \\
\tilde{\omega} &= 171^\circ 32'55''.14 + 5343''.958T_u + 0''.8539T_u^2 - 0''.00218T_u^3, \\
\Omega &= 73^\circ 28'37''.55 + 1795''.204T_u + 4''.722T_u^2, \\
e &= 0.0463444 - 0.00002658T_u + 0.000000077T_u^2, \\
i &= 0^\circ 46'20''.87 + 2''.251T_u + 0''.1422T_u^2, \\
a &= 19.21814.
\end{aligned} \tag{1.9}
$$

NEPTUNE

$$
\begin{aligned}
L &= 84^\circ 27'28''.78 + 791589''.291T_u + 1''.15374T_u^2 - 0''.002176T_u^3, \\
\tilde{\omega} &= 46^\circ 43'38''.37 + 5128''.468T_u + 1''.40694T_u^2 - 0''.002176T_u^3, \\
\Omega &= 130^\circ 40'52''.89 + 3956''.166T_u + 0''.89952T_u^2 - 0''.016984T_u^3, \\
e &= 0.00899704 + 0.000006330T_u - 0.000000002T_u^2, \\
i &= 1^\circ 46'45''.27 - 34''.357T_u - 0''.0328T_u^2, \\
a &= 30.10957.
\end{aligned} \tag{1.10}
$$

PLUTO

$$
\begin{aligned}
T &= 1989 \text{ October } 0.0344 \text{ day}, \\
\tilde{\omega} &= 113^\circ 31'17''.72, \\
\Omega &= 108^\circ 57'16''.18, \\
e &= 0.2486438, \\
i &= 17^\circ 8'48''.40, \\
a &= 39.517738.
\end{aligned} \tag{1.11}
$$

For Pluto the mean anomaly may be obtained directly from $M = n(t - T)$, where $n = k\sqrt{\mu}\, a^{-3/2}$. The orientation elements are referred to the equinox of 1900.0. The elements for this planet are osculating as opposed to mean.

1.3.2 The Minor Planets

Between the orbit of Mars and Jupiter exists a ring of minor planets whose origin is probably due to the decomposition of a major planet. These minor planets are small in size, ranging from a few inches to about 300 miles in diameter. An extended list of these members of the solar system, their approximate characteristics, and some of their possible uses in futuristic missions can be found in [34]. The approximate elements, referred to the ecliptic and mean equinox of 1950.0, for a few of the more important minor planets are listed as follows.

	Ceres	Pallas	Juno	Vesta
	(1957 June 11 at 0^{hr} E.T.)			
L	$72^\circ.247$	$34^\circ.549$	$25^\circ.907$	$332^\circ.903$
$\tilde{\omega}$	$152^\circ.367$	$122^\circ.734$	$56^\circ.571$	$253^\circ.236$
Ω	$80^\circ.514$	$172^\circ.975$	$170^\circ.438$	$104^\circ.102$
e	0.07590	0.23402	0.25848	0.08888
i	$10^\circ.607$	$34^\circ.798$	$12^\circ.993$	$7^\circ.132$
a	2.7675	2.7718	2.6683	2.3617
n	$0^\circ.21408$	$0^\circ.21358$	$0^\circ.22612$	$0^\circ.27157$

1.4 APPROXIMATE LUNAR THEORY

On a clear winter evening the Earth's closest neighbor appears to hang silently suspended, fixed forever against a deep blue sky. The motion of the Moon, however, is a complex balance and counterbalance of gravitational forces. The analysis of the motion of the Moon is very complicated, and it has taxed the minds of many of the giants in the field of dynamical astronomy. Newton, in his analytical attempts, is recorded as having said that the complex wobblings and librations of the lunar motion caused him to lose much sleep and to develop headaches. Euler, too, was intrigued by the lone satellite of our planet. Delaunay, Hansen, and Hill also attacked the problem of predicting the Moon's position. It was not until the commencement of the twentieth century that a suitable theory was developed by E. W. Brown [9]. The foundations of this theory were due to Hill [8] and others, but the patience of Brown is to be admired and remembered. At the beginning of his monumental work, Brown estimates that his own contributions to the computations required over eight thousand hours.

In this section lunar theory is not discussed; only the results are stated. The interested reader is certainly directed to Brown's original works [9]. Furthermore, only a truncated form of Brown's theory is given. The algorithm predicts the position of the Moon within about 30 seconds of arc. Additional coefficients may be added if desired; they are clearly tabulated at the conclusion of Brown's overpowering work on lunar theory. This truncated form of the lunar theory is useful in analytic studies of the Moon, where a rapid, self-contained prediction method is desired.

All constants and mean expansions of the required angles are consistent with the *Explanatory Supplement to the American Ephemeris and Nautical Almanac* [2]. The coefficients, however, as previously mentioned, are due to Brown. Slight correction terms in the longitude expressions have been omitted owing to the approximate nature of the algorithm.

1.4.1 Determination of Required Mean Angles

A requirement for the prediction of the position of the Moon is the determination of the following angles. These orientation parameters are specified in terms of Julian centuries measured since 1900 January 0.5 E.T., that is,

$$T_u = \frac{\text{J.D.} - 2415020.0}{36525.0}, \tag{1.12}$$

where J.D. is the Julian date at the instant of time. By utilizing T_u it is possible to determine

☾ ≡ the mean longitude of the Moon, measured in the ecliptic from the mean equinox of date to the mean ascending node of the lunar orbit, and then along the orbit,

Γ ≡ the Sun's mean longitude of perigee,

Γ' ≡ the mean longitude of the lunar perigee, measured in the ecliptic from the mean equinox of date to the mean ascending node of the lunar orbit, and then along the orbit,

Ω ≡ the longitude of the mean ascending node of the lunar orbit on the ecliptic measured from the mean equinox of date,

D ≡ the mean elongation of the Moon from the Sun;

and the auxiliary angles,

$$l \equiv \text{☾} - \Gamma', \qquad L \equiv \text{☾} - D, \qquad l' \equiv L - \Gamma, \qquad F \equiv \text{☾} - \Omega$$

from

$$\begin{aligned}
\text{☾} &= 270^\circ 26' 02''.99 + 1336^r 307^\circ 52' 59''.31 T_u - 4''.08 T_u^2 + 0''.0068 T_u^3, \\
\Gamma &\equiv 281^\circ 13' 15''.00 + 6189''.03 T_u + 1''.63 T_u^2 + 0''.012 T_u^3, \\
\Gamma' &\equiv 334^\circ 19' 46''.40 + 11^r 109^\circ 02' 02''.52 T_u - 37''.17 T_u^2 - 0''.045 T_u^3, \\
\Omega &\equiv 259^\circ 10' 59''.79 - 5^r 134^\circ 08' 31''.23 T_u + 7''.48 T_u^2 + 0''.008 T_u^3, \\
D &\equiv 350^\circ 44' 14''.95 + 1236^r 307^\circ 06' 51''.18 T_u - 5''.17 T_u^2 + 0''.0068 T_u^3.
\end{aligned} \tag{1.13}$$

1.4.2 Computation of the Moon's Ecliptic Longitude, Declination, and Parallax

Retaining coefficients of the trigonometric expansions to the order of 20 seconds of arc in magnitude for longitude and latitude terms, and $\frac{1}{2}$ second for the parallax term, from [9], it is possible to determine the Moon's longitude as[2]

$$\begin{aligned}
☾_\epsilon = ☾ &+ 22639.580 \sin (l) - 4586.438 \sin (l - 2D) \\
&+ 2369.899 \sin (2D) + 769.021 \sin (2l) \\
&- 668.944 \sin (l') - 411.614 \sin (2F) \\
&- 211.658 \sin (2l - 2D) - 206.219 \sin (l + l' - 2D) \\
&+ 191.954 \sin (l + 2D) - 165.351 \sin (l' - 2D) \\
&+ 147.878 \sin (l - l') \\
&- 124.785 \sin (D) - 109.804 \sin (l + l') - 55.174 \sin (2F - 2D) \\
&- 45.100 \sin (l + 2F) + 39.532 \sin (l - 2F) \\
&- 38.428 \sin (l - 4D) + 36.124 \sin (3l) - 30.773 \sin (2l - 4D) \\
&- 28.511 \sin (l - l' - 2D) - 24.451 \sin (l' + 2D); \qquad (1.14)
\end{aligned}$$

to obtain the Moon's declination from

$$\begin{aligned}
\delta_\epsilon = &\ 18461.480 \sin (F) + 1010.180 \sin (l + F) \\
&- 999.695 \sin (F - l) - 623.658 \sin (F - 2D) \\
&+ 199.485 \sin (F + 2D - l) - 166.577 \sin (l + F - 2D) \\
&+ 117.262 \sin (F + 2D) + 61.913 \sin (2l + F) \\
&- 33.359 \sin (F - 2D - l) \\
&- 31.763 \sin (F - 2l) - 29.689 \sin (l' + F - 2D); \qquad (1.15)
\end{aligned}$$

and to obtain the Moon's parallax as

$$\begin{aligned}
\pi_\epsilon = \pi_☾ &+ 186.5398 \cos (l) + 34.3117 \cos (l - 2D) \\
&+ 28.2333 \cos (2D) + 10.1657 \cos (2l) \\
&+ 3.0861 \cos (l + 2D) + 1.9202 \cos (l' - 2D) \\
&+ 1.4455 \cos (l + l' - 2D) + 1.1542 \cos (l - l') \\
&- 0.9752 \cos (D) - 0.9502 \cos (l + l') \\
&- 0.7136 \cos (l - 2F) + 0.6215 \cos (3l) \\
&+ 0.6008 \cos (l - 4D). \qquad (1.16)
\end{aligned}$$

[2] In a machine subroutine to determine the position and velocity of the Moon the repeated use of trigonometric functions can be avoided by direct expansion of composite arguments. Therefore, if the trigonometric arguments of the variables l, l', D, F, that is, eight trigonometric functions, are known, the computation of $☾_\epsilon$, δ_ϵ, π_ϵ can be achieved as a series of multiplications and additions with the aid of these trigonometric identities: $\sin 2\theta = 2 \cos \theta \sin \theta$, $\cos 2\theta = \cos^2 \theta - \sin^2 \theta$; $\sin 3\theta = 3 \sin \theta - 4 \sin^3 \theta$, $\cos 3\theta = 4 \cos^3 \theta - 3 \cos \theta$.

Table 1.4 Approximate Lunar Constants

Orbital Parameters and Physical Constants	Value
Orbital eccentricity ($e_☾$)	0.054900489
Sine of one-half inclination to the ecliptic (sin $\frac{1}{2}i_\epsilon$)	0.044886967
Sine parallax ($\pi_☾$)	3422″.5400
Ratio of mass of Earth to mass of Moon ($R_☾$)	81.53
Equatorial radius of Earth (a_e)	6378.388 km
Horizontal parallax at 60.2665 Earth radii ($\bar{\pi}_\epsilon$)	57′02″.70

The remaining lunar elements and physical constants are tabulated in Table 1.4. The constants adopted in this table are consistent, though they are not the very last known values of the parameters. In lunar theory the orbital parameters listed in Table 1.4 are true constants.

1.4.3 Determination of the Moon's Position

To determine the position of the Moon it is necessary first to determine the lunar position from the horizontal parallax as

$$r_☾ = \frac{a_e}{\pi_\epsilon}, \qquad (1.17)$$

where π_ϵ is obtained from (1.16). Examination of Figure 1.1 permits the direct position mappings from spherical to rectangular ecliptic coordinates

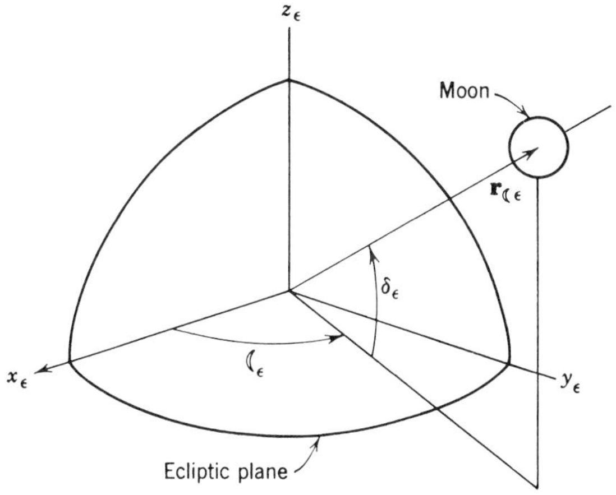

FIGURE 1.1 Coordinates of the Moon.

to be written at sight as

$$\mathbf{r}_{\leftmoon\epsilon} = \begin{bmatrix} x_{\leftmoon} \\ y_{\leftmoon} \\ z_{\leftmoon} \end{bmatrix}_{\epsilon} = \begin{bmatrix} r_{\leftmoon} \cos \delta_{\epsilon} \cos \leftmoon_{\epsilon} \\ r_{\leftmoon} \cos \delta_{\epsilon} \sin \leftmoon_{\epsilon} \\ r_{\leftmoon} \sin \delta_{\epsilon} \end{bmatrix}, \tag{1.18}$$

where $r_{\leftmoon}$, δ_{ϵ}, and $\leftmoon_{\epsilon}$ are obtained, respectively, from (1.17), (1.15), and (1.14). Velocity components, that is, derivatives of position with respect to $\tau \equiv k(t - t_0)$, may be accurately determined by direct differentiation of (1.14), (1.15), and (1.17) with the aid of

$$\dot{\mathbf{r}}_{\leftmoon\epsilon} \equiv \begin{bmatrix} \dot{x}_{\leftmoon} \\ \dot{y}_{\leftmoon} \\ \dot{z}_{\leftmoon} \end{bmatrix}_{\epsilon} = \begin{bmatrix} \dot{r}_{\leftmoon} \cos \delta_{\epsilon} \cos \leftmoon_{\epsilon} - r_{\leftmoon}\dot{\delta}_{\epsilon} \sin \delta_{\epsilon} \cos \leftmoon_{\epsilon} - r_{\leftmoon}\dot{\leftmoon}_{\epsilon} \cos \delta_{\epsilon} \sin \leftmoon_{\epsilon} \\ \dot{r}_{\leftmoon} \cos \delta_{\epsilon} \sin \leftmoon_{\epsilon} - r_{\leftmoon}\dot{\delta}_{\epsilon} \sin \delta_{\epsilon} \sin \leftmoon_{\epsilon} + r_{\leftmoon}\dot{\leftmoon}_{\epsilon} \cos \delta_{\epsilon} \cos \leftmoon_{\epsilon} \\ \dot{r}_{\leftmoon} \sin \delta_{\epsilon} + r_{\leftmoon}\dot{\delta}_{\epsilon} \cos \delta_{\epsilon} \cos \delta_{\epsilon} \end{bmatrix}. \tag{1.19}$$

If velocities are required then differentiation of the analytic expressions for π_{ϵ}, δ_{ϵ}, and $\leftmoon_{\epsilon}$ must be undertaken. It is well to note that in the geocentric right ascension-declination coordinate system ([1], Chapter 4)

$$\tau = k_e(32525.0 \times 1440.0)T_u, \tag{1.20}$$

which implies $\dot{\mathbf{r}}_{\leftmoon} \equiv d\mathbf{r}_{\leftmoon}/d\tau \not\equiv d\mathbf{r}_{\leftmoon}/dT_u$ and

$$\frac{dT_u}{d\tau} \equiv \frac{1}{k_e}(32525.0 \times 1440.0)^{-1} = 0.255666824 \times 10^{-6}. \tag{1.21}$$

1.4.4 Rotation into the Right Ascension-Declination Geocentric Coordinate System

Using the mean obliquity of the ecliptic,[3] ϵ, extracted from the *Explanatory Supplement* [2] as

$$\epsilon = 23°27'08''.26 - 46''.845T_u - 0''.0059T_u^2 + 0''.00181T_u^3, \tag{1.22}$$

the standard rotation to the right ascension-declination coordinate system ([1], Chapter 4) is given by the matrix M; that is,

$$\mathbf{r}_{\leftmoon} = \begin{bmatrix} x_{\leftmoon} \\ y_{\leftmoon} \\ z_{\leftmoon} \end{bmatrix} = [M]\begin{bmatrix} x_{\epsilon} \\ y_{\epsilon} \\ z_{\leftmoon} \end{bmatrix} = \begin{bmatrix} 1 & 0 & 0 \\ 0 & \cos \epsilon & -\sin \epsilon \\ 0 & \sin \epsilon & \cos \epsilon \end{bmatrix}\begin{bmatrix} x_{\epsilon} \\ y_{\epsilon} \\ z_{\leftmoon} \end{bmatrix}. \tag{1.23}$$

[3] For higher accuracy the effects of nutation in the obliquity should be considered.

In similar fashion, with a high degree of accuracy,

$$\dot{\mathbf{r}}_{☾} = \begin{bmatrix} \dot{x}_{☾} \\ \dot{y}_{☾} \\ \dot{z}_{☾} \end{bmatrix} = [M] \begin{bmatrix} \dot{x}_{\epsilon} \\ \dot{y}_{\epsilon} \\ \dot{z}_{\epsilon} \end{bmatrix}. \tag{1.24}$$

1.4.5 Differential Correction of the Lunar Coefficients

The coefficients in (1.14), (1.15), and (1.16) were developed with very good data but not the most current astrodynamical constants. As a matter of fact, all the constants of Table 1.4 have been determined to much greater precision. If better constants are to be used for the values given in Table 1.4, the coefficients of expansions (1.14), (1.15), and (1.16) should also be improved. Differential correction can be used to accomplish the desired improvement. The dynamical deterministic procedure for this improvement is now outlined.

Consider that the position of the Moon is a function of $☾_{\epsilon}$, δ_{ϵ}, π_{ϵ}, that is,

$$\mathbf{r}_{☾\epsilon} = \mathbf{r}_{☾\epsilon}(☾_{\epsilon}, \delta_{\epsilon}, \pi_{\epsilon}),$$

or of the a_i coefficients of the trigonometric terms, so that

$$\mathbf{r}_{☾} = \mathbf{r}_{☾\epsilon}(a_1, a_2, \ldots, a_q), \tag{1.25}$$

where q is the total number of terms carried in the lunar expansion, in this case 45 terms, and where a_q is the nominal value of the final coefficient of the expansion. Hence, by forming the differential

$$\Delta\mathbf{r}_{☾\epsilon} = \sum_{1}^{q} \frac{\partial \mathbf{r}_{☾\epsilon}}{\partial a_i} \Delta a_i \tag{1.26}$$

or omitting the ecliptic subscript, ϵ, it is possible to write

$$\Delta\mathbf{r}_{☾j} = \frac{\partial \mathbf{r}_{☾j}}{\partial a_1} \Delta a_1 + \frac{\partial \mathbf{r}_{☾j}}{\partial a_2} \Delta a_2 + \cdots + \frac{\partial \mathbf{r}_{☾j}}{\partial a_q} \Delta a_q. \tag{1.27}$$

Equation 1.27, written for j different times under the assumption that the partial derivative coefficients can be determined, forms a linear algebraic system of j equations in the a_q unknowns. If the system is written for q times, that is, $j = q$, the linear system becomes determined and can be solved for the Δa_i. It should be noticed that $\Delta\mathbf{r}_{☾j}$ can always be determined numerically from $\Delta\mathbf{r}_{☾j} \equiv \bar{\mathbf{r}}_{☾j} - \mathbf{r}_{☾j}$, where $\bar{\mathbf{r}}_{☾j}$ is the position of the Moon obtained subsequentially from the standard ephemeris tapes at times t_j, Section 1.5, and $\mathbf{r}_{☾j}$ is the position of the Moon obtained from the analytic computation of Section 1.4.3. Successive solution of (1.27) for Δa_i, $j = q$, drives the error or residual $\Delta \mathbf{r}_{☾ j}$ to a minimum; optimally the residual would vanish on adding the Δa corrections to the a coefficients. The partial derivatives of the differential system can be obtained by either variant calculations ([1], Chapter 9) or by analytic differentiation of (1.18) with respect to the a_j. Exact determination of the partial derivatives is not required for convergence of the correction process.

1.5 MORE EXACT ELEMENT DETERMINATION

1.5.1 Position and Velocities from Magnetic Tapes

Computer programs make much use of precise prediction techniques for representation of planetary position and velocity. The purpose of this section is to review quickly the available sources for the planetary (lunar) ephemerides and to outline a procedure that could be used to transform the basic data, that is, the position and velocity vectors $\mathbf{r}$, $\dot{\mathbf{r}}$ to standard auxiliary elements.

The Jet Propulsion Laboratories ephemeris tapes [4], [5], [6] contain highly accurate *ephemeral data*, that is, position and velocities of the Moon and of the planets—Mercury, Venus, the Earth-Moon barycenter, Mars, Jupiter, Saturn, Uranus, Neptune, and Pluto. Planetary and lunar coordinates are tabulated for the period December 30.0, 1949 (J.D. = 2433280.5) through January 5.0, 2000 (J.D. = 2451548.5). Planetary positions and velocities are referred to the heliocentric equatorial rectangular reference frame of the mean equator and equinox of 1950.0 (J.D. = 2433282.423) and are expressed in units of a.u. and a.u./mean solar day, respectively. Lunar positions and velocities are referred to the geocentric equatorial rectangular reference frame of 1950.0 and are expressed in units of Earth radii and Earth radii-mean solar day [4], [5].

It may be helpful to indicate the reduction of the fundamental rectangular ephemerides to the basic classical elements,

$$a,\ e,\ M_0,\ i,\ \Omega,\ \omega, \tag{1.28}$$

or, alternatively,

$$a,\ S_e,\ C_e,\ \mathbf{U}_0,\ \mathbf{V}_0, \tag{1.29}$$

where, for any orbit, $a \equiv$ the semimajor axis, $e \equiv$ the eccentricity, $i \equiv$ inclination to the fundamental plane, $\Omega \equiv$ longitude of the ascending node, $\omega \equiv$ argument of periapsis, $M_0 \equiv$ mean anomaly at time t_0, $S_e \equiv e \sin E_0$, $C_e \equiv e \cos E_0$, $\mathbf{U}_0 \equiv$ unit vector from the Sun pointing at the objective planet at time t_0, $\mathbf{V}_0 \equiv$ unit vector advanced to $\mathbf{U}_0$ by a right angle in the plane and direction of motion. In these elements, E_0 is the eccentric anomaly of the objective planet at epoch time t_0.

1.5.2 Reduction of Elements from Position and Velocity Vectors

The following chain of equations ([1], Chapter 3), where μ is the sum of the masses of the Sun and objective planet, can be used to determine the adopted elements from position and velocity vectors at a given epoch. Hence, given $\mathbf{r}_0$ and $\dot{\mathbf{r}}_0$, commence calculating with

$$r_0 = (x_0^2 + y_0^2 + z_0^2)^{1/2},$$

$$D_0 = \frac{x_0\dot{x}_0 + y_0\dot{y}_0 + z_0\dot{z}_0}{\mu^{1/2}},$$

$$\Delta = \frac{\dot{x}_0^2 + \dot{y}_0^2 + \dot{z}_0^2}{\mu},$$

$$a = \frac{1}{2/r_0 - \Delta},$$

$$C_e = 1 - \frac{r_0}{a},$$

$$S_e = \frac{D_0}{a^{1/2}},$$

$$e^2 = C_e^2 + S_e^2,$$

$$p = a(1 - e^2),$$

$$\mathbf{U}_0 = \frac{\mathbf{r}_0}{r_0},$$

$$\mathbf{V}_0 = \frac{r_0\dot{\mathbf{r}}_0 - \dot{r}_0\mathbf{r}_0}{(\mu p)^{1/2}},$$

$$\nabla_1 = U_{x0} + V_{y0},$$

$$\nabla_2 = U_{y0} - V_{x0},$$

$$\nabla_3 = (\nabla_1^2 + \nabla_2^2)^{1/2},$$

$$\sin i = (U_{z0}^2 + V_{z0}^2)^{1/2},$$

$$\cos i = \nabla_3 - 1,$$

$$\sin l_0 = \frac{\nabla_2}{\nabla_3},$$

$$\cos l_0 = \frac{\nabla_1}{\nabla_3}.$$

At this point, if $i \neq 0$ or π continue calculating with

$$\sin u_0 = \frac{U_{z0}}{\sin i},$$

$$\cos u_0 = \frac{V_{z0}}{\sin i},$$

$$\Omega = l_0 - u_0,$$

$$\dot{r}_0 = \frac{x_0\dot{x}_0 + y_0\dot{y}_0 + z_0\dot{z}_0}{r_0}.$$

Finally, if $e \neq 0$, obtain

$$\sin v_0 = \frac{1}{e}\dot{r}_0\left(\frac{p}{\mu}\right)^{1/2},$$

$$\cos v_0 = \frac{1}{e}\left(\frac{p}{r_0} - 1\right),$$

$$\omega = u_0 - v_0,$$

$$\sin E_0 = \frac{S_e}{e},$$

$$\cos E_0 = \frac{C_e}{e},$$

$$M_0 = E_0 - e \sin E_0. \tag{1.30}$$

In closing this section, it is well to note that set (1.29) is always valid for low eccentricity and inclination orbits. In opposite fashion, set (1.28) could conceivably yield poor results for low e and i orbits. The preceding algorithm, however, yields both sets of elements.

1.6 ATMOSPHERIC MODELS

A satellite in an eccentric orbit dips into and out of a planetary atmosphere; it has just passed periapsis and experienced retarding drag forces. A booster moves slowly upward from the ground, fighting both gravitation and drag forces. A manned reentry vehicle is decelerated and sustained by combined lift and drag forces. These are examples of the importance of predicting lift and drag forces within accurate bounds. The two atmosphere-dependent parameters that enter into the determination of these forces are the distribution of density, ρ, within the atmosphere and the rotational properties of the atmosphere. In Section 1.6 a discussion of various density models is undertaken.

1.6.1 Perfect Gas Law

The properties of gases are usually studied with simple rules called *perfect gas laws*. At one time these laws were formulations of the best experimental data, but accurate experiments have shown that they are only approximations. It does, however, become expedient to assume that there exist certain gases, called *perfect gases*, which obey these laws exactly. One of the most common gas laws which expresses the relationship between density, pressure, and temperature [30] is given by

$$\rho = \frac{pM}{RT}, \tag{1.31}$$

where $\rho \equiv$ the mass density of the gas, $p \equiv$ the absolute pressure of the gas, $M \equiv$ the mean molecular weight of the atmospheric constituents, $T \equiv$ the absolute temperature, and $R \equiv$ the universal gas constant.

For common gases the pertinent properties and gas constants are given in Table 1.5. Combinations of the gases in Table 1.5 have molecular weights

Table 1.5 Gas Properties
(R = 1545 ft-lb/lb-mole °R)

Gas	Molecular Weight M (lb/lb-mole)	Specific Heat Ratio (Standard Temperature and Pressure—STP) γ
Air	28.97	1.40
Monatomic gases		
Argon (A)	39.94	1.67
Helium (He)	4.003	1.66
Diatomic gases		
Carbon monoxide (CO)	28.01	1.40
Hydrogen (H_2)	2.016	1.41
Nitrogen (N_2)	28.02	1.40
Oxygen (O_2)	32.00	1.40
Triatomic gases		
Carbon dioxide (CO_2)	44.01	1.30
Sulfur dioxide (SO_2)	64.07	1.26
Water vapor (H_2O)	18.016	1.33
Hydrocarbons		
Acetylene (C_2H_2)	26.04	1.26
Methane (CH_4)	16.04	1.32
Ethane (C_2H_6)	30.07	1.22
Isobutane (C_4H_{10})	58.12	1.11

which are obtained as a weighted mean on a volumetric basis of the constituents. Assuming ideal mixing for a mixture of, for example, three gases, a, b, and c, it follows that

$$M = M_a X_a + M_b X_b + M_c X_c, \tag{1.32}$$

where X is the volumetric percent of each species.

Use of the gas law requires a change of temperature units from either Fahrenheit or centigrade as follows:

$$\begin{aligned} {}^\circ\mathrm{F} &= 1.8^\circ\mathrm{C} + 32^\circ, \\ {}^\circ\mathrm{R}\ \text{(degrees Rankine)} &= t_\mathrm{F} + 460.0, \\ {}^\circ\mathrm{K}\ \text{(degrees Kelvin)} &= t_\mathrm{C} + 273.0, \end{aligned} \tag{1.33}$$

where the symbols F and C, respectively, denote Fahrenheit and centigrade. With these gas constants the units of the so called equation of state are

$$\rho\,(\text{slugs/ft}^3) = \frac{p(\text{lb/ft}^2)M(\text{lb/lb-mole})}{g_0(\text{lbm/slug})R(\text{ft-lb/lb-mole }^\circ\text{R})T(^\circ\text{R})}.$$

Illustrative Example

Assuming that the atmosphere of Mars is composed of 95% nitrogen and 5% oxygen (by volume), what is the drag force on a vertically rising rocket with a frontal area of 25 ft² ($C_D = 0.2$)? The velocity of the rocket is 2000 fps, whereas the absolute pressure is 300 lb/ft² with a corresponding temperature of 30°F.

In order to obtain the drag force, the perfect gas law must be applied to determine the density of the Martian atmosphere at the point in question. First, however, the molecular weight of the Martian atmosphere is obtained from Table 1.5 and the use of (1.32) as follows:

$$M_{♂} = 0.95 \times 28.02 + 0.05 \times 32.0 = 28.219.$$

The density can be obtained from (1.31) as

$$\rho = \frac{pM}{g_0 RT} = \frac{28.219}{32.174 \times 1545} \times \frac{300}{(30 + 460)} = 0.000348 \text{ slug/ft}^3.$$

At this point the familiar expression for the computation of the drag force magnitude ([1], Chapter 2), that is,

$$D = \tfrac{1}{2} C_D A \rho V^2 = 0.5 \times 0.2 \times 25.0 \times 0.000348 \times 2000^2 = 3.48 \times 10^3 \text{ lb}$$

provides the final answer.

1.6.2 Equation of Hydrostatic Equilibrium

A second fundamental relationship in the construction of planetary atmospheric models is called the equation of hydrostatic equilibrium. Consider an infinitesimal segment of a rectangular atmospheric column with unit cross-sectional area as illustrated in Figure 1.2. Since the area of the top and bottom of the column is unity, the pressure p is simply equal to the force acting vertically on the column, and the weight of the differential rectangular column is equal to $\rho \bar{g}\, dh$, where $\bar{g}$ is the average local acceleration of gravity. Hence for equilibrium a vertical summation of forces yields

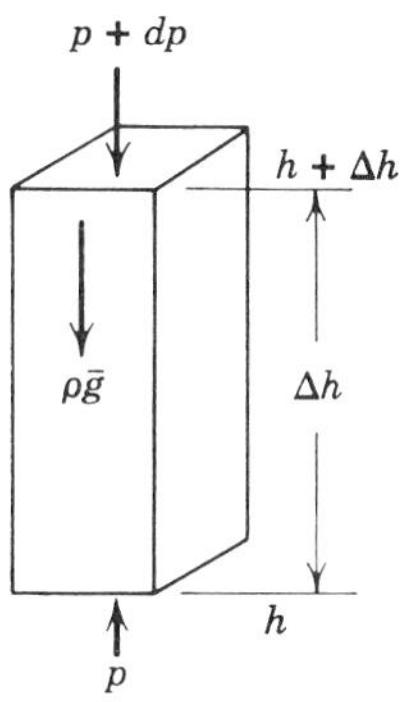

FIGURE 1.2 Differential column of atmosphere.

$$p - (p + \Delta p) = \rho \bar{g}\, \Delta h,$$

which reduces to

$$\Delta p = -\rho \bar{g}\, \Delta h,$$

or on passing to the limit it follows that

$$dp = -\rho g\, dh, \tag{1.34}$$

where $\bar{g}$ becomes the local acceleration of gravity, g, and h is the altitude.

1.6.3 Construction of a Static Atmospheric Model

The equation of state, that is, the perfect gas law and the equation of hydrostatic equilibrium, may be combined and utilized to define static atmospheric models. Consider that the local gravitational field is taken to act according to an inverse square force field so that the altitude variation of g can be described by

$$g = \frac{Gm}{(r_c + h)^2}, \tag{1.35}$$

where $G \equiv$ the universal gravitational constant, $m \equiv$ the central planetary mass, $r_c \equiv$ planetary radius at subpoint latitude and longitude, $h \equiv$ altitude measured along r_c above the datum or reference point on r_c. Exact expressions for r_c are derived in ([1], Chapter 1). Usually, however, it suffices to assume $r_c = a_p$, where a_p is the equatorial radius of the central planet.

Consider further that a new variable, commonly termed *molecular-scale temperature*, denoted T_m, is introduced by the relationship

$$T_m = \frac{M_0}{M} T, \tag{1.36}$$

where M_0 is defined by the surface or datum value of the atmospheric molecular weight. The molecular weight of a given planetary atmosphere is not constant for increasing altitude because of several factors such as photodissociation and diffusive separation.

Having the preceding relationships, it is possible to integrate the equation of hydrostatic equilibrium (1.34) by means of (1.31), that is,

$$dp = -\rho g\, dh = -\frac{pgM}{RT}\, dh. \tag{1.37}$$

Hence, by use of the molecular-scale temperature concept (1.36) and the local gravity relationship (1.35), it follows that

$$\int_{p_0}^{p_a} \frac{dp}{p} = -\int_{h_0}^{h_a} \frac{M_0 Gm}{RT_m(r_c + h)^2}\, dh = -\frac{M_0 Gm}{R} \int_{h_0}^{h_a} \frac{dh}{T_m(r_c + h)^2}. \tag{1.38}$$

The integrand between the limits of datum altitude h_0 and arbitrary altitude h_a is dependent on the molecular-scale temperature at altitude. To enable the integration of (1.38) to be performed, a temperature distribution must be determined or assumed.

The 1962 United States Standard Atmosphere and the 1959 ARDC Atmosphere are generated by assuming the following segmented linear distribution:

$$T_m = T_{m0j} + T_j^*(h - h_{0j}), \qquad j = 1, 2, \ldots, q, \tag{1.39}$$

where T_{m0j} = molecular-scale temperature at base altitude h_{0j}, T_j^* = constant temperature lapse rate or slope, and h_{0j} = initial base height of linear profile in question. The corresponding slopes and domain of the linear segments between altitudes $h_0, h_1, \ldots, h_q$ may be obtained from [15], [22].

Equation 1.38 can, therefore, be written

$$\int_{p_0}^{p_a} \frac{dp}{p} = \omega^* \int_{h_0}^{h_a} \frac{dh}{[T_{m0j} + T_j^*(h - h_{0j})][r_c + h]^2}, \tag{1.40}$$

which on integration becomes

$$p = p_0 \lambda^\alpha e^{-\beta}, \tag{1.41}$$

where

$$\omega^* \equiv \frac{M_0 Gm}{R},$$

$$\lambda \equiv \left[\frac{T_{m0j}}{r_c + h_{0j}}\right]\left[\frac{r_c + h}{T_{m0j} + T_j^*(h + h_{0j})}\right],$$

$$\alpha \equiv \frac{T_j^* \omega^*}{[T_{m0j} - T_j^*(r_c + h_{0j})]^2},$$

$$\beta \equiv \frac{\omega^*(h - h_{0j})}{[T_{m0j} - T_j^*(r_c + h_{0j})][r_c + h_{0j}][r_c + h]}.$$

If, under the preceding analysis, a polygonal molecular-scale temperature profile is assumed, it is possible to determine the atmospheric pressure directly as a function of h or altitude above the planetary surface in question. Several of these profiles for planetary bodies of interest are presented graphically in [11], [12].

At certain times it is also of interest to determine the auxiliary derivatives

$$\frac{dp}{dh} = -\frac{p\omega^*}{[r_c + h]^2[T_{m0j} + T^*(h - h_{0j})]}, \tag{1.42}$$

$$\frac{d\rho}{dh} = -\frac{pM_0}{R[T_{m0j} + T^*(h - h_{0j})]^2}\left[\frac{\omega^*}{[r_c + h]^2} + T_j^*\right], \tag{1.43}$$

and to compute the auxiliary parameter a, the speed of sound, which is approximately given by

$$a = \left[\gamma \frac{R}{M} T\right]^{1/2}, \tag{1.44}$$

where $T = (M/M_0)T_m$ with γ, the specific heat ratio of the planetary atmosphere, obtained from Table 1.5.

1.6.4 Time-Varying Models

Atmospheric variations are caused by many factors. Deviations from models, such as those of Section 1.6.3, have been pointed out by Gabbard [14].

LATITUDINAL VARIATION. A latitude variation enters through (1.35) in that r_c is actually a function of latitude. This effect, however, may be essentially eliminated by assuming a reference spheroid which accounts for the variation of r_c with latitude.

DIURNAL VARIATION. This variation is due to the heating effect of solar radiation. As the result of heating, the atmosphere bulges out in the general direction of the Sun but lags the Sun because of the planet's rotation.

SEMIDIURNAL VARIATION. This variation is due to solar and lunar tidal forces.

ANNUAL VARIATION. Even under the somewhat erroneous assumption that the planet moves about the Sun in a perfect circle, there would still be an annual variation due to the obliquity of the ecliptic, or relative inclination of the planet's equator and ecliptic planes. This is a latitude effect in that its magnitude is latitude dependent.

SEMIANNUAL VARIATION. Like the semiannual variation of geomagnetic activity, this variation probably is due to charged, corpuscular radiation.

MAGNETIC STORM VARIATION. This variation is a worldwide density increase which is correlated with magnetic storms, again indicative of a particular phenomenon.

SOLAR ACTIVITY VARIATION. This variation is strongly correlated with the solar flux in the decimeter (3 to 30 cm) wavelength range. This decimeter flux is thought not to be the cause but rather the index of the real cause which, in turn, is thought to be a variation in extreme ultraviolet solar radiation due to corona condensations.

SOLAR CYCLE VARIATION. Variations associated with solar activity show an 11-year solar cycle.

Most of these variations become important at altitudes generally beyond 100 nautical miles; they have little or no effect below this level. Undoubtedly, as time progresses, the list will be lengthened. For the moment these variations are sufficient to illustrate the types of phenomena that must be accounted for by proposed time-varying models. Model atmospheres are therefore of

Table 1.6 Model Atmospheres

ARDC, 1956	Static Model [15]
CIRA, 1961	Static Model [16]
U.S. Standard, 1962	Static Model [19]
Jacchia, 1960	Time-Varying Model [17]
Nicolet, 1961	Time-Varying Model [20]
Harris and Priester, 1962	Time-Varying Model [18]
Jacchia, 1964	Time-Varying Model [28]

two distinct types, static or time-varying, depending on the degree to which they attempt to represent variations in the atmosphere. The static models, such as those of Section 1.6.3, attempt to describe an "average" atmosphere which represents a mean of the diurnal, seasonal, solar activity, etc., variations described earlier. Inclusive of these variations, the time-varying models attempt to predict, to within a matter of minutes of time, the structure of the atmosphere. Present attempts to do so, however, result in densities accurate to only about a factor of 2 or so in describing order of magnitude changes of density at altitudes. For more detailed information about the concepts so far introduced see [17], [18], [20]. A collection of atmospheric models is tabulated in [26]. Some of the more popular models are listed in Table 1.6.

In the following sections two time-varying models for the Earth's atmosphere are presented. Because much of the original work in this field was pioneered by Jacchia [17], [27], [28], his original simplified model is presented first and a more exact version is described at a later stage.

Before these time-varying, or as they are also called *dynamic models*, are analyzed, an important parameter which is required as input is discussed. The parameter F_{10} is called the *solar flux*. Apparently the smoothed monthly means of the decimetric flux are correlated with the long term 11-year solar cycle. These smoothed monthly means, denoted $\bar{F}_{10}$, are used as an index of the actual variation. When the solar cycle is active, $\bar{F}_{10} \simeq 220$, in contrast to $\bar{F}_{10} \simeq 70$ for periods of low solar cycle activity. In order to determine the density of the atmosphere it is usually required to have the parameters F_{10} and $\bar{F}_{10}$ available. An estimate of the solar flux is usually provided by an examination of the solar flux variation, that is, the maxima and minima of F_{10}, over the previous year. Usually an extrapolation can be made to fair accuracy, or an average value $\bar{F}_{10}$ is used if no other information is available.[4]

1.6.5 Jacchia 1960 Model Atmosphere

The search for an analytic model which could be used to predict the density of the atmosphere and thus represent high altitude artificial satellite data led to the development of the Jacchia 1960 model. The two main effects outlined in Section 1.6.4 are included in the simplified model. These effects are the *diurnal* or daily variation and the 27-day solar effect. If it is assumed that the diurnal atmospheric bulge is axially symmetric, has the same geocentric latitude as the subsolar coordinates, and lags the Sun in longitude by a constant phase angle, it is possible to determine the atmospheric density by the following procedure. Compute the base density ρ_0 (gm/cm³) as a function of H_c (km), the altitude above the surface of the Earth, from the empirical relationship

$$\log_{10} \rho_0 = -16.021 - 0.001985 H_c + 6.363 \exp(-0.0026 H_c). \quad (1.45)$$

Obtain the parameter of the model ψ', the angle between the maximum of the diurnal bulge and the point in the atmosphere at which the density is desired, by means of spherical trigonometry:

$$\cos \psi' = \sin \delta \sin \delta_{\odot} + \cos \delta \cos \delta_{\odot} \cos (\alpha - \alpha_{\odot} - \lambda), \quad (1.46)$$

where $\alpha \equiv$ right ascension of point where the density is to be determined, $\delta \equiv$ declination of point where the density is to be determined, $\alpha_{\odot} \equiv$ right ascension of the Sun, $\delta_{\odot} \equiv$ declination of the Sun, and $\lambda \equiv$ constant atmospheric lag angle. It usually suffices to take λ between 25° and 30° [26].

The right ascension and declination of the Sun can be obtained adequately from the mean planetary coordinates of the Earth, Section 1.3, as

$$\sin \delta_{\odot} = \frac{-z_{\oplus}}{[x_{\oplus}^2 + y_{\oplus}^2 + z_{\oplus}^2]^{1/2}}, \qquad -\frac{\pi}{2} \leq \delta_{\odot} \leq \frac{\pi}{2}, \quad (1.47)$$

$$\cos \alpha_{\odot} = \frac{-x_{\oplus}}{[x_{\oplus}^2 + y_{\oplus}^2]^{1/2}}, \qquad \sin \alpha_{\odot} = \frac{-y_{\oplus}}{[x_{\oplus}^2 + y_{\oplus}^2]^{1/2}}, \quad (1.48)$$

where $0 \leq \alpha_{\odot} \leq 2\pi$.

[4] See the last paragraph of Section 1.6.6 for a source of these parameters.

An auxiliary function of ψ' must now be determined from empirical considerations, that is,

$$f(\psi') = \cos^n \frac{\psi'}{2} = \left(\frac{1 + \cos \psi'}{2}\right)^{n/2}, \tag{1.49}$$

where satellite data have been analyzed to obtain $n = 6$ [26].

Finally the actual density is obtained from

$$\rho = \rho_0 \frac{F_{10}}{100} \{1 + 0.19[\exp (0.0055H_c) - 1.9]f(\psi')\}, \tag{1.50}$$

where $\rho \equiv$ atmospheric density, gm/cm³, and $F_{10} \equiv$ solar flux, 10^{-22} W/m²/cps.

The Jacchia 1960 model, derived by semiempirical methods, has been used with a fair amount of success for the prediction of upper atmospheric densities. It should be used at altitudes above 100 km instead of the previously discussed static model of Section 1.6.3.

1.6.6 Jacchia 1964 Model

A more sophisticated model for the determination of atmospheric density was derived by Jacchia in 1964. This model attempts to account for the previously included effects, along with the semiannual and geomagnetic effects discussed in Section 1.6.4. According to this empirical model, the nighttime temperature T_0 is first obtained from the empirical relationship

$$\begin{aligned} T_0 = {} & 418° + 3°.60\bar{F}_{10} + 1°.8(F_{10} - \bar{F}_{10}) \\ & + \left[0.37 + 0.14 \sin \left(2\pi \frac{t - 151}{365}\right)\right]\bar{F}_{10} \sin \left(4\pi \frac{t - 59}{365}\right), \end{aligned} \tag{1.51}$$

where $\bar{F}_{10}$ = average solar flux, 10^{-22} W/m²/cps, F_{10} = solar flux, 10^{-22} W/m²/cps, and t = days elapsed from January 1. Next, in terms of the auxiliary parameters

$$\begin{aligned} \eta &\equiv \tfrac{1}{2}(\phi' - \delta_\odot), \\ \theta &\equiv \tfrac{1}{2}(\phi' - \delta_\odot), \\ \tau &\equiv HA - 45° + 12° \sin (HA + 45°), \qquad -\pi \le \tau \le \pi, \end{aligned} \tag{1.52}$$

where $\phi' \equiv$ geocentric latitude of point where the density is to be determined, $\delta_\odot \equiv$ declination of the Sun [see (1.47), (1.48)], and $HA \equiv$ hour angle of the Sun[5] ($HA \equiv 0$ at local noon), compute the *exospheric* temperature from

$$\begin{aligned} T_\infty = {} & T_0(1 + 0.28 \sin^{2.5} \theta)\left[1 + 0.28 \frac{\cos^{2.5} \eta - \sin^{2.5} \theta}{1 + 0.28 \sin^{2.5} \theta} \cos^{2.5} \frac{\tau}{2}\right] \\ & + 1.0a_p + 125[1 - \exp (-0.08a_p)] \end{aligned} \tag{1.53}$$

[5] See ([1], Chapter 4) for a further discussion of the hour angle.

as a function of T_0, θ, η, τ, and $a_p \equiv$ 3-hour planetary amplitude magnetic storm effect.

The average daily value of a_p is denoted A_p and is obtained by averaging the data from twelve observatories for different latitudes at 3-hour intervals. The value of A_p is expressed in units of 2γ ($\gamma = 10^{-5}$ gauss) and ranges between 10 and 20 during periods of average magnetic activity. During a period of magnetic activity, A_p can rise drastically. In fact, during a magnetic storm A_p can exceed 200.

In terms of T_∞ it is possible to determine the temperature T from

$$T = T_\infty - (T_\infty - 120) \exp(-s[H_c - 120]) \tag{1.54}$$

with

$$s \equiv 0.0291 \exp\left(-\frac{x^2}{2}\right),$$

$$x \equiv \frac{T_\infty - 800}{750 + 1.722 \times 10^{-4}(T_\infty - 800)^2}.$$

Two routes are now available for the computation of density, either entry into prepared temperature tables or numerical solution of the diffusion equation [26]:

$$\frac{dn_i}{n_i} = -\frac{dH_c}{H_i} - \frac{dT}{T}(1.0 + \alpha T), \tag{1.55}$$

where

$$H_i \equiv \frac{kT}{m_i g},$$

with $n_i \equiv$ concentration of each constituent of the atmosphere, $H_i \equiv$ scale height of individual constituent, $m_i \equiv$ molecular or atomic weight of each constituent, $H_c \equiv$ geometric altitude above planetary surface, $\alpha \equiv$ thermal diffusion constant, $k \equiv$ Boltzmann constant, and $g \equiv$ local acceleration of gravity.

Equations 1.55 can be solved with a set of adopted boundary conditions at, for example, 150 km, for the defined variables and the density obtained.

Other models representing the current state of the art are also available [26]. The two models discussed, however, are good representations of the time-varying atmosphere. For real-time simulation, current values of F_{10} and A_p may be obtained from the North Atlantic Radio Warning Service, Fort Belvoir, Virginia.

1.7 SUMMARY

A brief description of the planetary members of the solar system was introduced in narrative fashion. Approximate tables of the characteristics of the planets and natural satellites of the solar system were presented. Polynomial expansions of the mean elements of the major planets were tabulated along with a truncated form of Brown's lunar theory. These approximate techniques for obtaining positions of the planets and the Moon find much use in rendezvous and space targeting studies. The techniques may be very useful for self-contained position and velocity generators. Differential correction of the lunar coefficients in Brown's lunar theory was outlined. More exact planetary and lunar ephemeris determinations require the use of specially prepared magnetic tapes.

Knowledge of the position and velocity or trajectory state of a given planet leads naturally to the question concerned with planetary atmospheric models for the initial and final stages of space flight wherein a vehicle must pass through a dense medium. A general discussion of a typical static planetary atmospheric model was undertaken. The fluctuations and variances of this model from the true atmospheric state were described, and two dynamical atmospheric models for high altitude use were presented. The solar flux and geomagnetic indexes were discussed and their sources of availability for real-time simulation mentioned.

EXERCISES

1. Is the sum of all the masses of the solar system planets, exclusive of Jupiter, greater than, less than, or equal to the mass of Jupiter?
2. What is the approximate central angle between Mars and Earth on August 24, 1976?
3. How can the use of repeated trigonometric functions be avoided in predicting the position and velocity of the Moon?
4. Develop a differential correction scheme for correcting the coefficients of the classical Brown's lunar theory expansions in longitude, ecliptic declination, and parallax, based on angular observations of the Moon.
5. Spaceship Omega X-2 is moving through space ($V = 40{,}000$ fps) in the proximity of the planet Jupiter. An emergency landing must be made as soon as possible on Callisto, which is now on the opposite side of Jupiter and is not visible to the spaceship navigator. Since the rescue operation of spaceship Omega X-1, which has crashed on the surface of Callisto, must be accomplished in minimum time, a nearly straight line trajectory must be flown from the present position of Omega X-2 to Callisto. This orbit will unfortunately traverse the Jovian atmosphere. If at the point of closest approach to Jupiter the external atmospheric temperature is obtained from the navigator's tables as $-200°$F with an absolute pressure of 5000 lb/ft^2 and the maximum allowable axial drag force that Omega X-2 can safely experience is 20,000 lb, can the minimum time dip

trajectory be executed? The reference area of Omega X-2 is 100 ft^2 with an associated $C_D = 0.2$. The navigator's tables show that the Jovian atmosphere is composed of 62% methane and 38% hydrogen.

6. How long does it take light to travel from Earth to Saturn on January 1, 1984? Assume that c, the velocity of light, is 299,729.5 km/sec and that 1 a.u. = 149,597,850 km.

7. What is the atmospheric temperature on Earth at 10,000 ft if the sea level temperature is 59°F and the temperature lapse rate $T^* = 4°/1000$ ft? Assume a static atmospheric model. What is the speed of sound at this altitude?

8. Fort Belvoir, Virginia, reports that the solar flux $F_{10} = 100$ on midnight August 24, 1970. Assuming the atmospheric lag angle $\lambda = 28°$, what is the atmospheric density at this time at an altitude of 100 km? Use the Jacchia 1960 model.

9. What was the temperature at an altitude of 200 km if the values of the solar flux are $F_{10} = 90$ and $\bar{F}_{10} = 122$ on midnight, January 22, 1968?

10. If Kepler's equation is analytically inverted into a series expansion in the orbital eccentricity of a given planet, to what order of the eccentricity must the expansion be carried to permit all the planet's eccentric anomalies to be determined in closed form to eight decimal places?

REFERENCES

[1] P. R. Escobal, *Methods of Orbit Determination*, John Wiley and Sons, New York, 1965.

[2] *Explanatory Supplement to the Astronomical Ephemeris and the American Ephemeris and Nautical Almanac*, Her Majesty's Stationery Office, London, 1961.

[3] *Connaissance des Temps*, Le Bureau des Longitudes, Paris, France, 1965.

[4] *Jet Propulsion Laboratories Ephemeris Tapes*, E9510, E9511, and E9512.

[5] *Users' Description of Jet Propulsion Laboratories Ephemeris Tapes*, Jet Propulsion Laboratories Technical Report 32–580, March 2, 1964.

[6] Jet Propulsion Laboratories Memorandum 33–167, March 2, 1964.

[7] R. M. L. Baker, Jr., and M. W. Makemson, *An Introduction to Astrodynamics*, Academic Press, New York, 1960.

[8] G. W. Hill, *Collected Mathematical Works*, Carnegie Institute of Washington, Publication 9, 1964 reprint.

[9] E. W. Brown, "Motion of the Moon," *Memoirs of the Royal Astronomical Society*, Vol. 57, pp. 129–145.

[10] W. Ley and C. Bonestell, *The Conquest of Space*, Viking Press, New York, 1950.

[11] S. I. Rasool, "Structure of Planetary Atmospheres," *American Institute of Aeronautics and Astronautics Journal*, January 1963.

[12] D. P. Le Galley and A. Rosen, *Space Physics*, John Wiley and Sons, New York, 1964.

[13] G. P. Kuiper, *The Atmospheres of the Earth and Planets*, University of Chicago Press, 1952.

[14] T. P. Gabbard, *Earth-Atmosphere Models*, Lockheed Technical Memo, LTM 50275, Lockheed California Co., December 1962.

[15] R. A. Minzner and W. S. Ripley, *The ARDC Model Atmosphere, 1956*, AFSG 86, GRD, AFCRC, December 1956.
[16] H. Kallmann-Bijl et al., *CIRA, 1961*, North-Holland Publishing Co., Amsterdam, 1961.
[17] L. G. Jacchia, "A Variable Atmospheric Density Model from Satellite Accelerations," *Journal of Geophysical Research*, Vol. 65, 1960.
[18] I. Harris and W. Priester, *Time Dependent Structure of the Upper Atmosphere*, Goddard Space Flight Center, NASA, 1962.
[19] N. Sissenwine, H. Wexler, and M. Dubin, *A Preview*, "*The U.S. Standard Atmosphere, 1962*," ARS Reprint 2678-62; Paper presented at the 17th annual meeting of the American Rocket Society, Los Angeles, California, November 1962.
[20] M. Nicolet, *Density of the Heterosphere Related to Temperature*, Special Report 75, Smithsonian Astrophysical Observatory, Cambridge, Mass.
[21] H. A. Everett, *Thermodynamics* D. Van Nostrand, New York, 1941.
[22] R. A. Minzner, K. S. W. Champion, and H. L. Pond, *The ARDC Model Atmosphere, 1959*, AFSG 115, GRD, AFCRC, August 1959.
[23] M. Nicolet, "Structure of the Thermosphere," *Planetary and Space Science*, Vol. 5, 1961.
[24] L. G. Jacchia, "A Working Model for the Upper Atmosphere," *Nature*, Vol. 192, 1961.
[25] M. W. Reed, *The Stars for Sam*, Harcourt, Brace, and World, New York, 1931.
[26] R. W. Bruce, *A Survey of Model Atmospheres Used in the Analysis of Satellite Orbits*, Aerospace Corporation TDR-409(5540-10)-2, April 1965.
[27] L. G. Jacchia, *Variations in the Earth's Upper Atmosphere as Revealed by Satellite Drag*, Smithsonian Astrophysical Observatory, Cambridge, Mass., December 1962.
[28] L. G. Jacchia, *The Temperature above the Thermopause*, Special Report 150, Smithsonian Astrophysical Observatory, Cambridge, Mass., April 1964.
[29] M. P. Francis, private communication, TRW Systems, 1965.
[30] D. A. Mooney, *Introduction to Thermodynamics and Heat Transfer*, Prentice-Hall, Englewood Cliffs, N.J., 1955.
[31] N. Block, Jet Propulsion Laboratories, private communication, 1965.
[32] F. E. Ross, *Astronomical Papers*, Vol. IV, Part II.
[33] E. C. Bower, Orbit XIX, *Lick Observatory Bulletin*, No. 437.
[34] D. M. Cole and D. W. Cox, *Islands in Space*, Chilton Books, Philadelphia, 1964.

2 Optimization Techniques

Mathematics is the Queen of the Sciences.

GAUSS [6]

2.1 OPTIMIZATION PROBLEMS OF TRAJECTORY MECHANICS

Suppose that the life support system of a space vehicle is faltering. What is the fastest trajectory that will return the space vehicle to a safe base? How will it be possible to avoid over-the-Sun interplanetary one-impulse trajectories via the addition of a secondary impulse? Will it be possible to determine an optimum one- or two-impulse minimum fuel reentry trajectory? What is the point of closest approach of two orbits? These and other important questions require the use of specialized mathematical techniques. Specifically, from an impulse point of view these problems can be solved by use of the theory of maxima and minima of ordinary functions.

The purpose of this chapter is to discuss the techniques available for the determination of the *stationary values* of ordinary algebraic functions. By stationary values is meant that the *objective function* or function under investigation attains a maximum, minimum, or saddle point for these values.

In order to be complete, the theory includes the well-known results for obtaining the stationary values of functions of one variable and is later extended to include several variables subject to equational and inequality constraints. The problems chosen to illustrate the techniques are simplified from the astrodynamical point of view in order not to cloud the mathematical manipulations. Physically realistic trajectory problems, that is, some of the problems previously mentioned, are treated in more detail in later chapters.

2.2 MAXIMA AND MINIMA OF ORDINARY FUNCTIONS

Consider a mythical kingdom with the following simple terrain. The boundaries of the kingdom contain two hills and a valley. In the mathematical sense, hills refer to maxima; valleys, to minima. Of the two hills contained in this hypothetical kingdom one is probably of higher elevation than the other. Within the country's boundaries the higher hill is referred to as the *global maximum*; the lower hill, as the *local maximum*; and the valley as the *global minimum*. Between the hill and the valley exists a plateau or flattened region. In the mapping from the physical to the abstract world of mathematics, plateaus refer to saddle points or regions for which no true maximum or minimum is evident. The similar characteristic of the phenomena of maxima, minima, and inflexion, as may have been noted, is the horizontal position attained by a plane sliding over the domain of the kingdom. If the boundaries of the kingdom are extended, it is possible that other local maxima, minima, and inflexions will be found; perhaps a new global maxima or minima will be found.

In the language of mathematics the characteristic of a function f of n variables $(x_1, x_2, x_3, \ldots, x_n)$, achieving a maximum on the position specified n-tuple $(\xi_1, \xi_2, \xi_3, \ldots, \xi_n)$, can be defined for a sufficiently small neighborhood about $(\xi_1, \xi_2, \xi_3, \ldots, \xi_n)$ by

$$f(\xi_1, \xi_2, \xi_3, \ldots, \xi_n) > f(x_1, x_2, x_3, \ldots, x_n), \tag{2.1}$$

for all ξ_i in a neighborhood of x_i, $i = 1, 2, \ldots, n$. A minimum is, in opposite fashion, characterized by

$$f(\xi_1, \xi_2, \xi_3, \ldots, \xi_n) < f(x_1, x_2, x_3, \ldots, x_n), \tag{2.2}$$

for all ξ_i in a neighborhood of x_i, $i = 1, 2, \ldots, n$. If equality holds in either (2.1) or (2.2) for a sufficiently small region, the saddle point situation is existent.

Maxima, minima, and inflexions of a function f are referred to collectively as the *stationary values* of this function. Maxima and minima alone are denoted collectively as the *extreme values* of this function.

The theory about to be developed assumes that within the domain in question the partial derivatives of f are *continuous*. By continuity, roughly speaking,[1] is implied that, for *all* points $x_1, x_2, x_3, \ldots, x_n$ near $\xi_1, \xi_2, \xi_3, \ldots, \xi_n$, the value of a function, $g(x_1, x_2, x_3, \ldots, x_n)$, in this case the partial derivatives of f, differs little from $g(\xi_1, \xi_2, \xi_3, \ldots, \xi_n)$, that is, $|g(\xi_i) - g(x_i)|$ is smaller than any arbitrary positive number ϵ.

2.2.1 Functions of a Single Variable

Consideration of the extreme values of a function of one variable has certain merits and also allows determination of the necessary conditions for obtaining

[1] For an exact definition see [2].

the extreme values of functions of n variables. By inspection, examination of Figure 2.1 provides the necessary condition

$$\left.\frac{d}{dx} f(x)\right|_{x=\xi} = 0 \qquad (2.3)$$

for the occurrence of extreme values. This certainly follows if it is remembered that the geometric definition of the derivative of a function, that is, $(d/dx) f(x)$ defines the slope of the curve $y = f(x)$. Because the slope of the curve changes sign in going over a crest, it is intuitively appealing to accept (2.3) as the definition for the extrema of a function.

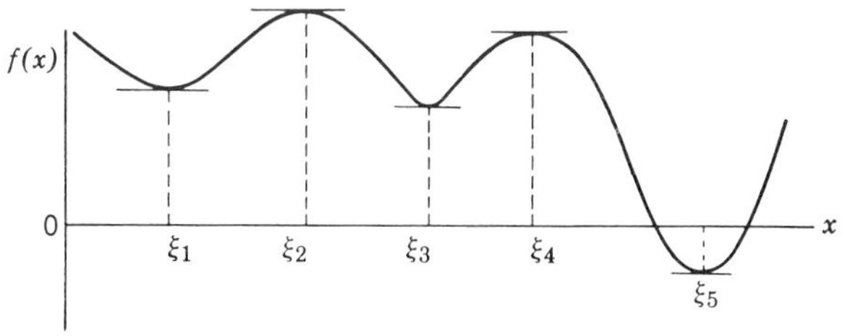

FIGURE 2.1 The maxima and minima of a function of one variable.

To be more explicit, consider a function f which is continuous and has a continuous derivative df/dx which vanishes at a finite number of points. Note that in Figure 2.1, to the left and right of $\xi = \xi_2$, intervals exist extending to the nearest points at which $(d/dx)f(x) = 0$, namely, $\xi_1 < x < \xi$ and $\xi < x < \xi_3$, in each of which $(d/dx)f(x)$ has but one sign. Assuming that the signs of $(d/dx)f(x)$ are different in these two intervals, it follows from the mean value theorem [2], that is,

$$f(\xi + \Delta\xi) - f(\xi) = \Delta\xi \frac{d}{dx} f(\xi + \theta\,\Delta\xi), \qquad (2.4)$$

where θ is a number between 0 and 1, that (2.4) has the same sign for all numerically small values of $\Delta\xi$, regardless of whether $\Delta\xi$ is positive or negative, so that $f(\xi)$ is an extreme value. When $(d/dx)f(x)$ is possessed of the same sign in both intervals, the right side of (2.4) changes sign when $\Delta\xi$ does; this states that $f(\xi + \Delta\xi) > f(\xi)$ on one side and $f(\xi + \Delta\xi) < f(\xi)$ on the other side. Hence there is no extreme value.

It is now possible to state that $f(x)$ has an extreme value at $x = \xi_2$ if and only if $(d/dx)f(x)$ changes sign on passing over this point; furthermore, if the derivative is positive to the left of ξ_2 and negative to the right, the function has a relative maximum, and a reversal of this case connotes a relative minimum.

Illustrative Example

An observer on Earth observes comet Encke in a near parabolic orbit sufficiently defined by the equation

$$y = ax^2 + bx + c \tag{2.5}$$

relative to the coordinate system of Figure 2.2. Assuming that the Earth is fixed, what is the point of closest approach of comet Encke to the Earth?

The point of closest approach is defined by the minimum of the distance between the comet trajectory and the Earth, that is, the minimum of

$$r^2 = x^2 + y^2. \tag{2.6}$$

Minimizing the square of the distance is equivalent to minimizing the

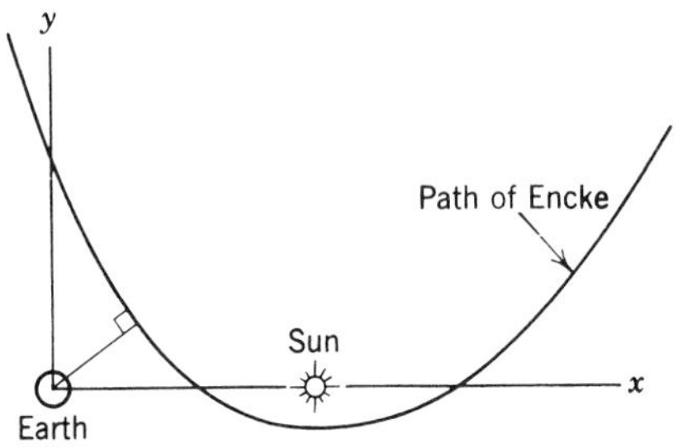

FIGURE 2.2 A parabolic orbit.

distance. Hence it is convenient to minimize r^2 instead of r. Since (2.5) defines y, (2.6) becomes

$$r^2 = x^2 + (ax^2 + bx + c)^2$$

or

$$r^2 = a^2x^4 + 2abx^3 + (b^2 + 2ac + 1)x^2 + 2bcx + c^2. \tag{2.7}$$

Hence differentiation and application of (2.3) yields the cubic form

$$\frac{dr^2}{dx} = 4a^2x^3 + 6abx^2 + 2(b^2 + 2ac + 1)x + 2bc = 0$$

or

$$2a^2\xi^3 + 3ab\xi^2 + (b^2 + 2ac + 1)\xi + bc = 0, \tag{2.8}$$

where $x = \xi$ is the value of x yielding an extreme situation. It follows that solution of the cubic form (2.8) ([1], Appendix 3) yields ξ and that substitution of ξ into (2.5) determines $y(\xi)$, so that an extreme point is $p = p[\xi, y(\xi)]$. The physics of the situation implies that the cubic has only one finite real root. In this case the determination of the true nature of p, that is, whether it is a maximum or minimum, can be very easily gleaned from the physics of the situation because a maximum occurs at infinity.[2] Hence p is a minimum. The actual test, that is, checking the sign change of the derivative, however, can still be applied to prove rigorously that p is a minimum. Furthermore, r^2 itself can be computed for small variations of x about ξ and criteria (2.2) applied as a secondary or direct check also to prove that ξ is the true global minimum of the problem. This particular example is examined in more detail in Section 2.3.

The process of deciding whether a maximum or minimum has been attained on the position $x = \xi$ in the interval $\xi_1 < \xi < \xi_2$ can sometimes be conveniently aided by the following alternative device. Assuming that both ξ and $\xi + \Delta\xi$ are contained in the interval in question, the mean value theorem of the differential calculus provides the relationship

$$\frac{d}{dx} f(\xi + \Delta\xi) - \frac{d}{dx} f(\xi) = \Delta\xi \frac{d^2}{dx^2} f(\xi + \theta\,\Delta\xi), \tag{2.9}$$

under the assumption that f is twice differentiable. Apparently, at the point $x = \xi + \Delta\xi$, $(d/dx)f(x)$ is possessed of the same sign as $\Delta\xi$ if $(d^2/dx^2)f(x) > 0$, and, in opposite fashion, when $(d^2/dx^2)f(x) < 0$ the derivative has the opposite sign. It therefore follows from this and the preceding reasoning that $(d^2/dx^2)f(x) > 0$ at the point of extremum corresponds to a minimum value and $(d^2/dx^2)f(x) < 0$ corresponds to a maximum value. The conditions for the maximum or minimum of a function of one variable are collected in Table 2.1.

Table 2.1 The Conditions Characterizing Maxima and Minima of a Function of One Variable

	Necessary Condition	Sufficient Condition	Sufficient Condition
Maxima	$\left.\frac{d}{dx} f(x)\right\|_{x=\xi} = 0$	$\left.\frac{d^2}{dx^2} f(x)\right\|_{x=\xi} < 0$	$\frac{d}{dx} f(x) > 0, \quad x < \xi$ $\frac{d}{dx} f(x) < 0, \quad x > \xi$
Minima	$\left.\frac{d}{dx} f(x)\right\|_{x=\xi} = 0$	$\left.\frac{d^2}{dx^2} f(x)\right\|_{x=\xi} > 0$	$\frac{d}{dx} f(x) < 0, \quad x < \xi$ $\frac{d}{dx} f(x) > 0, \quad x > \xi$

[2] Solutions occur at both $+\infty$ and $-\infty$.

2.2.2 Functions of *n* Variables

It is possible to use the knowledge gained from a study of the extreme values of a function of one variable to determine the procedure which should be used in determining the extreme values of a function of n variables. Consider that a function f of n variables is to be extremized, for example,

$$u = f(x_1, x_2, x_3, \ldots, x_n). \tag{2.10}$$

The variables x_1 through x_n can be parametrized by expanding them about the extreme point $(x_1, x_2, x_3, \ldots, x_n) = (\xi_1, \xi_2, \xi_3, \ldots, \xi_n)$ as

$$x_i = \xi_i + h_i t, \qquad i = 1, 2, \ldots, n, \tag{2.11}$$

where the h_i are arbitrary constants. It now follows that a function $U(t)$ can be written as a function of the single continuous parameter t, namely,

$$U(t) = f(\xi_1 + h_1 t, \xi_2 + h_2 t, \ldots, \xi_n + h_n t), \tag{2.12}$$

which for $t = 0$ has an extreme at the position $(\xi_1, \xi_2, \ldots, \xi_n)$. In other words, if $u = f(x_1, \ldots, x_n)$ is an extreme at the position $(\xi_1, \xi_2, \ldots, \xi_n)$, then $U(t)$ is extreme at the position $t = 0$.

On taking the derivative of U with respect to t it is possible to write

$$\frac{dU}{dt} = \frac{\partial f}{\partial x_1}\frac{dx_1}{dt} + \frac{\partial f}{\partial x_2}\frac{dx_2}{dt} + \cdots + \frac{\partial f}{\partial x_n}\frac{dx_n}{dt},$$

but

$$\frac{dx_i}{dt} = \frac{d}{dt}(\xi_i + h_i t) = h_i$$

so that

$$\frac{dU}{dt} = \frac{\partial f}{\partial x_1} h_1 + \frac{\partial f}{\partial x_2} h_2 + \cdots + \frac{\partial f}{\partial x_n} h_n,$$

where $\partial f/\partial x_i$ is evaluated at $x_i = \xi_i$. From the construction and the knowledge gained in the procedure for determining the extreme value of a function of a single variable it must be that, at $t = 0$,

$$\begin{aligned}\frac{d}{dt}U(0) = {} & \frac{\partial}{\partial x_1} f(\xi_1, \xi_2, \xi_3, \ldots, \xi_n) h_1 \\ & + \frac{\partial}{\partial x_2} f(\xi_1, \xi_2, \xi_3, \ldots, \xi_n) h_2 + \cdots \\ & + \frac{\partial}{\partial x_n} f(\xi_1, \xi_2, \xi_3, \ldots, \xi_n) h_n = 0.\end{aligned} \tag{2.13}$$

In order that (2.13) vanish identically for all arbitrary values of $h_1, h_2, \ldots, h_n$

it is necessary that

$$
\begin{aligned}
\frac{\partial}{\partial x_1} f(\xi_1, \xi_2, \xi_3, \ldots, \xi_n) &= 0, \\
\frac{\partial}{\partial x_2} f(\xi_1, \xi_2, \xi_3, \ldots, \xi_n) &= 0, \\
\vdots \qquad\qquad & \quad \vdots \\
\frac{\partial}{\partial x_n} f(\xi_1, \xi_2, \xi_3, \ldots, \xi_n) &= 0.
\end{aligned} \tag{2.14}
$$

Hence for functions of n variables a necessary condition for the extreme position is obtained by equating to zero the partial derivatives of the desired or objective function. In three space this corresponds to obtaining a horizontal position of a plane, for example, on the top of a hill, at the bottom of a valley, or at a saddle point.

The criterion for deciding whether the function is a maximum or minimum is provided directly by (2.1) and (2.2) or by the evaluation of the second derivative of (2.12), that is,

$$
\begin{aligned}
\frac{d^2}{dt^2} U(t) = \frac{\partial^2}{\partial x_1^{\,2}} f(x_1, x_2, x_3, \ldots, x_n) h_1^{\,2} + \frac{\partial^2}{\partial x_2^{\,2}} f(x_1, x_2, x_3, \ldots, x_n) h_2^{\,2} \\
+ 2 \frac{\partial^2}{\partial x_1\, \partial x_2} f(x_1, x_2, x_3, \ldots, x_n) h_1 h_2 + \cdots,
\end{aligned} \tag{2.15}
$$

which at $t = 0$, for all $h_1, h_2, \ldots, h_n$, by the rule for the extrema of one variable, must be negative for a maximum and positive for a minimum.

For the important case of a function of two variables, (2.15) can be abbreviated as

$$
\frac{d^2}{dt^2} U(t) = A h_1^{\,2} + 2B h_1 h_2 + C h_2^{\,2}, \tag{2.16}
$$

where the A, B, C are the appropriate partial derivatives. If h_1 is taken as αh_2 with arbitrary α and (2.16) is equated to zero, it follows that

$$
A\alpha^2 + 2B\alpha + C = 0.
$$

Hence, for the quadratic form (2.16) to change sign, $B^2 - AC < 0$. More explicitly, $[x_1, x_2] = [\xi_1, \xi_2]$ at the extremal producing point, if

$$
\left(\frac{\partial^2}{\partial x_1\, \partial x_2} f(\xi_1, \xi_2)\right)^2 - \frac{\partial^2}{\partial x_1^{\,2}} f(\xi_1, \xi_2) \frac{\partial^2}{\partial x_2^{\,2}} f(\xi_1, \xi_2) < 0; \tag{2.17}
$$

while a maximum exists if

$$\frac{\partial^2}{\partial x_1^{\,2}} f(\xi_1, \xi_2) < 0, \qquad \frac{\partial^2}{\partial x_2^{\,2}} f(\xi_1, \xi_2) < 0,$$

and a minimum exists if

$$\frac{\partial^2}{\partial x_1^{\,2}} f(\xi_1, \xi_2) > 0, \qquad \frac{\partial^2}{\partial x_2^{\,2}} f(\xi_1, \xi_2) > 0.$$

Failure of condition (2.17) [that is, if inequality (2.17) is greater than zero],

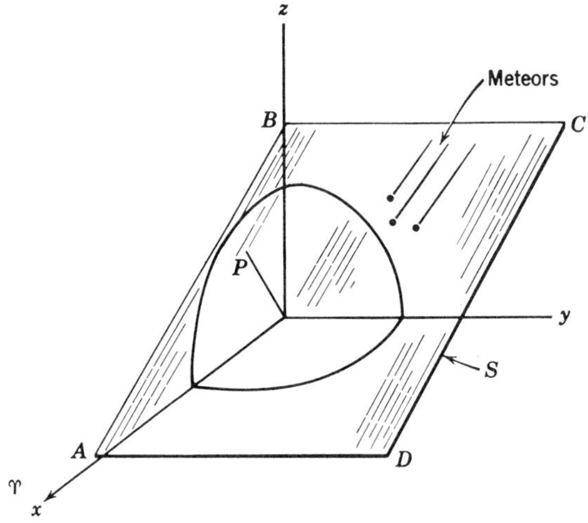

FIGURE 2.3 A meteoric swarm approaching Earth.

indicates that an extreme value has not been achieved. The case in which (2.17) vanishes is indeterminate.

Illustrative Example

A swarm of meteorites in the geocentric right ascension-declination system is moving in plane $ABCD$ of Figure 2.3. The plane of motion is defined by

$$ax + by + cz = d, \tag{2.18}$$

where the coefficients a, b, c are related to the classical angles i and Ω ([1],

Chapter 3, Exercise 14) for $i \leq \pi/2$ by

$$\Delta_1 = +[a^2 + b^2 + c^2]^{1/2}, \qquad \Delta_2 = +[a^2 + b^2]^{1/2}$$

and

$$\cos i = \frac{c}{\Delta_1}, \qquad \sin i = \frac{\Delta_2}{\Delta_1},$$

$$\cos \Omega = -\frac{b}{\Delta_2}, \qquad \sin \Omega = \frac{a}{\Delta_2}.$$

Will a meteoric shower be visible as the meteorites approach the Earth?

The visual effects of a meteoric shower are produced by the entrance of meteors into the atmosphere which, through friction, causes them to glow. The question of a metoric shower is therefore equivalent to determining the minimum distance r_m between the center of the Earth and the plane of motion of the meteorites. Certainly, if $r_m < a_e + H_a$, where a_e is the radius of the Earth and H_a is the depth of the atmosphere, a meteoric fireworks display is imminent. The square of the distance from the center of the Earth to arbitrary point P in plane S of Figure 2.3 is given by

$$r^2 = x^2 + y^2 + z^2. \tag{2.19}$$

Hence we seek to minimize r^2 subject to condition (2.18). Consider solution of (2.18) for one of the variables, z for example, such that

$$z = \frac{d}{c} - \frac{a}{c}x - \frac{b}{c}y. \tag{2.20}$$

It is now possible to make r^2 a function of two variables, x and y, by substituting (2.20) into (2.19) as follows:

$$u \equiv r^2 = x^2 + y^2 + \left(\frac{d}{c} - \frac{a}{c}x - \frac{b}{c}y\right)^2.$$

By applying (2.14) in two dimensions, the extreme values may be obtained through

$$\frac{\partial u}{\partial x} = 2x - 2\frac{a}{c}\left(\frac{d}{c} - \frac{a}{c}x - \frac{b}{c}y\right) = 0,$$

$$\frac{\partial u}{\partial y} = 2y - 2\frac{b}{c}\left(\frac{d}{c} - \frac{a}{c}x - \frac{b}{c}y\right) = 0, \tag{2.21}$$

or on simplifying:

$$\begin{aligned} (a^2 + c^2)x \qquad &+ aby = ad, \\ abx + (b^2 + c^2)y &= bd. \end{aligned} \tag{2.22}$$

Here x and y should really be denoted by ξ_1 and ξ_2, the extreme values of u, as soon as the partial derivatives are equated to zero. No confusion results, however, if this is kept in mind. The determinant of this linear system (2.22) becomes

$$\Delta = (a^2 + c^2)(b^2 + c^2) - a^2b^2 = (a^2 + b^2 + c^2)c^2,$$

and so, by Cramer's rule [5],

$$x = \frac{ad}{a^2 + b^2 + c^2}, \qquad y = \frac{bd}{a^2 + b^2 + c^2}$$

with z obtained from (2.20) as

$$z = \frac{d}{c} - \frac{dc}{\Delta}(a^2 + b^2) = \frac{cd}{a^2 + b^2 + c^2}.$$

The magnitude of r^2 can now be determined as the sum of squares of these coordinates and the meteoric shower function S written as

$$S = \frac{(ad)^2 + (bd)^2 + (cd)^2}{(a^2 + b^2 + c^2)^2} - (a_e + H_a)^2, \tag{2.23}$$

where $S > 0$ implies no visible meteoric shower, $S < 0$ implies a meteoric shower, and $S = 0$ denotes the limiting case. The physics of the situation implies that $\partial u/\partial x$, $\partial u/\partial y$ yield a minimum r, because the maximum is at infinity. The sufficient condition for a minimum can be established by a slow variance of x and y about the extreme values ξ_1 and ξ_2. This shows that, for all $[x, y] \neq [\xi_1, \xi_2]$, r^2 increases and hence $x = \xi_1$, $y = \xi_2$ is the true global minimum. An auxiliary check is also provided by applying condition (2.17), that is, a minimum will exist if and only if

$$\frac{\partial^2 u}{\partial x^2} = 2\left[1 + \left(\frac{a}{c}\right)^2\right] > 0,$$

$$\frac{\partial^2 u}{\partial y^2} = 2\left[1 + \left(\frac{b}{c}\right)^2\right] > 0,$$

$$\left(\frac{\partial^2 u}{\partial x\,\partial y}\right)^2 - \frac{\partial^2 u}{\partial x^2}\frac{\partial^2 u}{\partial y^2} = 4\left\{\frac{a^2b^2}{c^4} - \left[1 + \left(\frac{a}{c}\right)^2\right]\left[1 + \left(\frac{b}{c}\right)^2\right]\right\}$$

$$= -4\left\{1 + \left(\frac{b}{c}\right)^2 + \left(\frac{a}{c}\right)^2\right\} < 0.$$

The conditions characterizing this minimum are always true for all values of a, b, c, and d.

2.3 MAXIMA AND MINIMA SUBJECT TO AUXILIARY CONSTRAINTS

Another way to express a necessary condition for an extreme value of the function $f(x_1, x_2, x_3, \ldots, x_n)$ to occur on the position $\xi_1, \xi_2, \xi_3, \ldots, \xi_n$ is to say that the linear differential df must vanish; that is, since

$$df \equiv \frac{df}{dt}\,dt = 0,$$

it follows that

$$df = \frac{\partial f}{\partial x_1}\,dx_1 + \frac{\partial f}{\partial x_2}\,dx_2 + \frac{\partial f}{\partial x_3}\,dx_3 + \cdots + \frac{\partial f}{\partial x_n}\,dx_n = 0. \qquad (2.24)$$

This equation explicitly assumes that the variables $x_1, x_2, x_3, \ldots, x_n$ are independent. Suppose, however, that the variables $x_1, x_2, x_3, \ldots, x_n$ are related by the auxiliary equations

$$g_i(x_1, x_2, x_3, \ldots, x_k, \ldots, x_n) = 0, \qquad i = 1, 2, 3, \ldots, p \qquad (2.25)$$

so that the variables x_k $(k = p, \ldots, n)$ are actually related. The easy way to attempt a solution of this problem is to express $p < n$ of the variables by means of (2.25) in terms of the remaining n–p variables and to substitute their values into the objective function $f(x_1, x_2, x_3, \ldots, x_n)$. The function f is now reduced to

$$f(x_{p+1}, x_{p+2}, \ldots, x_n), \qquad (2.26)$$

where the variables $x_{p+1}, x_{p+2}, \ldots, x_n$ are now independent variables. The extreme values of f can now be obtained by the methods of Section 2.2.2.

Illustrative Example

In the illustrative example of Section 2.2.1 it was desired to obtain the minimum distance between the fictitious orbit of comet Encke and Earth. Explain explicitly how this was achieved.

The function that is to be minimized is

$$f \equiv r^2 = x^2 + y^2,$$

which is a function of the two independent variables x and y. Comet Encke, however, has the orbit defined by

$$g_1 \equiv y - ax^2 + bx + c = 0,$$

which is a relationship between x and y. The function g_1 is therefore the auxiliary constraint to which f must be subjected. Accordingly, g_1 is solved for one of the variables; in this case the simplest policy is to solve for y, so that

$$y = ax^2 + bx + c.$$

Substitution of y into the objective function f yields

$$f \equiv x^2 + (ax^2 + bx + c)^2;$$

thus f becomes a function of a single independent variable. Finally, f can be extremized as in Section 2.2.1. The procedure above, which was intuitively accepted in the illustrative example of Section 2.2.1, should be carefully examined. If the constraining equation, the g_1 functional constraint, can be solved for one of the variables, then f can be reduced to a function of independent variables by substitution. Also, in this example f is a function of the two variables x and y. Hence it is possible to impose only *one constraint*, that is, one substitution. If a second constraining equation g_2 were imposed, it would be impossible to minimize f because both variables would be predetermined; that is, in the example $n = 2$ and $p = 1$, so that $p < n$.

2.3.1 Equational Constraints

The procedure of reducing a given objective function to a relationship of strictly independent variables at times becomes intractable. This is said in light of the fact that there are many equations in which it is not possible or convenient explicitly to eliminate the dependent variables. To circumvent this problem the method of *undetermined multipliers* can be introduced. This method was developed by Lagrange and can be verified as follows. Consider a function of two variables $f(x_1, x_2)$ which is to be extremized subject to the constraint $g(x_1, x_2) = 0$. If it is assumed that $\partial g/\partial x_1$ and $\partial g/\partial x_2$ do not vanish at the extreme point, $[x_1, x_2] = [\xi_1, \xi_2]$, it follows that by the implicit function theorem it is possible to solve $g(x_1, x_2) = 0$ in the neighborhood of the extreme point to obtain $x_2 = h(x_1)$. With this understanding, $f(x_1, x_2)$ can be written $f\{x_1, h(x_1)\}$. Hence on differentiation, f must have an extremum at the point $x_1 = \xi_1$. However, evaluating total derivatives yields

$$\frac{df}{dx_1} = \frac{\partial f}{\partial x_1} + \frac{\partial f}{\partial x_2}\frac{dh}{dx_1} = 0,$$

and, from $g(x_1, x_2) \equiv 0$,

$$\frac{\partial g}{\partial x_1} + \frac{\partial g}{\partial x_2}\frac{dh}{dx_1} = 0,$$

which is satisfied by $x_2 = h(x_1)$. If the last equation is multiplied by

$$\lambda \equiv \frac{-\partial f/\partial x_2}{\partial g/\partial x_2}$$

and added to df/dx_1, that is, added to

$$\frac{\partial f}{\partial x_1} + \frac{\partial f}{\partial x_2}\frac{dh}{dx_1} = 0,$$

it is easy to verify that the one condition that must hold is

$$\frac{\partial f}{\partial x_1} + \lambda \frac{\partial g}{\partial x_1} = 0; \tag{2.27}$$

the second condition, the definition of λ, is

$$\frac{\partial f}{\partial x_2} + \lambda \frac{\partial g}{\partial x_2} = 0. \tag{2.28}$$

These two conditions are called Lagrange's rule for two variables. The use of this rule for obtaining the extreme values of a given function f, subject to the constraint $g = 0$, is to be discussed presently; however, it is instructive to discuss Lagrange's rule from a different point of view which will lend itself naturally for extension to more than two dimensions.

Consider the differentiation of constraining equations of the form (2.25) to express directly the differentials of the variables involved, that is

$$dg_i = \frac{\partial g_i}{\partial x_1} dx_1 + \frac{\partial g_i}{\partial x_2} dx_2 + \frac{\partial g_i}{\partial x_3} dx_3 + \cdots + \frac{\partial g_i}{\partial x_n} dx_n = 0, \qquad i = 1, 2, \ldots, p. \tag{2.29}$$

These differentials of the constraining equations relate the connection between the individual differentials involved in the problem at hand. For the sake of clarity, following the preceding discussion, let a function of two variables be considered, subject to a single constraining equation. From (2.24) and (2.29) it follows that

$$df = \frac{\partial f}{\partial x_1} dx_1 + \frac{\partial f}{\partial x_2} dx_2 = 0,$$

$$dg_1 = \frac{\partial g_1}{\partial x_1} dx_1 + \frac{\partial g_1}{\partial x_2} dx_2 = 0. \tag{2.30}$$

If the second equation, $dg_1 \equiv 0$, is multiplied by an arbitrary constant λ and added to $df = 0$, that is, to the differential of the objective function, the condition for optimality obviously remains unchanged. More specifically, by use of (2.30),

$$df + \lambda\, dg_1 = \left(\frac{\partial f}{\partial x_1} + \lambda \frac{\partial g_1}{\partial x_1}\right) dx_1 + \left(\frac{\partial f}{\partial x_2} + \lambda \frac{\partial g_1}{\partial x_2}\right) dx_2 = 0. \tag{2.31}$$

Hence, for any arbitrary values of dx_1 and dx_2, (2.31) holds true if

$$\frac{\partial f}{\partial x_1} + \lambda \frac{\partial g_1}{\partial x_1} = 0, \qquad \frac{\partial f}{\partial x_2} + \lambda \frac{\partial g_1}{\partial x_2} = 0. \tag{2.32}$$

The left side of (2.31) expresses the fact that conditions $df = 0$ and $dg_1 = 0$ can both be satisfied if an unknown value λ is found. Equations (2.32) are identical with the previously obtained relationships, (2.27) and (2.28), defining Lagrange's rule except that they were derived by using differential notation. A completely rigorous proof of this principle can be found in [2]. In an analogous manner, if two constraining equations are involved in the problem at hand,

$$df + \lambda_1\, dg_1 + \lambda_2\, dg_2 = 0, \tag{2.33}$$

which imply

$$\frac{\partial f}{\partial x_1} + \lambda_1 \frac{\partial g_1}{\partial x_1} + \lambda_2 \frac{\partial g_2}{\partial x_1} = 0,$$

$$\frac{\partial f}{\partial x_2} + \lambda_1 \frac{\partial g_1}{\partial x_2} + \lambda_2 \frac{\partial g_2}{\partial x_2} = 0,$$

$$\frac{\partial f}{\partial x_3} + \lambda_1 \frac{\partial g_1}{\partial x_3} + \lambda_2 \frac{\partial g_2}{\partial x_3} = 0. \tag{2.34}$$

The reader may be puzzled by the appearance of a third variable in set (2.34). It must be emphasized that, if two constraining equations are involved in a given physical problem, f must be a function of at least three variables, that is, $f = f(x_1, x_2, x_3)$, or no actual optimization or extremal problem exists since the variables are uniquely determined. This process can be generalized to n variables. With the previous introduction, Lagrange's rule can now be stated as follows.

If the extreme value of a function of n variables is sought, that is, the extreme value of the function

$$f(x_1, x_2, x_3, \ldots, x_n), \tag{2.35}$$

where the n variables are not all independent but are related by the p subsidiary conditions,

$$g_i(x_1, x_2, x_3, \ldots, x_n) = 0, \qquad i = 1, 2, 3, \ldots, p, \qquad p < n, \tag{2.36}$$

it is possible to construct the new function

$$M \equiv f + \lambda_1 g_1 + \lambda_2 g_2 + \lambda_3 g_3 + \cdots + \lambda_p g_p \tag{2.37}$$

and equate the partial derivatives $\partial M/\partial x_1$, $\partial M/\partial x_2$, $\partial M/\partial x_3, \ldots, \partial M/\partial x_n$ to zero in order to obtain the relationships

$$\begin{aligned}
\frac{\partial f}{\partial x_1} + \lambda_1 \frac{\partial g_1}{\partial x_1} + \lambda_2 \frac{\partial g_2}{\partial x_1} + \lambda_3 \frac{\partial g_3}{\partial x_1} + \cdots + \lambda_p \frac{\partial g_p}{\partial x_1} &= 0, \\
\frac{\partial f}{\partial x_2} + \lambda_1 \frac{\partial g_1}{\partial x_2} + \lambda_2 \frac{\partial g_2}{\partial x_2} + \lambda_3 \frac{\partial g_3}{\partial x_2} + \cdots + \lambda_p \frac{\partial g_p}{\partial x_2} &= 0, \\
\vdots \qquad\qquad\qquad\qquad\qquad & \vdots \\
\frac{\partial f}{\partial x_n} + \lambda_1 \frac{\partial g_1}{\partial x_n} + \lambda_2 \frac{\partial g_2}{\partial x_n} + \lambda_3 \frac{\partial g_3}{\partial x_n} + \cdots + \lambda_p \frac{\partial g_p}{\partial x_n} &= 0,
\end{aligned} \tag{2.38}$$

which, with constraining equations

$$\begin{aligned}
g_1(x_1, x_2, x_3, \ldots, x_n) &= 0, \\
g_2(x_1, x_2, x_3, \ldots, x_n) &= 0, \\
g_3(x_1, x_2, x_3, \ldots, x_n) &= 0, \\
\vdots \qquad\qquad & \vdots \\
g_p(x_1, x_2, x_3, \ldots, x_n) &= 0,
\end{aligned} \tag{2.39}$$

provide a system of $n + p$ equations in $n + p$ unknowns which determine the extreme values of the objective function f.

Illustrative Example

A simple illustration of Lagrange's rule can be brought forth by solving the previously discussed problem of comet Encke. It should be remembered that the minimum of the function

$$f = r^2 = x^2 + y^2$$

was sought subject to the constraint that

$$y - ax^2 - bx - c = 0. \tag{2.40}$$

Hence, by Lagrange's rule, the new function

$$M \equiv f + \lambda(y - ax^2 - bx - c) \tag{2.41}$$

is formed and the partial derivatives of M with respect to x and y are equated to zero, that is,

$$\frac{\partial M}{\partial x} = 2x + \lambda(-2ax - b) = 0, \tag{2.42}$$

$$\frac{\partial M}{\partial y} = 2y + \lambda \qquad\qquad = 0. \tag{2.43}$$

From (2.43) it follows that substitution of $\lambda = -2y$ into (2.42) yields

$$2x + 2y(2ax + b) = 0$$

or, on utilizing the constraint $g \equiv 0$, that is, (2.40),

$$x + (ax^2 + bx + c)(2ax + b) = 0.$$

Combining terms the preceding equation reduces to

$$2a^2x^3 + 3abx^2 + (b^2 + 2ac + 1)x + bc = 0. \tag{2.44}$$

Direct comparison shows that (2.44) and (2.8) are identical for $x = \xi$.

The power of this technique becomes even more evident if it is desired to minimize f subject to a more complicated constraint, for example,

$$g = y - x \sin xy = 0. \tag{2.45}$$

In this case the straightforward substitution technique is not very satisfactory owing to the transcendental nature of the problem. But, by Lagrange's rule,

$$M = f + \lambda g, \tag{2.46}$$

$$\frac{\partial M}{\partial x} = \frac{\partial f}{\partial x} + \lambda \frac{\partial g}{\partial x} = 0, \tag{2.47}$$

$$\frac{\partial M}{\partial y} = \frac{\partial f}{\partial y} + \lambda \frac{\partial g}{\partial y} = 0, \tag{2.48}$$

so that solving the last equation for λ and substituting into (2.47) yields

$$\frac{\partial f}{\partial x}\frac{\partial g}{\partial y} - \frac{\partial f}{\partial y}\frac{\partial g}{\partial x} = 0 \tag{2.49}$$

or

$$x + x(y^2 - x^2)\cos xy + y \sin xy = 0. \tag{2.50}$$

The extreme values of f can now be determined by iterative solution of (2.45) and (2.50) for x and y. This method is further employed in Chapter 3.

2.3.2 Inequality Constraints

Many problems in optimization theory have physical constraints that are not equations. Suppose, for example, that it is desired to find the fastest possible trajectory between two points subject to the condition that the acceleration on the pilot does not exceed certain safety limits. Or perhaps it is desired to obtain the fastest trajectory from one terminal to another subject to avoiding some object, perhaps a planet. Thus, as illustrated in Figure 2.4, trajectory A may be the fastest from terminal 1 to 2, but it requires that the pilot or astronaut traverse the planet; an unhealthy situation. This compressing state of affairs can be remedied by imposing the constraint that the

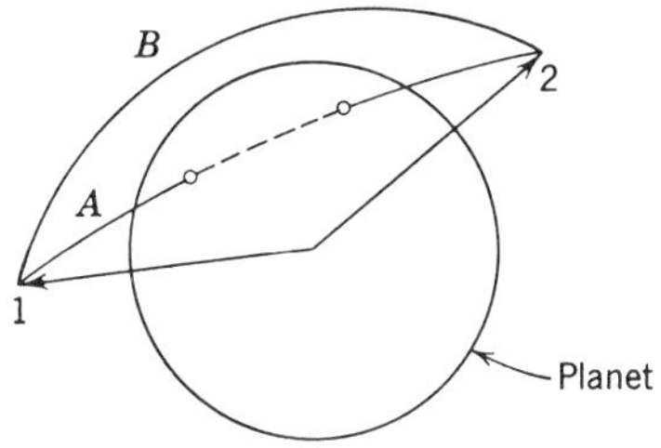

FIGURE 2.4 An invalid trajectory.

minimum radius of the trajectory be greater than some specified or limiting radius.

Therefore consider that

$$f(x_1, x_2, x_3, \ldots, x_n) \tag{2.51}$$

is to be maximized subject to the additional conditions

$$a_i < g_i^*(x_1, x_2, x_3, \ldots, x_n) < b_i, \tag{2.52}$$

where a_i and b_i are certain limits imposed by physical considerations. Since it is possible to express (2.52) equivalently as

$$\frac{g_i^* - a_i}{b_i - g_i^*} > 0, \qquad a_i < b_i, \tag{2.53}$$

it suffices to consider constraints of the form

$$g_i(x_1, x_2, x_3, \ldots, x_n) > 0,$$

where

$$g_i \equiv \frac{g_i^* - a_i}{b_i - g_i^*}, \qquad a_i < b_i. \tag{2.54}$$

Illustrative Example

If the objective function is

$$r^2 = x^2 + y^2 + z^2$$

and it is required that $r_1^2 < r^2 < r_2^2$, this condition can be represented by

$$\frac{r^2 - r_1^2}{r_2^2 - r^2} > 0$$

with $r_2^2 > r_1^2$. Notice that the two inequalities $r^2 > r_1^2$ and $r^2 < r_2^2$ have been reduced to one.

As it is difficult to deal with inequalities, the trick to finding the stationary values of (2.51), subject to inequality constraints of the form depicted by (2.54), is to convert the inequality constraints into equational constraints. This can be done at the expense of introducing more unknown variables into the physical problem. For example, if

$$x_1 > 0, \tag{2.55}$$

it certainly follows that

$$x_1 = \epsilon^2 > 0 \tag{2.56}$$

expresses the identical constraint; ϵ is now an unknown, however. In the same fashion, for some given number a, if

$$x_1 > a, \tag{2.57}$$

it follows that

$$x_1 - a = \delta^2 \tag{2.58}$$

is valid for all unknown values of δ.

In general, if certain inequality constraints are given, the inequality constraint can be written

$$g_i(x_1, x_2, x_3, \ldots, x_n) - a_i = \delta_i^2, \qquad i < n, \tag{2.59}$$

where the a_i are stated numbers and the δ_i are unknowns. In fact, it is possible to write

$$\begin{aligned} g_1(x_1, x_2, x_3, \ldots, x_n) - a_1 &= x_{n+1}^2, \\ g_2(x_1, x_2, x_3, \ldots, x_n) - a_2 &= x_{n+2}^2, \\ &\vdots \\ g_m(x_1, x_2, x_3, \ldots, x_n) - a_m &= x_{n+m}^2, \end{aligned} \tag{2.60}$$

or in the physical problem at hand the inequality constraints g_i can be replaced by equational constraints $\tilde{g}_i$ where

$$\tilde{g}_i(x_1, x_2, x_3, \ldots, x_n, x_{n+1}, x_{n+2}, x_{n+3}, \ldots, x_{n+i}) = 0 \tag{2.61}$$

with

$$\tilde{g}_i \equiv g_i - a_i - x_{n+i}^2, \qquad i = 1, 2, \ldots, m. \tag{2.62}$$

Illustrative Example

Derive the fundamental form of the equations to minimize the time interval τ between a given radius vector and a second radius vector terminal with a limited amount of fuel or velocity increment.

Suppose an expression for the time interval τ between two radius vectors is available from [1], Chapter 3. The function to be minimized in this case is the time interval. If the fuel or velocity increment that can be applied at the initial terminal is propulsion limited by velocity budget, ΔV_0, so that

$$\Delta V^2 < \Delta V_0^2, \tag{2.63}$$

it follows from the discussion of this section that

$$\Delta V^2 - \Delta V_0^2 = \epsilon^2, \tag{2.64}$$

where ϵ is an unknown variable. Hence inequality (2.63) has the counterpart equation given by (2.64).

It is now possible to form the objective M according to Lagrange's rule as follows:

$$M = \tau + \lambda(\Delta V^2 - \Delta V_0^2 - \epsilon^2). \tag{2.65}$$

The equations defining the stationary or minimum time trajectory become

$$\frac{\partial M}{\partial x_i} = \frac{\partial \tau}{\partial x_i} + \lambda \frac{\partial \Delta V^2}{\partial x_i} = 0, \tag{2.66}$$

$$\frac{\partial M}{\partial \epsilon} = -2\lambda\epsilon = 0, \tag{2.67}$$

where x_i are the parameters of the problem. The second equation states that either $\lambda = 0$, which must be rejected as this would eliminate the constraint, or $\epsilon = 0$, which implies that

$$\Delta V^2 = \Delta V_0^2. \tag{2.68}$$

Therefore, for a minimum time one-impulse trajectory, all the available impulse must be applied at the initiation of the maneuver. Notice the importance of this constraint. If this problem were solved without applying (2.63), it should be intuitively obvious that the minimum time trajectory would be a straight line trajectory with an infinite impulse applied at the first terminal. Equation (2.66) cannot actually be used because two-position vectors and (2.68) account for the seven degrees of freedom inherent in this problem.

2.4 APPLICATION: THE POINT OF CLOSEST APPROACH OF TWO ORBITS

2.4.1 Discussion of the Problem

The problem of determining the point of closest approach of two orbits has two important applications. In planetary theory the point of closest approach is of interest for visual observations of planetary surfaces. On the other hand, for planetocentric orbits the point of nearest approach has interest not only for visual observation but also for rendezvous considerations. In this section are developed the formulas which determine the point of closest approach between the Earth and an asteroid with known elements. The analysis is, of course, valid for any two bodies whose orbital elements are known.

2.4.2 Minimization Process

If the orbit plane coordinate system is employed ([1], Chapter 4), the rectangular coordinates of the Earth can be written as follows:

$$x_{\omega\oplus} = a_{\oplus}(\cos E_{\oplus} - e_{\oplus}), \qquad y_{\omega\oplus} = a_{\oplus}[1 - e_{\oplus}{}^2]^{1/2} \sin E_{\oplus}, \tag{2.69}$$

where, referred to the ecliptic, a denotes the semimajor axis, e denotes the eccentricity, and E is the eccentric anomaly. Similarly for any given body or asteroid, denoted by the subscript b,

$$x_{\omega b} = a_b(\cos E_b - e_b); \qquad y_{\omega b} = a_b(1 - e_b{}^2)^{1/2} \sin E_b. \tag{2.70}$$

Transformation of the orbit plane coordinates to the standard ecliptic coordinate system [1] is provided by the well-known vector mappings

$$\mathbf{r}_{\oplus} = x_{\omega\oplus}\mathbf{P}_{\oplus} + y_{\omega\oplus}\mathbf{Q}_{\oplus}, \qquad \mathbf{r}_b = x_{\omega b}\mathbf{P}_b + y_{\omega b}\mathbf{Q}_b, \tag{2.71}$$

where the ecliptic unit vector **P** points at perihelion and **Q** is advanced to **P** in the direction of motion by a right angle in the plane of motion. Expressions in terms of the classical orbital elements for **P** and **Q** can be found in [1] and Section 6.3.2.

The problem is to find the minimum of the difference between $\mathbf{r}_{\oplus}$ and $\mathbf{r}_b$. More specifically, the objective is to minimize the equivalent expression

$$(\Delta r)^2 \equiv (\mathbf{r}_{\oplus} - \mathbf{r}_b) \cdot (\mathbf{r}_{\oplus} - \mathbf{r}_b). \tag{2.72}$$

By direct expansion (2.72) can be written

$$f \equiv \Delta r^2 = x_{\omega\oplus}{}^2 + y_{\omega\oplus}{}^2 + x_{\omega b}{}^2 + y_{\omega b}{}^2 + \alpha x_{\omega\oplus}x_{\omega b} + \beta x_{\omega\oplus}y_{\omega b} + \gamma x_{\omega b}y_{\omega\oplus} + \zeta y_{\omega b}y_{\omega\oplus}, \tag{2.73}$$

where

$$\alpha \equiv -2\mathbf{P}_{\oplus} \cdot \mathbf{P}_b, \qquad \beta \equiv -2\mathbf{P}_{\oplus} \cdot \mathbf{Q}_b,$$
$$\gamma \equiv -2\mathbf{Q}_{\oplus} \cdot \mathbf{P}_b, \qquad \zeta \equiv -2\mathbf{Q}_{\oplus} \cdot \mathbf{Q}_b.$$

Equation 2.73 is the objective function which is to be minimized. Since the problem involves the dynamics of two bodies, it must also be that for any given universal time t, Kepler's equation [1] must be satisfied, namely,

$$\frac{1}{n_{\oplus}}[E_{\oplus} - e_{\oplus} \sin E_{\oplus}] + T_{\oplus} = t = \frac{1}{n_b}[E_b - e_b \sin E_b] + T_b, \tag{2.74}$$

where n is the mean motion and T is the latest time of perifocal passage. The constraint equation of the problem, g, can there be stated as

$$g = \frac{1}{n_\oplus}[E_\oplus - e_\oplus \sin E_\oplus] - \frac{1}{n_b}[E_b - e_b \sin E_b] + T_\oplus - T_b = 0. \quad (2.75)$$

To incorporate the constraint into the minimization process the new function to be written is

$$M \equiv f + \lambda g. \quad (2.76)$$

Proceeding formally,

$$\frac{\partial M}{\partial E_\oplus} = \frac{\partial f}{\partial E_\oplus} + \lambda \frac{\partial g}{\partial E_\oplus} = 0, \quad (2.77)$$

$$\frac{\partial M}{\partial E_b} = \frac{\partial f}{\partial E_b} + \lambda \frac{\partial g}{\partial E_b} = 0, \quad (2.78)$$

and elimination of λ yields

$$\frac{\partial f}{\partial E_b}\frac{\partial g}{\partial E_\oplus} - \frac{\partial f}{\partial E_\oplus}\frac{\partial g}{\partial E_b} = 0. \quad (2.79)$$

The pertinent partial derivatives can be obtained by differentiation of (2.73) and (2.75) as

$$\frac{\partial f}{\partial E_\oplus} = -2a_\oplus^2 e_\oplus^2 \sin E_\oplus \cos E_\oplus + a_\oplus[2a_\oplus e_\oplus - \alpha x_{\omega b} - \beta y_{\omega b}] \sin E_\oplus$$

$$+ a_\oplus(1 - e_\oplus^2)^{1/2}(\gamma x_{\omega b} + \zeta y_{\omega b}) \cos E_\oplus,$$

$$\frac{\partial f}{\partial E_b} = -2a_b^2 e_b^2 \sin E_b \cos E_b + a_b[2a_b e_b - \alpha x_{\omega\oplus} - \gamma y_{\omega\oplus}] \sin E_b$$

$$+ a_b(1 - e_b^2)^{1/2}[\beta x_{\omega\oplus} + \zeta y_{\omega\oplus}] \cos E_b,$$

$$\frac{\partial g}{\partial E_\oplus} = \frac{1}{n_\oplus}(1 - e_\oplus \cos E_\oplus),$$

$$\frac{\partial g}{\partial E_b} = -\frac{1}{n_b}(1 - e_b \cos E_b). \quad (2.80)$$

Equations 2.79 and 2.75 form a system of two algebraic equations in the two unknown $E_\oplus$ and E_b. Iterative solution [7] of the system for elliptic orbits yields two values of $E_\oplus$ and two values of E_b corresponding to the minimum and maximum excursions of the Earth and the asteroid with respect to each other.

2.5 APPLICATION: MULTISTAGE ROCKET PROBLEM

2.5.1 Discussion of the Problem

Propellant calculations based on the physical properties of available fuels show that the energy levels required to place satellites in orbit are not possible for single-stage rockets. In order to overcome this problem the simple technique of staging has been adopted. Because the final velocity of each stage of a rocket just before it is jettisoned becomes the initial velocity of the next stage, it is possible to attain orbital speeds with relatively low-energy propellant-oxidizer combinations. In this section the problem of determining the optimum mass ratios, R, or ratios of weight before, W_b, to weight after burnout, W_a, for each stage to attain a desired terminal or injection velocity, is analyzed. Hence the problem treated in this section deals with the best overall stage weights to reach a specified injection velocity.

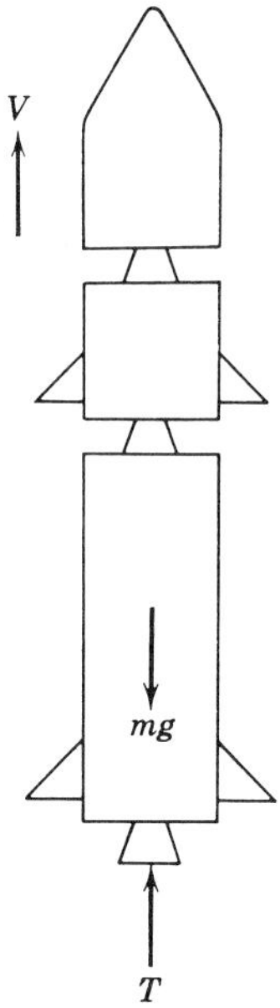

FIGURE 2.5 Vertically rising staged rocket and associated forces.

2.5.2 The Rocket Equation

In order to analyze this problem a relationship is required that connects the velocity increment, ΔV, attained by a given stage for a given mass ratio, that is, for a specified amount of fuel expenditure. This relationship is known in the literature as the *rocket equation*; it can be derived as follows. Consider the forces on the vertically rising rocket stage illustrated in Fig. 2.5. A vertical summation of forces yields

$$\frac{dV}{dt} = \frac{T}{m} - g, \tag{2.81}$$

where T = thrust, m = stage mass at time t, g = acceleration of gravity, and V = stage velocity. Assuming constant gravity acceleration, and introducing the thrust equation

$$T = -gI_{sp}\frac{dm}{dt}, \tag{2.82}$$

where I_{sp} = specific impulse, dm/dt = negative rate of propellant depletion, and g = exact weight to mass conversion factor, it follows that

$$dV = -gI_{\text{sp}}\frac{dm}{m} - g\,dt.$$

Integration yields

$$\Delta V = gI_{\text{sp}} \log \frac{m_b}{m_a} - g\,\Delta t = gI_{\text{sp}} \log R - g\,\Delta t. \tag{2.83}$$

Equation 2.83 states that the magnitude of the velocity increment, ΔV, gained by a rocket in vertical flight during Δt seconds is equal to a constant times the logarithm of the mass ratio, R, minus the gravity loss.

2.5.3 Rocket-Stage Optimization

Because of the staging process the overall mass ratio of the rocket, R^*, that is, the ratio of the takeoff or initial weight to the final payload weight, W_{b1} to W_{ap}, can be written as a function of the individual stages as follows:

$$R^* \equiv \frac{W_{b1}}{W_{ap}} = \left(\frac{W_{b1}}{W_{b1} - W_{p1} - W_{s1}}\right)\left(\frac{W_{b2}}{W_{b2} - W_{p2} - W_{s2}}\right) \cdots \frac{W_{bn}}{W_{ap}}, \tag{2.84}$$

where W_{bi} = initial weight before burning of ith stage, W_{pi} = propellant weight contained in the ith stage, and $W_{si} \equiv$ the structural weight of the ith stage. The general term of (2.84) can be manipulated algebraically to yield the following identity:

$$\frac{W_{bi}}{W_{bi} - W_{pi} - W_{si}} \equiv \frac{R_i[1 - \sigma_i]}{1 - R_i\sigma_i}, \qquad i = 1, 2, \ldots, n, \tag{2.85}$$

where

$$R_i \equiv \frac{W_{bi}}{W_{bi} - W_{pi}} = \frac{W_{bi}}{W_{ai}},$$

$$\sigma_i \equiv \frac{W_{si}}{W_{pi} + W_{si}}.$$

The factor σ is referred to as the stage structural factor; it is a useful parameter in design studies. Use of the parameters R_i and σ_i permits the following transformation of (2.84) to be made:

$$R^* = \left[\frac{R_1(1 - \sigma_1)}{1 - R_1\sigma_1}\right]\left[\frac{R_2(1 - \sigma_2)}{1 - R_2\sigma_2}\right] \cdots \left[\frac{R_n(1 - \sigma_n)}{1 - R_n\sigma_n}\right], \tag{2.86}$$

or taking the logarithm of both sides

$$\log R^* = \sum_{i=1}^{i=n} \log\left(\frac{R_i[1 - \sigma_i]}{1 - R_i\sigma_i}\right).$$

Further use of the properties of logarithms yields the relationship

$$\log R^* = \sum_{i=1}^{i=n} (\log R_i + \log [1 - \sigma_i] - \log [1 - R_i\sigma_i]). \qquad (2.87)$$

In order to optimize the design of any multistage rocket, for a specified maximum velocity of injection, $V_m = V_n - V_1$, where V_n is the final-stage velocity and V_1 is the first-stage velocity, it is desired to minimize R^* or equivalently $\log R^*$. If V_m is a specified value, it follows from Section 2.5.2 that the final velocity is the sum of the velocity increments of all n stages; namely,

$$V_m = \sum_{i=1}^{i=n} gI_{spi} \log R_i, \qquad (2.88)$$

where for purposes of simplicity gravitational losses have been neglected. The analysis therefore requires the minimization of (2.87) subject to constraint (2.88). In order to perform the desired optimization appeal is made to Lagrange's rule. Hence the objective function M is introduced as

$$M = \log R^* + \lambda\left(V_m - \sum_{i=1}^{i=n} gI_{spi} \log R_i\right),$$

where λ is an unknown multiplier. By differentiating M and equating to zero, it follows that

$$\frac{\partial M}{\partial R_i} = \frac{1}{R_i} + \frac{\sigma_i}{1 - R_i\sigma_i} - \lambda \frac{gI_{spi}}{R_i} = 0. \qquad (2.89)$$

Solving (2.89) for R_i results in

$$R_i = -\frac{1 - \lambda gI_{spi}}{\lambda gI_{spi}\sigma_i}, \qquad (2.90)$$

which on substitution into (2.88) yields

$$V_m = \sum_{i=1}^{i=n} gI_{spi} \log \left(\frac{\lambda gI_{spi} - 1}{\lambda gI_{spi}\sigma_i}\right). \qquad (2.91)$$

Equation 2.91 for specified values of V_m, I_{spi} and σ_i is a relationship that permits the determination of the unknown multiplier λ. Hence, after solving (2.91) iteratively, the sought-for mass ratios are directly obtained from (2.90).

2.6 SUMMARY

In this chapter the necessary mathematical tools and techniques required for the determination of the extreme values of functions related to current astrodynamic problems have been developed. The simple case of determining the maxima and minima of functions of one variable was discussed. The theory was extended to account for more than one variable. Optimization of functions with equation constraints was discussed and a convenient technique, the method of Lagrange multipliers, was introduced as an aid in the solution of such problems. Inequality constraints were then discussed, that is, the determination of the extreme values of a function subject to bounds on the variables of the problem. Some problems to illustrate the theory were discussed, namely, the determination of the point of closest approach of two celestial bodies and the optimum staging of space vehicles. The mathematical techniques of this chapter provide the reader with the necessary background material for the discussion that follows in Chapters 3 and 4.

EXERCISES

1. At what sidereal time θ will the speed V of a spaceship with eccentricity, e, be greatest if

$$V = e\cos\theta + (1 - e^2)^2 \sin\theta\,?$$

2. A spaceship is proceeding in the equatorial plane along a rectilinear path defined by

$$x = at + ae, \qquad y = 2at + e,$$

where a and e are constants. At what time t will the ship be nearest the Earth if the Earth is taken to be a sphere of radius a_e with equatorial form defined by

$$x^2 + y^2 = a_e^{\,2}\,?$$

3. Heliocentric probe Zeta-Z1 has been transferred into the orbital plane of comet Biela. Zeta-Z1 is to obtain photographs of Biela from as close a distance as possible without endangering the crew. In the orbit plane system of Biela the equation of orbital laboratory Zeta-Z1 is

$$\frac{(x_\omega + ae)^2}{a^2} + \frac{y_\omega^{\,2}}{a^2(1 - e^2)} = 1,$$

and the orbit of Biela is approximately

$$y_\omega^{\,2} = -4qx_\omega + 4q^2.$$

If the orbital time phasing is assumed correct and the analysis is restricted to nonintersecting orbits, what is the point of closest approach of Zeta-Z1 to Biela?

4. What is the distance of closest approach of geocentric satellites Beta X1 and Beta X2 if the position vector of Beta X1 is

$$\mathbf{r}_1 = x_{\omega 1}\mathbf{P}_1 + y_{\omega 1}\mathbf{Q}_1$$

and the position vector of Beta X2 is

$$\mathbf{r}_2 = x_{\omega 2}\mathbf{P}_2 + x_{\omega 2}\mathbf{Q}_2?$$

Consider that

$$x_{\omega i} = a_i(\cos E_i - e_i), \qquad y_{\omega i} = a_i[1 - e_i^2]^{1/2} \sin E_i, \qquad i = 1, 2,$$

with $\mathbf{P}_i$ defined as a unit vector from the dynamical center pointing at perigee of the respective orbit while $\mathbf{Q}_i$ is a unit vector advanced to $\mathbf{P}_i$ by a right angle in the plane and direction of motion. Assume near circularity of the orbits; that is, only linear terms are carried in the orbital eccentricities.

5. Solve the meteoric shower problem of Section 2.2.2 by the method of Lagrange multipliers.

6. When is it not possible to employ directly the Lagrange rule for obtaining the maxima and minima of a given function f subject to the constraints g_i, $i = 1, 2, \ldots, n$?

REFERENCES

[1] P. R. Escobal, *Methods of Orbit Determination*, John Wiley and Sons, New York, 1965.

[2] R. Courant, *Differential and Integral Calculus*, Interscience, New York, 1955, Volumes 1 and 2.

[3] H. Hancock, *Theory of Maxima and Minima*, Dover Publications, New York, 1960.

[4] G. Leitmann, *Optimization Techniques*, Academic Press, New York, 1962, Chapter 1.

[5] J. V. Uspensky, *Theory of Equations*, McGraw-Hill, Book Company, New York, 1948.

[6] E. T. Bell, *Men of Mathematics*, Simon & Schuster, New York, 1937.

[7] F. B. Hildebrand, *Introduction to Numerical Analysis*, McGraw-Hill Book Company, New York, 1965, Chapter 10.

[8] W. T. Thomson, *Introduction to Space Dynamics*, John Wiley and Sons, New York, 1961.

3 Optimum Planetocentric Orbital Maneuvers

Newton lectured now and then to the few students who chose to hear him; and it is recorded that very frequently he came to the lecture-room and found it empty. On such occasions he would remain fifteen minutes, and then, if no one came, return to his apartment.

JAMES PARTON [15]

3.1 PRELIMINARY ASPECTS OF OPTIMUM ORBITAL MANEUVERS

The purpose of this chapter is to develop analytic techniques that permit the determination of approximate orbits to satisfy certain optimization criteria. Apart from ordinary transfer methods, two main trajectory modes are discussed. These modes are minimum fuel and minimum time of transfer planetocentric trajectories. Chapters 4 and 5 discuss the interplanetary and lunar transfer problems.

In this chapter *minimum fuel trajectories are considered to be in equivalence with minimum velocity increment trajectories*; that is, if the minimum amount of velocity increment is utilized to perform a specified transfer, it is also assumed that this trajectory corresponds to a transfer process requiring the least amount of fuel.

A second concept which simplifies the trajectory calculations by a considerable amount is the idea of an *impulsive change of state*. This process is characterized by an instantaneous application of thrust in order to accomplish a change of state. In real life the inertia of physical systems does not permit a trajectory state to change instantaneously. Hence, in order to perform a specified maneuver, a vehicle must undergo a thrusting arc of finite time duration. For high thrust levels, however, impulsive thrusting is an excellent approximation.

These two concepts, velocity increment equivalence to fuel expenditure and impulsive change of state, permit the solution of some very difficult problems in optimum orbit determination to be accomplished. In most cases the orbits thus determined are quite fair approximations to actual real life trajectory transfers. For greater accuracy, or when the two previously introduced concepts are not acceptable, either perturbation of the Keplerian equations or direct numerical methods must be used. Even here, however, the theory to be developed provides a set of first estimates or initial conditions to the direct numerical techniques, that is, suitable reference orbits or starting values for special perturbations procedures. These numerical procedures are examined in Chapter 7.

3.2 COPLANAR TRANSFERS BETWEEN CIRCULAR ORBITS

3.2.1 Hohmann Transfers

It will perhaps benefit the reader to become acquainted with some simple transfer mechanics. The simplest of all transfer modes is called the Hohmann transfer [2]. This type of transfer has been investigated in detail by Hoelker and Silber [3], Roth [4], and Lawden [5]. Basically, the mode of transfer discovered by Hohmann is as follows. From a circular orbit I, a transfer is made to a final circular orbit F in the same plane as I. An impulsive increment is applied at point A tangent to, and in the direction of motion of, the vehicle in the initial circular orbit. This addition of velocity increment increases the energy of the orbit, and the transfer ellipse T, illustrated in Figure 3.1, is obtained. A coasting transfer is now made to point B where a second impulse in the direction of motion is applied at apogee of the intermediate elliptical trajectory. The energy of the orbit is again increased and circular orbit F is achieved. The process of going from B to A is a mirror

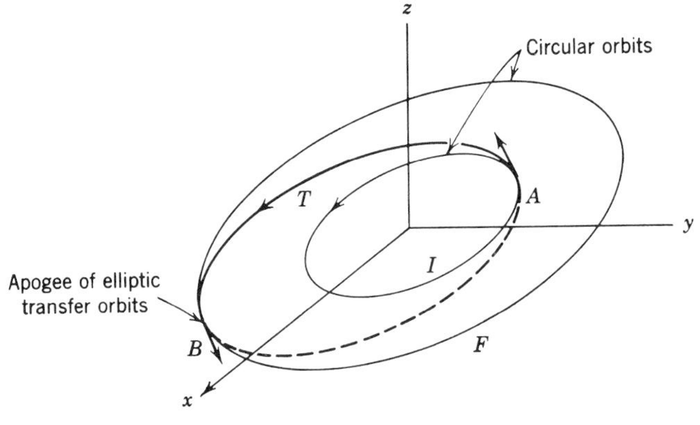

FIGURE 3.1 Hohmann transfer.

image of the A to B transfer illustrated in Figure 3.1, where the firing is now directed in opposition to the actual motion.

The velocity increment to escape orbit I is easily obtained from repeated application of the vis-viva equation ([1], Chapter 3). Using this equation, the velocity at point A on the circular orbit is

$$V_I = \left[\mu\left(\frac{2}{r} - \frac{1}{a_I}\right)\right]^{1/2} = \left[\frac{\mu}{a_I}\right]^{1/2}, \tag{3.1}$$

where a_I is the semimajor axis of orbit I, and μ is the sum of the masses of vehicle and the central planet. Furthermore, if the semimajor axis of orbit F is denoted by a_F, it follows that the semimajor axis of the transfer orbit, a_T, is given by

$$a_T = \tfrac{1}{2}(a_I + a_F).$$

The velocity at point A on the elliptical transfer orbit now can be determined:

$$V_{AT} = \left[\mu\left(\frac{2}{a_I} - \frac{2}{a_I + a_F}\right)\right]^{1/2} = \left[\frac{2\mu}{a_I}\left(\frac{a_F}{a_I + a_F}\right)\right]^{1/2}. \tag{3.2}$$

Therefore, in order to escape orbit I, the following impulse must be applied

$$\Delta V_A = \left[\frac{2\mu}{a_I}\left(\frac{a_F}{a_I + a_F}\right)\right]^{1/2} - \left[\frac{\mu}{a_I}\right]^{1/2}. \tag{3.3}$$

If no further impulse is applied to orbit T, the new orbit is an ellipse which becomes tangent to orbit F but always falls back to point A. By a second double application of the vis-viva equation it follows that the velocity increment required to achieve orbit F is given by

$$\Delta V_B = \left[\frac{\mu}{a_F}\right]^{1/2} - \left[\frac{2\mu}{a_F}\left(\frac{a_I}{a_I + a_F}\right)\right]^{1/2}, \tag{3.4}$$

so that the total Hohmann velocity expenditure ΔV_H can be written as

$$\Delta V_H = \Delta V_A + \Delta V_B. \tag{3.5}$$

A convenient nondimensional form for the velocity increment can be obtained by dividing by the circular velocity of the initial orbit, that is, by (3.1):

$$\frac{\Delta V_H}{V_I} = \frac{\Delta V_A}{V_I} + \frac{\Delta V_B}{V_I}. \tag{3.6}$$

In effect, this operation throws the Hohmann relationship into the form

$$\frac{\Delta V_H}{V_I} = \left(1 - \frac{1}{R}\right)\left(\frac{2R}{1 + R}\right)^{1/2} + \left(\frac{1}{R}\right)^{1/2} - 1, \tag{3.7}$$

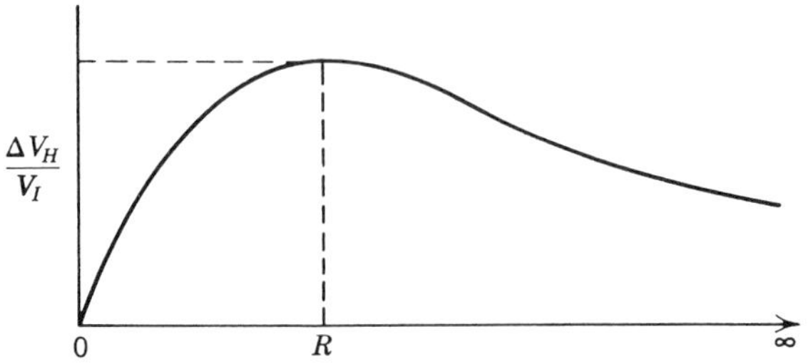

FIGURE 3.2 Graphical representation of Hohmann transfer velocity increment requirement.

where for convenience $R \equiv a_F/a_I$. The graph of (3.7) is illustrated in Figure 3.2. Equation 3.7 possesses a maximum at point R. To obtain this maximum it is possible to differentiate and obtain

$$\frac{d(\Delta V_H/V_I)}{dR} = \frac{1}{R^2}\left[\frac{2R}{1+R}\right]^{1/2} + \frac{R-1}{R(1+R)^2}\left[\frac{2R}{1+R}\right]^{-1/2} - \frac{1}{2R^{3/2}}. \tag{3.8}$$

On equating to zero, after a little reduction, (3.8) reduces to the cubic

$$R^3 - 15R^2 - 9R - 1 = 0, \tag{3.9}$$

which has a root at

$$R = R_M \equiv 15.58176.$$

Because, for all values of $R > R_M$, the velocity expenditure diminishes, the idea of ascending via a Hohmann transfer to a very high orbit, that is, $a_T \gg a_F$, and then returning to orbit F sounds feasible. In fact, consider

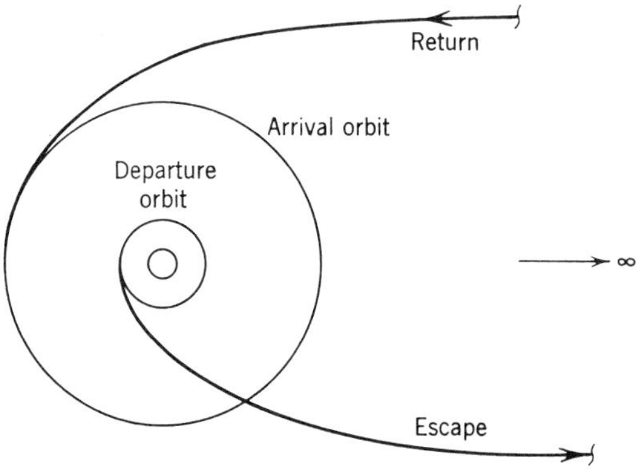

FIGURE 3.3 Escape to a very distant point and return.

the process of transferring to infinity or, in more precise terms, to a very large value of a_T as illustrated in Figure 3.3. From (3.7),

$$\lim_{R\to\infty}\frac{\Delta V_H}{V_I} = \lim_{R\to\infty}\left(1-\frac{1}{R}\right)\lim_{R\to\infty}\left(\frac{2R}{1+R}\right)^{1/2} + \lim_{R\to\infty}\left(\frac{1}{R}\right)^{1/2} - 1 = \sqrt{2}-1. \tag{3.10}$$

Hence the velocity increment to escape to infinity is given by

$$\Delta V_{\infty e} = (\sqrt{2}-1)V_I, \tag{3.11}$$

and the velocity to return from infinity, or equivalently to escape to infinity from orbit F, is given by

$$\Delta V_{\infty r} = (\sqrt{2}-1)V_F. \tag{3.12}$$

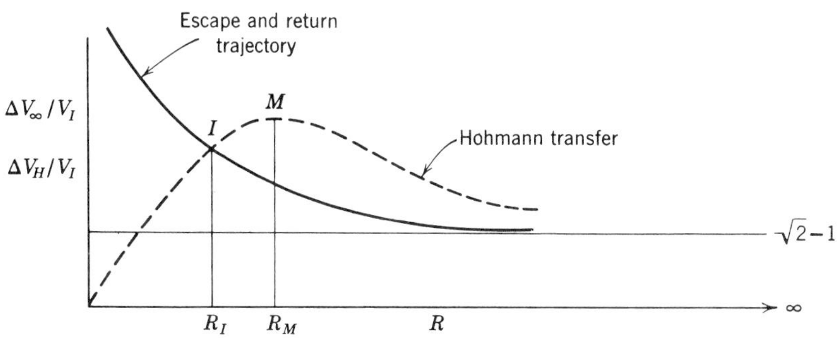

FIGURE 3.4 Superposition of escape and return trajectory and Hohmann transfer.

The nondimensionalized velocity increment for the total maneuver can be obtained by addition as

$$\frac{\Delta V_\infty}{V_I} = \frac{\Delta V_{\infty e}}{V_I} + \frac{\Delta V_{\infty r}}{V_I} = (\sqrt{2}-1)\left(1+\left[\frac{1}{R}\right]^{1/2}\right). \tag{3.13}$$

Let (3.13) be investigated graphically as illustrated in Figure 3.4. As the figure shows, the curves of ΔV_H and ΔV_∞ intersect. To seek the point of intersection, it is possible to equate expressions (3.7) and (3.13):

$$\left(1-\frac{1}{R}\right)\left(\frac{2R}{1+R}\right)^{1/2} - 1 = (\sqrt{2}-1)\left(1+\left[\frac{1}{R}\right]^{1/2}\right) - \left[\frac{1}{R}\right]^{1/2}, \tag{3.14}$$

which after reduction reduces to

$$R^3 - (7+4\sqrt{2})R^2 + (3+4\sqrt{2})R - 1 = 0 \tag{3.15}$$

with a real root given by

$$R = R_I \equiv 11.93876.$$

3.2.2 Hohmann Transfer Limits

The preceding section obtained two critical ratios of final to initial circular orbit radii, that is,

$$R_M = 15.58176, \qquad R_I = 11.93876.$$

The following conclusions can evidently be inferred as limit rules for a Hohmann transfer. For $0 < R \leq 11.93876$ a standard Hohmann transfer is the best possible transfer; for $11.93876 < R \leq 15.58176$ a Hohmann transfer to a very removed point and then a return to the desired orbit is feasible; the case $R > 15.58176$ is yet to be investigated in full detail. A new transfer mode called the bielliptic transfer mode is superior in the case $R > 15.58176$ and partially superior in the case $11.93876 \leq R \leq 15.58176$. The exact transfer rules are listed in Table 3.1.

3.2.3 Bielliptic Transfers

Consider the following transfer mode. From a circular orbit an impulse is applied and transfer along an elliptical orbit is initiated. The elliptical orbit ascends to an apoapsis farther removed from the dynamical center than the final circular orbit which is to be achieved. At apoapsis a second impulse is applied, and the return leg of a new elliptic orbit is initiated, which tangentially returns to the final orbit desired. This process is schematically represented in Figure 3.5. The bielliptic maneuver is a three-impulse transfer technique. By application of the vis-viva integral, the nondimensionalized

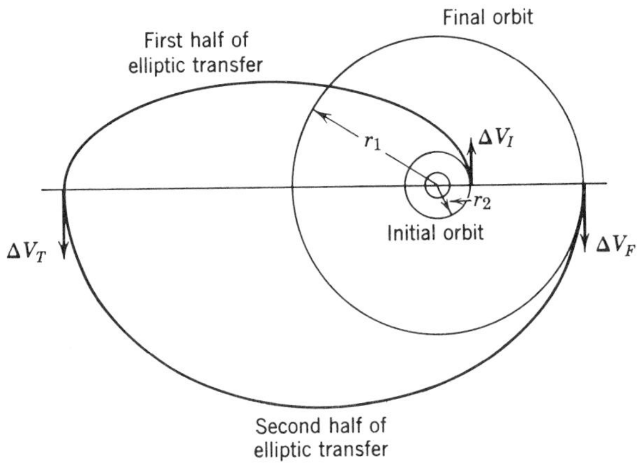

FIGURE 3.5 The bielliptic maneuver.

impulses are given as

$$\frac{\Delta V_I}{V_I} = \left| \left[\frac{2r_T}{r_I + r_T} \right]^{1/2} - 1 \right|, \tag{3.16}$$

$$\frac{\Delta V_T}{V_I} = \left| r_F \frac{V_F}{V_I} \left[\frac{2}{r_T(r_T + r_F)} \right]^{1/2} - r_I \left[\frac{2}{r_T(r_I + r_T)} \right]^{1/2} \right|, \tag{3.17}$$

$$\frac{\Delta V_F}{V_I} = \left| \frac{V_F}{V_I} \left(\left[\frac{2r_T}{r_F + r_T} \right]^{1/2} - 1 \right) \right|, \tag{3.18}$$

where $r_I = a_I$, $r_F = a_F$, and $r_T = 2a_T - a_F$.

For the case $r_T > r_F > r_I$ the differences can be oriented correctly, and it is possible to write the total velocity increment as

$$\frac{\Delta V_B}{V_I} = \frac{\Delta V_I}{V_I} + \frac{\Delta V_T}{V_I} + \frac{\Delta V_F}{V_I} \tag{3.19}$$

or

$$\frac{\Delta V_B}{V_I} = \left[\frac{2r_T/r_I}{1 + r_T/r_I} \right]^{1/2} - 1 + \left[\frac{2}{r_T/r_I} \right]^{1/2} \left(\left[\frac{1}{1 + (r_T/r_I)/(r_F/r_I)} \right]^{1/2} - \left[\frac{1}{1 + r_T/r_I} \right]^{1/2} \right) + \left[\frac{1}{r_F/r_I} \right]^{1/2} \left(\left[\frac{2r_T/r_I}{r_F/r_I + r_T/r_I} \right]^{1/2} - 1 \right). \tag{3.20}$$

More compactly, by defining

$$R \equiv \frac{r_F}{r_I}, \qquad R^* \equiv \frac{r_T}{r_I}, \tag{3.21}$$

the total nondimensionalized bielliptic velocity increment can be written as

$$\frac{\Delta V_B}{V_I} = \left[\frac{2R^*}{1 + R^*} \right]^{1/2} - 1 + \left[\frac{2}{R^*} \right]^{1/2} \left(\left[\frac{1}{1 + R^*/R} \right]^{1/2} - \left[\frac{1}{1 + R^*} \right]^{1/2} \right) + \left[\frac{1}{R} \right]^{1/2} \left(\left[\frac{2R^*}{R + R^*} \right]^{1/2} - 1 \right). \tag{3.22}$$

Equation 3.22 contains two parameters, R and R^*, and for fixed R represents an equation in R^* which should be possessed of maxima and minima. To obtain the extreme parametric members of this family of curves, let (3.22) be differentiated with respect to R^* and equated to zero; that is,

$$\frac{d\, \Delta V_B/V_I}{dR^*} = 0.$$

The solution is a little lengthy, but it can be verified that

$$R^* = -\frac{3 + 1/R}{3(1 + 1/R) \pm 2(3 - 2/R)^{1/2}}. \tag{3.23}$$

Furthermore, since the numerator is negative, the negative sign in the denominator is necessary in order that $R^* > 0$. From Figure 3.5, since for the case under investigation,

$$r_T > r_F > r_I \Rightarrow \frac{r_T}{r_I} > \frac{r_F}{r_I} > 1$$

or

$$R^* > R > 1, \tag{3.24}$$

it follows that

$$-\frac{3 + 1/R}{3(1 + 1/R) - 2(3 - 2/R)^{1/2}} > R. \tag{3.25}$$

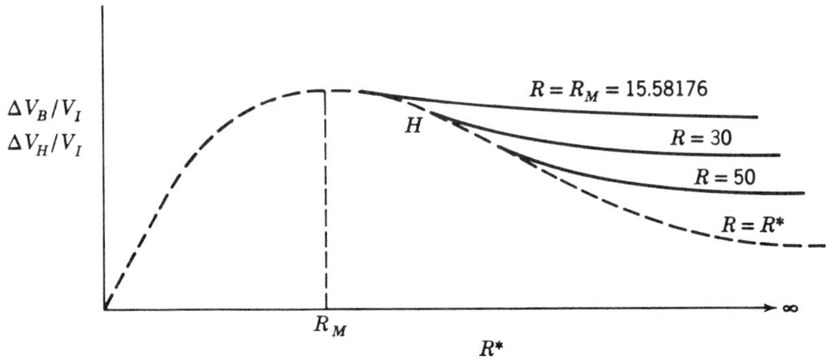

FIGURE 3.6 Hohmann and bielliptic transfer modes.

Rearranging the inequality results in

$$3R^4 - 44R^3 - 42R^2 - 12R - 1 < 0 \tag{3.26}$$

or

$$(3R + 1)(R^3 - 15R^2 - 9R - 1) < 0. \tag{3.27}$$

Hence the final limiting value of R can be obtained by setting

$$(3R + 1)(R^3 - 15R^2 - 9R - 1) = 0 \tag{3.28}$$

and rejecting the physically impossible root $R = -\frac{1}{3}$. It is striking to notice that by some mystical phenomenon

$$R^3 - 15R^2 - 9R - 1 = 0 \tag{3.29}$$

is the same polynomial obtained in the case of the Hohmann transfer! Therefore, from (3.9),

$$R = R_M = 15.58176. \tag{3.30}$$

The previous results are depicted graphically in Figure 3.6.

These facts should be noticed. For $R = R^*$, the bielliptic curves degenerate into the Hohmann curves by definition. For $R = R_M = 15.58176$, the upper limiting curve, starting at the crest of the Hohmann curve, is produced. As mentioned previously, R^* is the transfer radius ratio from the initial orbit to the intermediate high altitude orbit. Consider the curve labeled $R = 30$. The initial point of this curve, H, determines the velocity increment required to effect the orbital change using a Hohmann maneuver. Points farther to the right of H show the total increment needed to effect the same transfer by first proceeding to attain a high intermediate orbit and then returning to the desired orbit. As can be seen from the figure, the magnitude of $\Delta V_B/V_I$ decreases over the amount required by the Hohmann transfer for large values of R^*.

3.2.4 Superiority of Hohmann or Bielliptic Transfer

It is possible to obtain the absolute limit of applicability of the Hohmann or bielliptic transfer by noticing that (3.23), that is, the functional relation of R^* to R, becomes unbounded for the condition

$$3\left(1 + \frac{1}{R}\right) - 2\left(3 - \frac{2}{R}\right)^{1/2} = 0 \tag{3.31}$$

or

$$3R^2 - 26R - 9 = 0, \tag{3.32}$$

which has the real root

$$R = 9. \tag{3.33}$$

With this understanding the basic limit where the bielliptic maneuver should not be used is

$$1 < R < 9. \tag{3.34}$$

It should be recalled from Section 3.2.2 that two other critical ratios exist, namely, R_M and R_I. In terms of these ratios it is possible to write the inequalities

$$9 \leq R \leq R_I \tag{3.35}$$

and

$$R_I < R \leq R_M. \tag{3.36}$$

Finally, the remaining interval of interest is given by

$$R > R_M. \tag{3.37}$$

The discussion in the previous sections now can be summ arized as a function of these intervals as displayed in Table 3.1.

Table 3.1 Comparison of Hohmann and Bielliptic Transfers for the Case $r_T \geq r_F > r_I$

Limit of Final to Initial Orbit Radius Ratio R	Type of Transfer Resulting in Least Velocity Increment Expenditure
$1 < R < 9$	Hohmann
$9 \leq R \leq R_I$	Hohmann
$R_I < R \leq R_M$	Test required
$R_M < R$	Bielliptic

The test required to decide whether to initiate a Hohmann or bielliptic transfer in the situation $R_I < R \leq R_M$ can be obtained by equating the velocity increments required for each type of transfer, that is, (3.7) and (3.22). It follows that

$$\left(1 - \frac{1}{R}\right)\left[\frac{2R}{1 + R}\right]^{1/2} + 2\left[\frac{1}{R}\right]^{1/2} - \left[\frac{2R^*}{1 + R^*}\right]^{1/2} - \left[\frac{2}{R^*}\right]^{1/2}\left[\frac{1}{1 + R^*/R}\right]^{1/2}$$
$$+ \left[\frac{2}{R^*}\right]^{1/2}\left[\frac{1}{1 + R^*}\right]^{1/2} - \left[\frac{1}{R}\right]^{1/2}\left[\frac{2R^*}{R + R^*}\right]^{1/2} = 0. \quad (3.38)$$

Solving for the value of R^*, corresponding to the zero of (3.38), it can be shown that

$$R_0^* = \frac{A + B^{1/2}}{2}, \quad (3.39)$$

where

$$A \equiv \frac{1}{R}\left[\frac{(M - 3N)^2 + 1}{(M + N)^2 - 1}\right] + \frac{1}{R^2}\left[\frac{1}{1 - (M + N)^2}\right],$$

$$B \equiv A^2 - \frac{4}{R[(M + N)^2 - 1]},$$

$$M \equiv \frac{P^2 - 1/R}{2P},$$

$$N \equiv \frac{1}{2P},$$

$$P \equiv \frac{R - 1}{R}\left[\frac{R}{1 + R}\right]^{1/2} + \left[\frac{2}{R}\right]^{1/2}.$$

Because of the continuously decreasing nature of these curves, it is possible to state that for all values of $R^* > R_0^*$ the bielliptic transfer is superior to the Hohmann transfer. The interested reader is directed to [3] for a more detailed treatment of this problem. It is also shown in [3] that only the case

$$r_T \geq r_F > r_I \quad (3.40)$$

benefits from the bielliptic maneuver; that is, for interior transfers where

$$r_F \geq r_T \geq r_I, \tag{3.41}$$

and for the case in which

$$r_F > r_I \geq r_T, \tag{3.42}$$

the Hohmann maneuver is superior.

3.3.1 Cotangential Transfers

In this section the transfer of a vehicle from an inner to an outer elliptic orbit in the same plane is discussed. This mode of transfer is called the cotangential transfer. It has been investigated by Lawden from a geometric

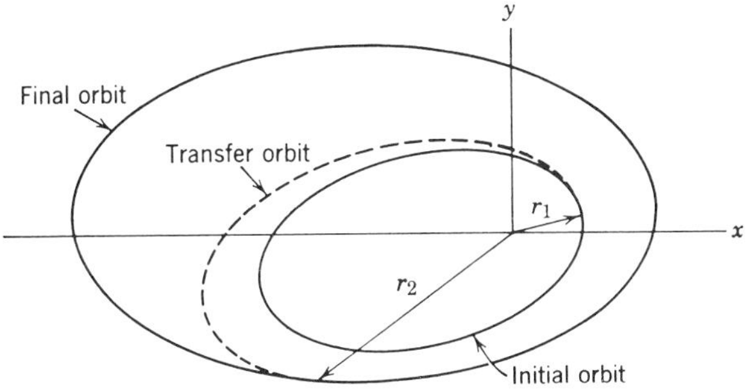

FIGURE 3.7 Cotangential transfer process.

point of view [6]. It appears intuitively feasible to accept the fact that a cotangential transfer, that is, the avoidance of velocity vector rotations at initial and final terminals, would be an optimum transfer mode or at any rate reasonably close to optimum from the point of view of minimum fuel. In fact, numerical studies have shown that this particular transfer mode is indeed always close to optimum. The exact optimum orbit determination process however, may, deviate slightly from the analysis proposed herein. The transfer process is illustrated in Figure 3.7. Intersecting initial and final orbits are not allowed in the following analysis.

3.3.2 Necessary Tangency Conditions

Apart from tangency considerations at the departure terminal r_1, it follows that continuity in position requires that $r_1 = r_{T1}$, where r_{T1} is the radius vector magnitude at terminal 1 on the transfer orbit. Therefore, in more explicit form,

$$\frac{p_1}{1 + e_1 \cos (v + \phi_1)} = \frac{p_T}{1 + e_T \cos (v + \phi_T)}, \tag{3.43}$$

where p = the orbital semiparameter, e = the orbital eccentricity, ϕ = the angle between arbitrary reference line and periapsis. If $\phi_1 = 0$, the phase angles are with respect to periapsis of the initial orbit. Equivalently, (3.43) can be written as follows:

$$\tilde{p}_1 + \tilde{q}_1 \cos (v + \phi_1) = \tilde{p}_T + \tilde{q}_T \cos (v + \phi_T), \tag{3.44}$$

where

$$\tilde{p} \equiv \frac{1}{p}, \qquad \tilde{q} \equiv \frac{e}{p}. \tag{3.45}$$

This condition, which expresses equality of the radii, can be written more symmetrically as

$$\tilde{q}_T \cos (v + \phi_T) - \tilde{q}_1 \cos (v + \phi_1) = -(\tilde{p}_T - \tilde{p}_1). \tag{3.46}$$

On squaring and collecting terms (3.46) becomes

$$\tilde{q}_T{}^2 + \tilde{q}_1{}^2 - 2\tilde{q}_T\tilde{q}_1 \cos (\phi_T - \phi_1) - [\tilde{q}_T \sin (v + \phi_T) - \tilde{q}_1 \sin (v + \phi_1)]^2 = [\tilde{p}_T - \tilde{p}_1]^2. \tag{3.47}$$

This equation can be simplified considerably by the tangency condition, namely,

$$\frac{1}{r_1}\frac{dr_1}{dv} = \frac{1}{r_T}\frac{dr_T}{dv}. \tag{3.48}$$

Hence, by differentiation of (3.46), it follows that

$$\tilde{q}_T \sin (v + \phi_T) - \tilde{q}_1 \sin (v + \phi_1) = 0, \tag{3.49}$$

and therefore (3.47) becomes

$$\tilde{q}_T{}^2 + \tilde{q}_1{}^2 - 2\tilde{q}_T\tilde{q}_1 \cos (\phi_T - \phi_1) = [\tilde{p}_T - \tilde{p}_1]^2. \tag{3.50}$$

In similar fashion, at the second or arrival terminal

$$\tilde{q}_T{}^2 + \tilde{q}_2{}^2 - 2\tilde{q}_T\tilde{q}_2 \cos (\phi_T - \phi_2) = [\tilde{p}_T - \tilde{p}_2]^2. \tag{3.51}$$

Equations 3.50 and 3.51 are a system of two equations in the three unknowns $\tilde{p}_T$, $\tilde{q}_T$, and ϕ_T. The parameters $\tilde{p}_i$, $\tilde{q}_i$, ϕ_i, $i = 1, 2$ are known from the characteristics of the initial and final orbits. These equations can be written in the alternative form

$$-2\tilde{q}_T\tilde{q}_1 \cos (\phi_T - \phi_1) = [\tilde{p}_T - \tilde{p}_1]^2 - [\tilde{q}_T{}^2 + \tilde{q}_1{}^2], \tag{3.52}$$

$$-2\tilde{q}_T\tilde{q}_2 \cos (\phi_T - \phi_2) = [\tilde{p}_T - \tilde{p}_2]^2 - [\tilde{q}_T{}^2 + \tilde{q}_2{}^2]. \tag{3.53}$$

Solving (3.52) for ϕ_T yields

$$\phi_T = \phi_1 + \cos^{-1}\left(\frac{[\tilde{q}_T^{\,2} + \tilde{q}_1^{\,2}] - [\tilde{p}_T - \tilde{p}_1]^2}{2\tilde{q}_T\tilde{q}_1}\right), \tag{3.54}$$

which on substitution into (3.53) and expansion results in

$$\pm\frac{\{4\tilde{q}_T^{\,2}\tilde{q}_1^{\,2} - ([\tilde{q}_T^{\,2} + \tilde{q}_1^{\,2}] - [\tilde{p}_T - \tilde{p}_1]^2)^2\}^{1/2}}{2\tilde{q}_T\tilde{q}_1}$$
$$= \frac{[\tilde{q}_T^{\,2} + \tilde{q}_2^{\,2}] - [\tilde{p}_T - \tilde{p}_2]^2}{2\tilde{q}_T\tilde{q}_2 \sin(\phi_2 - \phi_1)} - \frac{[\tilde{q}_T^{\,2} + \tilde{q}_1^{\,2}] - [\tilde{p}_T - \tilde{p}_1]^2}{2\tilde{q}_T\tilde{q}_1}\cdot\frac{\cos(\phi_2 - \phi_1)}{\sin(\phi_2 - \phi_1)}. \tag{3.55}$$

Squaring both sides of (3.55) and transposing produces the tangency condition F defined by

$$F \equiv \left(\frac{[\tilde{q}_T^{\,2} + \tilde{q}_2^{\,2}] - [\tilde{p}_T - \tilde{p}_2]^2}{\tilde{q}_2 \sin(\phi_2 - \phi_1)} - \frac{[\tilde{q}_T^{\,2} + \tilde{q}_1^{\,2}] - [\tilde{p}_T - \tilde{p}_1]^2}{\tilde{q}_1 \tan(\phi_2 - \phi_1)}\right)^2 - \left(\frac{4\tilde{q}_T^{\,2}\tilde{q}_1^{\,2} - ([\tilde{q}_T^{\,2} + \tilde{q}_1^{\,2}] + [\tilde{p}_T - \tilde{p}_1]^2)^2}{\tilde{q}_1^{\,2}}\right) = 0. \tag{3.56}$$

Note that $F = F(\tilde{p}_T, \tilde{q}_T)$, that is, the dimensional parameters of the transfer conic, are related by means of this tangency function. This condition is used presently.

3.3.3 Minimum Cotangential Velocity Increment Transfer

It is possible to determine the total velocity increment to accomplish the cotangential transfer from the vis-viva equation ([1], Chapter 3), written in terms of the auxiliary variables $\tilde{p}$, $\tilde{q}$ defined by (3.45); that is,

$$V^2 = \mu\left(\frac{2}{r} - \frac{\tilde{p}^2 - \tilde{q}^2}{\tilde{p}}\right). \tag{3.57}$$

In order to utilize (3.57) it is necessary to evaluate the radius vector in terms of the auxiliary variables. To do this, consider the square of (3.46), namely,

$$\tilde{q}_T^{\,2}\cos^2(v + \phi_T) + \tilde{q}_1^{\,2}\cos^2(v + \phi_1) - 2\tilde{q}_T\tilde{q}_1\cos(v + \phi_T)\cos(v + \phi_1) = [\tilde{p}_T - \tilde{p}_1]^2 \tag{3.58}$$

or equivalently, since $\cos^2(v + \phi) + \sin^2(v + \phi) = 1$,

$$\tilde{q}_T^{\,2} + \tilde{q}_1^{\,2} - \tilde{q}_T^{\,2}\sin^2(v + \phi_T) - \tilde{q}_1^{\,2}\sin^2(v + \phi_1) - 2\tilde{q}_T\tilde{q}_1\cos(v + \phi_T)\cos(v + \phi_1) = [\tilde{p}_T - \tilde{p}_1]^2. \tag{3.59}$$

By (3.49) this relationship reduces to

$$\tilde{q}_T^{\,2} + \tilde{q}_1^{\,2} - 2\tilde{q}_1^{\,2} \sin^2 (v + \phi_1)$$
$$- 2\tilde{q}_T\tilde{q}_1 \cos (v + \phi_T) \cos (v + \phi_1) = [\tilde{p}_T - \tilde{p}_1]^2, \qquad (3.60)$$

and on utilization of (3.46) a further reduction is obtained:

$$\tilde{q}_T^{\,2} - \tilde{q}_1^{\,2} - 2\tilde{q}_1(\tilde{p}_T - \tilde{p}_1) \cos (v + \phi_1) = [\tilde{p}_T - \tilde{p}_1]^2. \qquad (3.61)$$

Hence,

$$\tilde{q}_1 \cos (v + \phi_1) = \frac{\tilde{p}_T - \tilde{p}_1}{2} - \frac{q_T^{\,2} - q_1^{\,2}}{2(\tilde{p}_T - \tilde{p}_1)}, \qquad (3.62)$$

and, from the equation of conic,

$$\frac{1}{r_1} = \tilde{p}_1 + \tilde{q}_1 \cos (v + \phi_1) = \frac{\tilde{p}_T + \tilde{p}_1}{2} - \frac{\tilde{q}_T^{\,2} - \tilde{q}_1^{\,2}}{2(\tilde{p}_T - \tilde{p}_1)}, \qquad (3.63)$$

so that (3.57), evaluated at the first terminal, becomes

$$V_1^{\,2} = \mu\left(\tilde{p}_T + \frac{\tilde{p}_1\tilde{q}_T^{\,2} - \tilde{p}_T\tilde{q}_1^{\,2}}{\tilde{p}_1(\tilde{p}_1 - \tilde{p}_T)}\right). \qquad (3.64)$$

Similarly, the velocity at the first terminal of the transfer orbit can be obtained from

$$V_{T1}^{\,2} = \mu\left(\tilde{p}_1 + \frac{\tilde{p}_1\tilde{q}_T^{\,2} - \tilde{p}_T\tilde{q}_1^{\,2}}{\tilde{p}_T(\tilde{p}_1 - \tilde{p}_T)}\right). \qquad (3.65)$$

Furthermore, because the cotangential condition permits an algebraic addition of velocities, the total impulse for the complete two-impulse transfer becomes

$$\Delta V = |V_{T2} - V_2| + |V_{T1} - V_1|, \qquad (3.66)$$

or, by use of (3.64) and (3.65) and similar forms at the second terminal,

$$\Delta V = \mu^{1/2}\left[\frac{\tilde{p}_1^{1/2} - \tilde{p}_T^{1/2}}{\tilde{p}_1^{1/2} + \tilde{p}_T^{1/2}}\left\{\tilde{p}_1 - \tilde{p}_T + \frac{\tilde{q}_T^{\,2}}{\tilde{p}_T} - \frac{\tilde{q}_1^{\,2}}{\tilde{p}_1}\right\}\right]^{1/2}$$
$$+ \mu^{1/2}\left[\frac{\tilde{p}_2^{1/2} - \tilde{p}_T^{1/2}}{\tilde{p}_2^{1/2} + \tilde{p}_T^{1/2}}\left\{\tilde{p}_2 - \tilde{p}_T + \frac{\tilde{q}_T^{\,2}}{\tilde{p}_T} - \frac{\tilde{q}_2^{\,2}}{\tilde{p}_2}\right\}\right]^{1/2}. \qquad (3.67)$$

It should be noted that $\Delta V = \Delta V(\tilde{p}_T, \tilde{q}_T)$. To obtain a cotangential transfer orbit with a minimum amount of impulse, Lagrange's rule, discussed in Section 2.3, must be introduced because of the intractability of the equations involved. The problem is therefore the minimization of ΔV, (3.67), subject to constraint F (3.56). To accomplish this result construct the objective function M defined by

$$M \equiv \Delta V + \lambda F \qquad (3.68)$$

and equate the partials of M with respect to $\tilde{p}_T$ and $\tilde{q}_T$ to zero; that is

$$\frac{\partial M}{\partial \tilde{p}_T} = \frac{\partial \Delta V}{\partial \tilde{p}_T} + \lambda \frac{\partial F}{\partial \tilde{p}_T} = 0,$$

$$\frac{\partial M}{\partial \tilde{q}_T} = \frac{\partial \Delta V}{\partial \tilde{q}_T} + \lambda \frac{\partial F}{\partial \tilde{q}_T} = 0. \tag{3.69}$$

Eliminating λ yields

$$G(\tilde{p}_T, \tilde{q}_T) \equiv \frac{\partial \Delta V}{\partial \tilde{q}_T}\frac{\partial F}{\partial \tilde{p}_T} - \frac{\partial \Delta V}{\partial \tilde{p}_T}\frac{\partial F}{\partial \tilde{q}_T} = 0, \tag{3.70}$$

which, with

$$F(\tilde{p}_T, \tilde{q}_T) = 0, \tag{3.71}$$

provides a system of two algebraic equations in two unknowns whose zeros define the minimum velocity increment cotangential transfer conic. The partial derivatives of F and ΔV with respect to $\tilde{p}_T$ and $\tilde{q}_T$ can be obtained explicitly from (3.56) and (3.67).

3.4 NONCOPLANAR ORBITAL TRANSFERS

The analysis of the two preceding sections was restricted to the transfer of vehicles without plane changes; that is, the initial and final orbits were taken to be in the same plane. Consider, however, the general problem of transfer between two orbits inclined to each other by an angle θ_T. The geometry of the transfer process is illustrated in Figure 3.8. A departure is made from orbit plane 1 with plane change angle θ_1. At apoapsis on the resulting

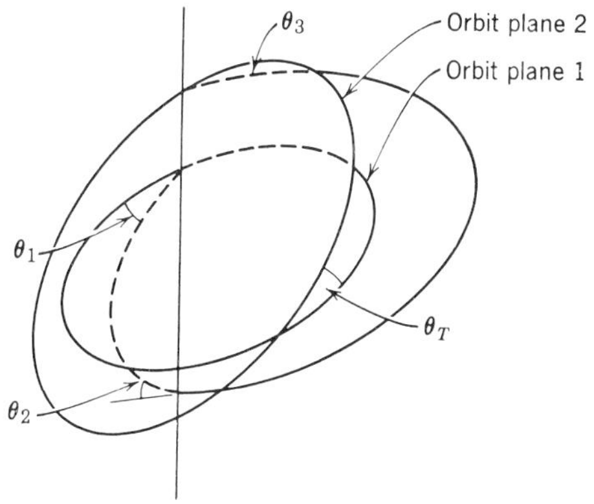

FIGURE 3.8 Bielliptic transfer with plane change.

elliptic orbit a second impulse is applied to change the orbit plane through an angle θ_2. Finally, at apoapsis of the resulting second ellipse a final plane change through an angle θ_3 places the vehicle in the desired final orbit, that is, orbit plane 2. The initial and final orbits are taken to be circular.

The resulting bielliptic maneuver just described is general enough that appropriate choice of the θ_i and radius magnitudes will include Hohmann transfers, if these modes of transfer are desired. Bielliptic transfers with plane changes have been discussed in detail by Roth [12].

3.4.1 Velocity Requirements with Plane Change

In any periapsis to apoapsis transfer, since the semiparameter of the transfer conic is constant, it follows from ([1], Chapter 3) that

$$\frac{r_p^4 \dot{v}_p^2}{\mu} = p = \frac{r_a^4 \dot{v}_a^2}{\mu}, \tag{3.72}$$

where the subscripts p and a, respectively, refer to periapsis and apoapsis. Since the rate of change of true anomaly $\dot{v}$ at periapsis and that at apoapsis is given, respectively, by

$$\dot{v}_p = \frac{V_p}{r_p}, \qquad \dot{v}_a = \frac{V_a}{r_a}, \tag{3.73}$$

it follows that, in terms of the speed V,

$$r_p V_p = r_a V_a. \tag{3.74}$$

Hence it is possible to write the relationships

$$V_a = \frac{r_p}{r_a} V_p, \qquad V_p = \frac{r_a}{r_p} V_a. \tag{3.75}$$

Furthermore, equating $1/a$ from the vis-viva integral at the two points yields

$$V_a^2 - \frac{2\mu}{r_a} = V_p^2 - \frac{2\mu}{r_p}, \tag{3.76}$$

which can be solved for V_p by virtue of relationship (3.75) as

$$V_p = \left[\frac{2\mu r_a}{r_p(r_a + r_p)}\right]^{1/2}. \tag{3.77}$$

Similarly it is possible to obtain

$$V_a = \left[\frac{2\mu r_p}{r_a(r_a + r_p)}\right]^{1/2}. \tag{3.78}$$

To derive the velocity increments at each plane change consider the vector diagram in Figure 3.9. This vector diagram gives the geometric relationships between, for example, the initial velocity V_x, the velocity on the transfer orbit, V_y, and the total velocity change ΔV. Adopting the previous notation, the velocity increment at all three plane changes can be obtained directly from the law of cosines:

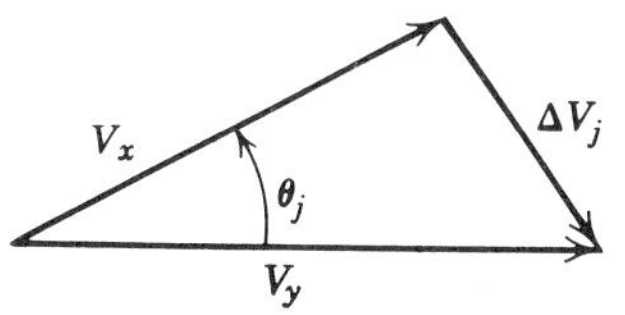

FIGURE 3.9 Velocity triangle.

$$\Delta V_j = [V_x^2 + V_y^2 - 2V_xV_y \cos \theta_j]^{1/2}, \qquad j = 1, 2, 3. \tag{3.79}$$

Proceeding formally for $j = 1$, since the initial orbit is circular and the vehicle is also at periapsis of the transfer leg, it follows that

$$V_x = \left[\frac{\mu}{r_p}\right]^{1/2}, \qquad V_y = \left[\frac{2\mu r_a}{r_p(r_a + r_p)}\right]^{1/2}.$$

Substituting the functional values of V_x and V_y into (3.79) yields

$$\Delta V_1 = \frac{\mu}{r_p}\left[1 + \frac{2r_a}{(r_a + r_p)} - 2\left(\frac{2r_a}{r_a + r_p}\right)^{1/2}\right]^{1/2}.$$

Velocity increments for ΔV_2 and ΔV_3 may be derived similarly by making the appropriate substitutions in (3.79). Finally, removing the initial circular velocity from ΔV_1, ΔV_2, and ΔV_3 and summing, the total velocity increment becomes

$$\frac{\Delta V}{V_I} = (1 + H_1^2 - 2H_1 \cos \theta_1)^{1/2} + H_2(1 + H_3^2 - 2H_3 \cos \theta_2)^{1/2}$$
$$+ H_4(1 + H_5^2 - 2H_5 \cos \theta_3)^{1/2}, \tag{3.80}$$

where, in terms of the radius magnitudes at the point of application of the first, second, and third impulses, that is, r_I, r_T, and r_F,

$$H_1 \equiv \left[\frac{2r_T}{r_I + r_T}\right]^{1/2}, \qquad H_2 \equiv \left[\frac{2r_Fr_I}{r_T(r_F + r_T)}\right]^{1/2},$$

$$H_3 \equiv \left[\frac{(r_T + r_F)r_I}{r_F(r_I + r_T)}\right]^{1/2}, \qquad H_4 \equiv \left[\frac{r_I}{r_F}\right]^{1/2}, \qquad H_5 \equiv \frac{H_1H_4}{H_3}.$$

At times a convenient parameter to specify is θ_T, the total plane change, that is,

$$\theta_T = \theta_1 + \theta_2 + \theta_3. \tag{3.81}$$

Owing to conditions of symmetry, (3.80) can be used equally well for outward as well as inward transfers, that is, $r_I > r_F$ or $r_F > r_I$.

3.4.2. Hohmann Transfers with Plane Change

The general velocity function, (3.80), can be amended readily to determine Hohmann transfers with plane changes. For example, the conditions

$$r_T = r_F, \qquad \theta_3 = 0 \tag{3.82}$$

imply an outward Hohmann transfer with arbitrary θ_1 and $\theta_2 = \theta_T - \theta_1$. Similarly, the conditions

$$r_T = r_I, \qquad \theta_1 = 0 \tag{3.83}$$

imply an inward Hohmann transfer with arbitrary θ_3 and $\theta_2 = \theta_T - \theta_3$.

An interesting and useful mode of transfer is the so-called modified Hohmann defined by the conditions

$$r_T = r_F, \qquad \theta_1 = \theta_2 = 0. \tag{3.84}$$

These conditions imply a regular Hohmann transfer in the initial orbit plane and then a final plane change. If $\theta_3 = \theta_T$, this would correspond to a final plane change at the line of intersection of both orbit planes.

Finally the case $\theta_1 = \theta_2 = \theta_3 = 0$ is the special coplanar case discussed in Section 3.2.

Illustrative Example

Obtain the velocity requirements for a simple plane change between any two orbital planes.

Using the general function defined by (3.80), it is possible to choose

$$r_T = r_I = r_F$$

and take any two of the θ_i, $i = 1, 2, 3$, equal to zero. Therefore, letting

$$\theta_2 = \theta_3 = 0,$$

it follows that $H_j = 1$, $j = 1, 2, 3, 4, 5$. Equation 3.80 therefore can be written

$$\frac{\Delta V}{V_I} = \sqrt{2}(1 - \cos\theta_1)^{\frac{1}{2}}$$

from which it follows that

$$\Delta V = 2V_I \sin\left(\frac{\theta_1}{2}\right). \tag{3.85}$$

3.4.3 Optimum Split of Plane Change

The manner in which to split the total plane change in the general transfer function defined by (3.80) is not at all obvious. Evidently it is desirable to minimize the total velocity requirement as a function of θ_1, θ_2, and θ_3 subject to constraint (3.81). Hence, by formal differentiation,

$$\frac{\partial(\Delta V_T/V_I)}{\partial\theta_1} = [1 + H_1^{\,2} - 2H_1 \cos \theta_1]^{-1/2} H_1 \sin \theta_1 - [1 + H_5^{\,2} - 2H_5 \cos (\theta_T - \theta_2 - \theta_1)]^{-1/2} \times H_4 H_5 \sin (\theta_T - \theta_2 - \theta_1), \tag{3.86}$$

$$\frac{\partial(\Delta V_T/V_I)}{\partial\theta_2} = [1 + H_3^{\,2} - 2H_3 \cos \theta_2]^{-1/2} H_3 \sin \theta_2 - [1 + H_5^{\,2} - 2H_5 \cos (\theta_T - \theta_2 - \theta_1)]^{-1/2} \times H_4 H_5 \sin (\theta_T - \theta_2 - \theta_1). \tag{3.87}$$

On equating the respective partial derivatives to zero, a system of two equations in two unknowns is obtained which, on solution, yields the appropriate optimum plane changes for a specified θ_T.

3.5 OPTIMUM INTERCEPT MANEUVERS

The process of departing from a state on a specified orbit and attaining a desired position is called interception. In the intercept maneuver no attempt is made to match velocity at the final terminal. Hence interception is quite different from rendezvous in which position, velocity, and phasing are matched at the final terminal. A typical intercept maneuver is illustrated in Figure 3.10.

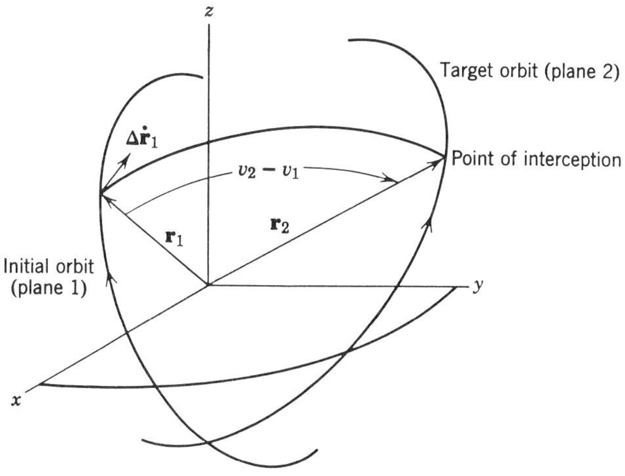

FIGURE 3.10 Intercept maneuver.

In this section two different intercept modes are considered, namely, the minimum velocity increment and minimum time intercept modes. The minimum velocity intercept mode was first investigated in detail by Roth [7] who showed that the solution can be obtained as a quartic equation in the flight path angle of the intercept trajectory. It is expedient, however, to switch to the semiparameter of the intercept orbit as a fundamental parameter in order to treat the intercept and two-impulse transfer problems in a coordinated fashion. Furthermore, as discussed by Escobal [9], this parameter also permits an easy derivation of the minimum time intercept mode.

3.5.1 Velocity Vector as a Function of Semiparameter

Consider the geometric situation depicted in Figure 3.11. The vehicle velocity vector $\dot{\mathbf{r}}_1$ at the first terminal can be written as a linear combination of the orthogonal set of unit vectors $\mathbf{V}_1$ and $\mathbf{U}_1$, where $\mathbf{U}_1$ is a unit vector from the dynamical center which points at the vehicle and $\mathbf{V}_1$ is advanced to $\mathbf{U}_1$ by a right angle in the direction and plane of motion, as follows:

$$\dot{\mathbf{r}}_1 = \dot{r}_1\mathbf{U}_1 + r_1\dot{v}_1\mathbf{V}_1. \tag{3.88}$$

Note that $\dot{r}$ is the radial rate of change along $\mathbf{U}$, whereas $r\dot{v}$ is the traverse rate of change in the $\mathbf{V}$ direction. Utilizing the true anomaly difference, $v_2 - v_1$, depicted in Figure 3.12, the following relationships can be written at sight:

$$\mathbf{U}_1 = \mathbf{U}_2 \cos(v_2 - v_1) - \mathbf{V}_2 \sin(v_2 - v_1), \tag{3.89}$$

$$\mathbf{V}_1 = \mathbf{U}_2 \sin(v_2 - v_1) + \mathbf{V}_2 \cos(v_2 - v_1). \tag{3.90}$$

Solving (3.89) for $\mathbf{V}_2$ and substituting into (3.90) yields

$$\mathbf{V}_1 \sin(v_2 - v_1) = \mathbf{U}_2 - \mathbf{U}_1 \cos(v_2 - v_1); \tag{3.91}$$

therefore elimination of $\mathbf{V}_1$ from (3.88) results in

$$\dot{\mathbf{r}}_1 = \left[\dot{r} - \frac{r_1\dot{v}_1}{\tan(v_2 - v_1)}\right]\mathbf{U}_1 + \left[\frac{r_1\dot{v}_1}{\sin(v_2 - v_1)}\right]\mathbf{U}_2. \tag{3.92}$$

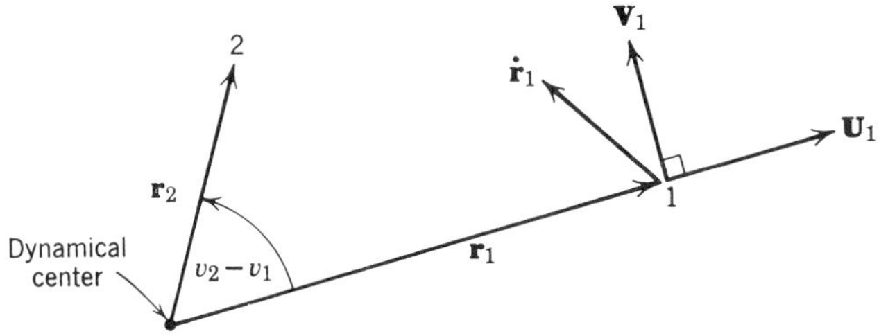

FIGURE 3.11 Orbit geometry.

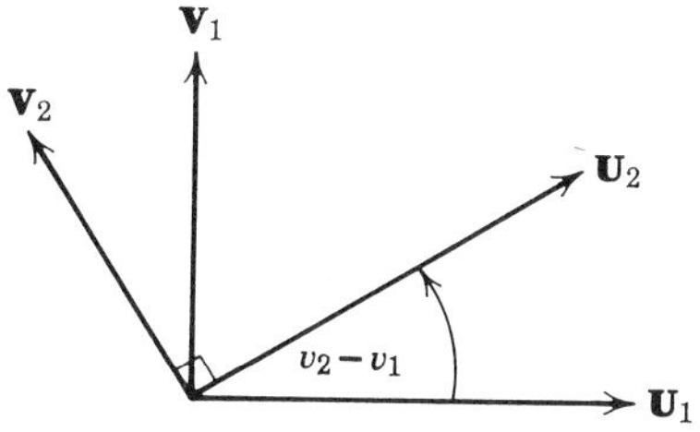

FIGURE 3.12 Displacement of vehicle through the angle $v_2 - v_1$.

By expanding the equation of a conic,

$$r_2 = \frac{p}{1 + e\cos(v_2 - v_1 + v_1)}, \tag{3.93}$$

and equating p, the semiparameter, at the two universal times t_1 and t_2 it follows, from ([1], Section 6.7) that

$$S_{v1} = \frac{\beta_{r1} + \beta_{c1}C_{v1}}{\beta_{s1}}, \tag{3.94}$$

where

$$\begin{aligned} \beta_{r1} &= r_2 - r_1, \\ \beta_{c1} &= r_2 \cos(v_2 - v_1) - r_1, \\ \beta_{s1} &= r_2 \sin(v_2 - v_1), \end{aligned} \tag{3.95}$$

with $S_{v1} \equiv e \sin v_1$ and $C_{v1} \equiv e \cos v_1$. Furthermore, since a well-known expression for the radial rate of change of the satellite radius vector magnitude is

$$\dot{r} = \left[\frac{\mu}{p}\right]^{1/2} e \sin v, \tag{3.96}$$

it is possible to write

$$\dot{r}_1 = \left[\frac{\mu}{p}\right]^{1/2} \frac{\beta_{r1} + \beta_{c1}C_{v1}}{\beta_{s1}}, \tag{3.97}$$

where $e \equiv$ orbital eccentricity, and $\mu \equiv$ the sum of the masses of vehicle and central body.

Finally, the radial and traverse velocity components at the first terminal can be written as a function of the semiparameter with the aid of $C_v = (p/r) - 1$ so that

$$\dot{r}_1 = \tilde{\omega}_1 p^{-1/2} + \tilde{\omega}_2 p^{1/2}, \tag{3.98}$$

$$r_1 \dot{v}_1 = \tilde{\omega}_3 p^{1/2}, \tag{3.99}$$

where

$$\tilde{\omega}_1 \equiv \frac{\sqrt{\mu}[\beta_{r1} - \beta_{c1}]}{\beta_{s1}},$$

$$\tilde{\omega}_2 \equiv \frac{\sqrt{\mu}\beta_{c1}}{r_1\beta_{s1}},$$

$$\tilde{\omega}_3 \equiv \frac{\sqrt{\mu}}{r_1}. \tag{3.100}$$

It follows at once that the velocity vector at the first terminal becomes

$$\dot{\mathbf{r}}_1 = (\omega_1 p^{-1/2} + \omega_2 p^{1/2})\mathbf{U}_1 + (\omega_3 p^{1/2})\mathbf{U}_2 \tag{3.101}$$

with

$$\omega_1 \equiv \tilde{\omega}_1 = \frac{\sqrt{\mu}[1 - \cos(v_2 - v_1)]}{\sin(v_2 - v_1)},$$

$$\omega_2 \equiv \tilde{\omega}_2 - \tilde{\omega}_3 \cot(v_2 - v_1) = -\frac{\sqrt{\mu}}{r_2 \sin(v_2 - v_1)},$$

$$\omega_3 \equiv \frac{\tilde{\omega}_3}{\sin(v_2 - v_1)} = \frac{\sqrt{\mu}}{r_1 \sin(v_2 - v_1)}.$$

3.5.2 Velocity Minimization Process

The velocity increment at the first terminal as illustrated in Figure 3.10 is directly given by

$$|\Delta\dot{\mathbf{r}}| = |\dot{\mathbf{r}}_1 - \dot{\mathbf{r}}_1^*|, \tag{3.102}$$

where $\dot{\mathbf{r}}_1^*$ is the known velocity vector before application of the targeting velocity increment $\Delta\dot{\mathbf{r}}_1$. Equation 3.102 can be written equivalently as the scalar relationship

$$\Delta V_1 = [(\dot{\mathbf{r}}_1 - \dot{\mathbf{r}}_1^*)\cdot(\dot{\mathbf{r}}_1 - \dot{\mathbf{r}}_1^*)]^{1/2}. \tag{3.103}$$

Utilizing (3.101), by direct substitution and expansion, it is possible to verify that

$$\begin{aligned}\Delta V_1^2 &= (\omega_1 p^{-1/2} + \omega_2 p^{1/2})^2 + (\omega_3 p^{1/2})^2 + \dot{\mathbf{r}}_1^* \cdot \dot{\mathbf{r}}_1^* \\ &\quad + 2(\omega_1 p^{-1/2} + \omega_2 p^{1/2})(\omega_3 p^{1/2})\mathbf{U}_1 \cdot \mathbf{U}_2 \\ &\quad - 2(\omega_1 p^{-1/2} + \omega_2 p^{1/2})\mathbf{U}_1 \cdot \dot{\mathbf{r}}_1^* - 2(\omega_3 p^{1/2})\mathbf{U}_2 \cdot \dot{\mathbf{r}}_1^*,\end{aligned} \tag{3.104}$$

which collapses to

$$\Delta V_1^2 = \nu_1 p + \nu_2 p^{1/2} + \nu_3 p^{-1/2} + \nu_4 p^{-1} + \nu_5, \tag{3.105}$$

where

$$\nu_1 \equiv \omega_2^{\,2} + \omega_3^{\,2} + 2\omega_2\omega_3\mathbf{U}_1 \cdot \mathbf{U}_2,$$
$$\nu_2 \equiv -2\omega_2\dot{\mathbf{r}}_1^* \cdot \mathbf{U}_1 - 2\omega_3\dot{\mathbf{r}}_1^* \cdot \mathbf{U}_2,$$
$$\nu_3 \equiv -2\omega_1\dot{\mathbf{r}}_1^* \cdot \mathbf{U}_1,$$
$$\nu_4 \equiv \omega_1^{\,2},$$
$$\nu_5 \equiv \dot{\mathbf{r}}_1^* \cdot \dot{\mathbf{r}}_1^* + 2\omega_1\omega_3\mathbf{U}_1 \cdot \mathbf{U}_2 + 2\omega_1\omega_2.$$

Differentiation of (3.105) yields the algebraic form

$$\frac{\partial\, \Delta V_1^{\,2}}{\partial p} = \nu_1 + \tfrac{1}{2}\nu_2 p^{-1/2} - \tfrac{1}{2}\nu_3 p^{-3/2} - \nu_4 p^{-2}, \tag{3.106}$$

which on equating to zero and removal of fractional powers yields the quartic resolvent

$$\nu_1 s^4 + \tfrac{1}{2}\nu_2 s^3 - \tfrac{1}{2}\nu_3 s - \nu_4 = 0, \tag{3.107}$$

where for convenience the substitution $p = s^2$ has been made. Solution of the quartic resolvent for the determination of the corresponding roots s_i and computation of

$$\Delta V_{1i} = |\dot{\mathbf{r}}_{1i} - \dot{\mathbf{r}}_1^*| \tag{3.108}$$

via (3.101), namely,

$$\dot{\mathbf{r}}_{1i} = (\omega_1 p_i^{-1/2} + \omega_2 p_i^{1/2})\mathbf{U}_1 + (\omega_3 p_i^{1/2})\mathbf{U}_2, \tag{3.109}$$

enables segregation of the p_i corresponding to the true minimum ΔV. Furthermore, since (3.109) has produced the appropriate velocity vector $\dot{\mathbf{r}}_1$, and the position vector $\mathbf{r}_1$ is already known, the minimum velocity increment intercept orbit has been determined.

3.5.3 Time Minimization Process

Suppose that it is desired to depart from a specified orbital position $\mathbf{r}_1$ as depicted in Figure 3.10 and intercept a vehicle at position $\mathbf{r}_2$ as quickly as possible. This procedure is called minimum time interception [9]. If the only method of accomplishing the intercept is with a single impulse applied at the first terminal, it would appear obvious that, the greater the impulse applied, the sooner will the interception occur. Usually, however, the amount of impulse available at the first terminal is limited by the fuel capacity of the vehicle. This is a very important constraint which must be included in the analysis in order to have physically realizable transfers.

To be more exact, it is desired to minimize τ, the modified time between $\mathbf{r}_1$ and $\mathbf{r}_2$, where, as illustrated in Figure 3.10, $\mathbf{r}_1$ is contained in orbit plane 1 and $\mathbf{r}_2$ is contained in orbit plane 2. For convenience τ is defined by

$$\tau \equiv k(t_2 - t_1), \tag{3.110}$$

with k = the central planet gravitational constant, t_1 = universal departure time from orbit 1, and t_2 = universal time of arrival at orbit 2.

The fuel or velocity increment constraint can be stated as

$$|\dot{\mathbf{r}}_1 - \dot{\mathbf{r}}_1^*| < \Delta V_{\max}, \tag{3.111}$$

where $\dot{\mathbf{r}}_1^*$ is the known velocity vector in orbit plane 1 before commencing the intercept maneuver. Equivalently, this inequality can be expressed as

$$[(\dot{\mathbf{r}}_1 - \dot{\mathbf{r}}_1^*) \cdot (\dot{\mathbf{r}}_1 - \dot{\mathbf{r}}_1^*)]^{1/2} < \Delta V_{\max}. \tag{3.112}$$

Proceeding as in Section 2.3.2, at the expense of introducing an unknown parameter, the inequality can be transformed into the equation

$$\Delta V_{\max} - [\Delta V_1{}^2]^{1/2} = \epsilon^2, \tag{3.113}$$

where ϵ is an unknown parameter and, by (3.105),

$$\Delta V_1{}^2 = \nu_1 p + \nu_2 p^{1/2} + \nu_3 p^{-1/2} + \nu_4 p^{-1} + \nu_5. \tag{3.114}$$

Applying Lagrange's rule, Section 2.3.1, it is possible to form the minimization function M as

$$M = \tau + \mu(\Delta V_{\max} - [\Delta V_1{}^2]^{1/2} - \epsilon^2). \tag{3.115}$$

In (3.115), μ is an unknown Lagrange multiplier. In order to minimize τ the standard procedure would demand that the partial derivatives of M with respect to the independent variables vanish. However, this would overdetermine the physical problem. Note that, if two position vectors are specified, six constants of the orbit are known and only one degree of freedom is available. This one degree of freedom is absorbed by (3.113). Hence the condition of optimality becomes $\partial M/\partial \epsilon = 0$, or, from (3.115),

$$-2\mu\epsilon = 0. \tag{3.116}$$

If $\mu = 0$, no constraint would be imposed. Therefore, as a consequence of (3.113), for the case $\epsilon = 0$, the controlling equation of the minimum time intercept problem becomes

$$[\Delta V_1{}^2]^{1/2} = \Delta V_{\max}. \tag{3.117}$$

Hence, on squaring both sides and using (3.114),

$$\nu_1 p + \nu_2 p^{1/2} + \nu_3 p^{-1/2} + \nu_4 p^{-1} + \nu_5 - \delta = 0, \tag{3.118}$$

where for convenience $\Delta V_{\max} \equiv \delta$. On clearing radicals, the following quartic equation is obtained:

$$\nu_1 s^4 + \nu_2 s^3 + (\nu_5 - \delta)s^2 + \nu_3 s + \nu_4 = 0, \tag{3.119}$$

with the understanding that $p = s^2$. This quartic resolvent determines the minimum time trajectory between the terminals $\mathbf{r}_1$ and $\mathbf{r}_2$ with the constraint that the impulse at $\mathbf{r}_1$ is less than some bounding impulse. Solution of the quartic and computation of the time interval τ_i from the p-iteration theory ([1], Section 6.4) allows segregation of the p_i corresponding to the true minimum time orbit. Note that, because p is numerically determined, the time interval obtained from the p-iteration theory is obtained in closed form; that is, no iteration is required.

3.5.4 Fixed Time of Flight Intercept

A mode of interception which is also of importance is that of fixed time of flight between the initial terminal in orbit 1, $\mathbf{r}_1$, and the final terminal in orbit 2, $\mathbf{r}_2$. In this case, depending on the given conditions and the angular radius vector spread, one of the six methods discussed in ([1], Chapter 6) can be used to determine the orbit. The procedure for solution is iterative. Once the fixed time transfer orbit has been determined, the velocity increment required to perform the intercept is directly given by

$$\Delta V_1 = |\dot{\mathbf{r}}_1 - \dot{\mathbf{r}}_1^*|, \tag{3.120}$$

where $\dot{\mathbf{r}}_1^*$ is the known velocity vector in orbit 1 at $\mathbf{r}_1$ as depicted in Figure 3.10.

3.6 OPTIMUM TWO-IMPULSE MANEUVERS

In the preceding section an examination of optimum intercept trajectories was undertaken. The two-impulse maneuver is defined as a departure from an initial orbit via a velocity impulse and attainment of a desired position and velocity vector at some future or past state. Hence, over and above the interception process, the two-impulse technique attempts to match final velocities. The transfer geometry is shown in Figure 3.13. As may be

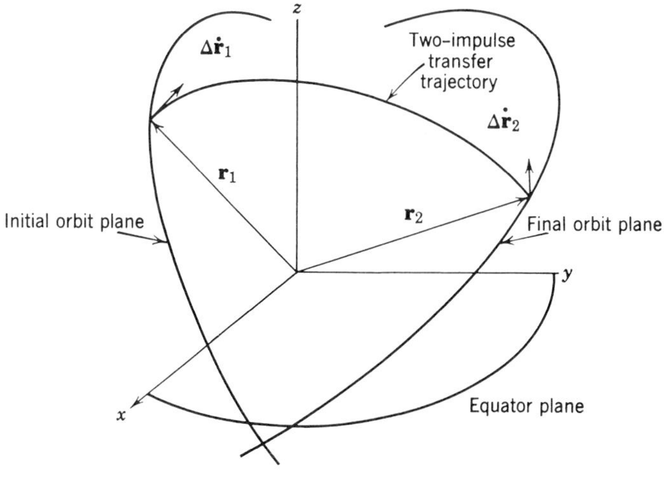

FIGURE 3.13 Two-impulse transfer.

obvious, the two-impulse maneuver is of fundamental importance in the supply of space stations or in rescue operations in space.

3.6.1. Minimum Velocity Increment Two-Impulse Transfer

The procedure developed in this section is constrained by the initial and final positions of the specified terminals and therefore solves the important case of optimum transfer modes between fixed terminals. If the absolute minimum velocity increment two-impulse transfer is desired, that is, the best departure and arrival points on the initial and final orbits, then a stepping procedure should be undertaken; that is, hold $\mathbf{r}_1$ fixed and vary $\mathbf{r}_2$ in steps through an angle of 2π, then increase $\mathbf{r}_1$ by a small amount and repeat the search on $\mathbf{r}_2$, and so on. For each step the optimum is determined by the analysis developed herein, and the absolute global minimum is then the minimum of all the local minima. The problem of determining the global minimum velocity increment trajectory is very difficult. The procedure described in this section has the advantage of determining both local and global minima, even though the global minimum must be sought by a direct search procedure. It is, however, believed that the attack on determination of the global minimum, if it is desired, is easier because this attack breaks down to the selection of the minimum of a set of local minima, whereas a direct search for the global minimum would involve a great number of unknowns and a very complicated iterative procedure. Furthermore, with the analysis contained here it is possible to locate the true global minimum, but with an iterative procedure to determine the global minimum directly it is not possible to obtain the desired local results for the very important two-impulse case that is of primary importance in present-day studies [6], [8], [9].

In order to be general and allow for different engine sizing or to reduce the two-impulse problem to an intercept problem, the minimum of the weighted sum of $|\Delta\dot{\mathbf{r}}_1|$ and $|\Delta\dot{\mathbf{r}}_2|$ is sought.

By reasoning analogous to that used in Section 3.5.1 it is possible to determine the velocity vector at the second terminal from

$$\dot{\mathbf{r}}_2 = (-\omega_1 p^{-1/2} + \omega_3 p^{1/2})\mathbf{U}_2 + (\omega_2 p^{1/2})\mathbf{U}_1, \tag{3.121}$$

where

$$\omega_1 \equiv \frac{\sqrt{\mu}[1 - \cos(v_2 - v_1)]}{\sin(v_2 - v_1)},$$

$$\omega_2 \equiv \frac{-\sqrt{\mu}}{r_2 \sin(v_2 - v_1)},$$

$$\omega_3 \equiv \frac{\sqrt{\mu}}{r_1 \sin(v_2 - v_1)}.$$

Equations 3.101 and 3.121 determine the velocity vectors at both terminals as linear combinations of the known $\mathbf{U}_1$ and $\mathbf{U}_2$ vectors and the unknown scalar p, the orbit semiparameter.

The impulsive velocity requirements for orbital transfer from the first to the second terminal are given as

$$\Delta \equiv |\dot{\mathbf{r}}_1 - \dot{\mathbf{r}}_1^*| + |\dot{\mathbf{r}}_2 - \dot{\mathbf{r}}_2^*|, \tag{3.122}$$

where $\dot{\mathbf{r}}_1^*$ and $\dot{\mathbf{r}}_2^*$ are known velocity vectors at the departure and arrival orbits at the times t_1 and t_2. Equivalently,

$$\Delta = \sqrt{\lambda_1}[(\dot{\mathbf{r}}_1 - \dot{\mathbf{r}}_1^*)\cdot(\dot{\mathbf{r}}_1 - \dot{\mathbf{r}}_1^*)]^{1/2} + \sqrt{\lambda_2}[(\dot{\mathbf{r}}_2 - \dot{\mathbf{r}}_2^*)\cdot(\dot{\mathbf{r}}_2 - \dot{\mathbf{r}}_2^*)]^{1/2} \tag{3.123}$$

defines the function whose minimum is sought, where λ_1 and λ_2 are arbitrary positive multipliers. As mentioned previously, by carrying the multipliers it is possible not only to minimize velocity requirements for a one-impulse transfer from the first or second terminal but also by the appropriate choice of the λ's, to account for variable engine propulsion systems.

More precisely, by computation of the dot products in (3.123) it is possible to write

$$\Delta = \sqrt{\lambda_1}[\Delta V_1^2(p)]^{1/2} + \sqrt{\lambda_2}[\Delta V_2^2(p)]^{1/2}, \tag{3.124}$$

where

$$\begin{aligned}\Delta V_1^2(p) = {} & (\omega_1 p^{-1/2} + \omega_2 p^{1/2})^2 + (\omega_3 p^{1/2})^2 + \dot{\mathbf{r}}_1^* \cdot \dot{\mathbf{r}}_1^* \\ & + 2(\omega_1 p^{-1/2} + \omega_2 p^{1/2})(\omega_3 p^{1/2})\mathbf{U}_1 \cdot \mathbf{U}_2 \\ & - 2(\omega_1 p^{-1/2} + \omega_2 p^{1/2})\mathbf{U}_1 \cdot \dot{\mathbf{r}}_1^* - 2(\omega_3 p^{1/2})\mathbf{U}_2 \cdot \dot{\mathbf{r}}_1^*,\end{aligned}$$

which collapses to

$$\Delta V_1^2(p) = \nu_1 p + \nu_2 p^{1/2} + \nu_3 p^{-1/2} + \nu_4 p^{-1} + \nu_5, \tag{3.125}$$

with

$$\begin{aligned}\nu_1 &\equiv \omega_2^2 + \omega_3^2 + 2\omega_2\omega_3 \mathbf{U}_1 \cdot \mathbf{U}_2, \\ \nu_2 &\equiv -2\omega_2 \dot{\mathbf{r}}_1^* \cdot \mathbf{U}_1 - 2\omega_3 \dot{\mathbf{r}}_1^* \cdot \mathbf{U}_2, \\ \nu_3 &\equiv -2\omega_1 \dot{\mathbf{r}}_1^* \cdot \mathbf{U}_1, \\ \nu_4 &\equiv \omega_1^2, \\ \nu_5 &\equiv \dot{\mathbf{r}}_1^* \cdot \dot{\mathbf{r}}_1^* + 2\omega_1\omega_3 \mathbf{U}_1 \cdot \mathbf{U}_2 + 2\omega_1\omega_2.\end{aligned}$$

Similarly,

$$\Delta V_2^2(p) = \nu_6 p + \nu_7 p^{1/2} + \nu_8 p^{-1/2} + \nu_9 p^{-1} + \nu_{10}, \tag{3.126}$$

where

$$\begin{aligned}\nu_6 &\equiv \omega_2^2 + \omega_3^2 + 2\omega_2\omega_3 \mathbf{U}_1 \cdot \mathbf{U}_2. \\ \nu_7 &\equiv -2\omega_3 \dot{\mathbf{r}}_2^* \cdot \mathbf{U}_2 - 2\omega_2 \dot{\mathbf{r}}_2^* \cdot \mathbf{U}_1, \\ \nu_8 &\equiv 2\omega_1 \dot{\mathbf{r}}_2^* \cdot \mathbf{U}_2, \\ \nu_9 &\equiv \omega_1^2, \\ \nu_{10} &\equiv \dot{\mathbf{r}}_2^* \cdot \dot{\mathbf{r}}_2^* - 2\omega_1\omega_2 \mathbf{U}_1 \cdot \mathbf{U}_2 - 2\omega_1\omega_3.\end{aligned}$$

3.6.2 Velocity Minimization Process

Equation 3.124 can be extremized by extracting the derivative of Δ with respect to p and equating to zero so that

$$\frac{\partial \Delta}{\partial p} = \tfrac{1}{2}\sqrt{\lambda_1}\,\frac{\partial\, \Delta V_1^2(p)/\partial p}{[\Delta V_1^2(p)]^{1/2}} + \tfrac{1}{2}\sqrt{\lambda_2}\,\frac{\partial\, \Delta V_2^2(p)/\partial p}{[\Delta V_2^2(p)]^{1/2}} = 0. \tag{3.127}$$

Hence the condition controlling the analysis is

$$-\frac{\sqrt{\lambda_2}[\partial\, \Delta V_2^2(p)/\partial p]}{\sqrt{\lambda_1}[\partial\, \Delta V_1^2(p)/\partial p]} = \frac{[\Delta V_2^2(p)]^{1/2}}{[\Delta V_1^2(p)]^{1/2}}, \tag{3.128}$$

or on removing radicals it follows that

$$F \equiv \lambda_2\left[\frac{\partial\, \Delta V_2^2(p)}{\partial p}\right]^2 \Delta V_1^2(p) - \lambda_1\left[\frac{\partial\, \Delta V_1^2(p)}{\partial p}\right]^2 \Delta V_2^2(p) = 0, \tag{3.129}$$

which determines the general impulse function F.

The velocity impulses $\Delta V_1^2(p)$ and $\Delta V_2^2(p)$ are given directly by (3.125) and (3.126), so that differentiation of these equations provides

$$\frac{\partial\, \Delta V_1^2(p)}{\partial p} = \nu_1 + \tfrac{1}{2}\nu_2 p^{-1/2} - \tfrac{1}{2}\nu_3 p^{-3/2} - \nu_4 p^{-2}, \tag{3.130}$$

$$\frac{\partial\, \Delta V_2^2(p)}{\partial p} = \nu_6 + \tfrac{1}{2}\nu_7 p^{-1/2} - \tfrac{1}{2}\nu_8 p^{-3/2} - \nu_9 p^{-2}. \tag{3.131}$$

Substitution of (3.125), (3.126), (3.130), and (3.131) into (3.129) yields

$$\begin{aligned} F^* = {} & \alpha_1 p + \alpha_2 p^{1/2} + \alpha_3 + \alpha_4 p^{-1/2} + \alpha_5 p^{-1} + \alpha_6 p^{-3/2} \\ & + \alpha_7 p^{-2} + \alpha_8 p^{-5/2} + \alpha_9 p^{-3} + \alpha_{10} p^{-7/2} + \alpha_{11} p^{-4} \\ & + \alpha_{12} p^{-9/2} + \alpha_{13} p^{-5}, \end{aligned} \tag{3.132}$$

where

$$\alpha_1 \equiv \lambda_2\nu_6^2\nu_1 - \lambda_1\nu_1^2\nu_6,$$

$$\alpha_2 \equiv \lambda_2\nu_6^2\nu_2 - \lambda_1\nu_1^2\nu_7 + \lambda_2\nu_6\nu_7\nu_1 - \lambda_1\nu_1\nu_2\nu_6,$$

$$\alpha_3 \equiv \tfrac{1}{4}\lambda_2\nu_7^2\nu_1 - \tfrac{1}{4}\lambda_1\nu_2^2\nu_6 + \lambda_2\nu_6^2\nu_5 - \lambda_1\nu_1^2\nu_{10} + \lambda_2\nu_6\nu_7\nu_2 - \lambda_1\nu_1\nu_2\nu_7,$$

$$\begin{aligned} \alpha_4 \equiv {} & \tfrac{1}{4}\lambda_2\nu_7^2\nu_2 - \tfrac{1}{4}\lambda_1\nu_2^2\nu_7 + \lambda_2\nu_6^2\nu_3 - \lambda_1\nu_1^2\nu_8 - \lambda_2\nu_6\nu_8\nu_1 + \lambda_1\nu_1\nu_3\nu_6 \\ & + \lambda_2\nu_6\nu_7\nu_5 - \lambda_1\nu_1\nu_2\nu_{10}, \end{aligned}$$

$$\begin{aligned} \alpha_5 \equiv {} & \tfrac{1}{4}\lambda_2\nu_7^2\nu_5 - \tfrac{1}{4}\lambda_1\nu_2^2\nu_{10} + \lambda_2\nu_6^2\nu_4 - \lambda_1\nu_1^2\nu_9 - \tfrac{1}{2}\lambda_2\nu_7\nu_8\nu_1 \\ & + \tfrac{1}{2}\lambda_1\nu_2\nu_3\nu_6 - \lambda_2\nu_6\nu_8\nu_2 + \lambda_1\nu_1\nu_3\nu_7 + \lambda_2\nu_6\nu_7\nu_3 \\ & - \lambda_1\nu_1\nu_2\nu_8 - 2\lambda_2\nu_6\nu_9\nu_1 + 2\lambda_1\nu_1\nu_4\nu_6, \end{aligned}$$

$$\alpha_6 \equiv \tfrac{1}{4}\lambda_2\nu_7^2\nu_3 - \tfrac{1}{4}\lambda_1\nu_2^2\nu_8 - \tfrac{1}{2}\lambda_2\nu_7\nu_8\nu_2 + \tfrac{1}{2}\lambda_1\nu_7\nu_3\nu_2 - \lambda_2\nu_7\nu_9\nu_1$$
$$+ \lambda_1\nu_2\nu_4\nu_6 + \lambda_2\nu_6\nu_7\nu_4 - \lambda_1\nu_1\nu_2\nu_9 - \lambda_2\nu_6\nu_8\nu_5$$
$$+ \lambda_1\nu_1\nu_3\nu_{10} - 2\lambda_2\nu_6\nu_9\nu_2 + 2\lambda_1\nu_1\nu_4\nu_7,$$
$$\alpha_7 \equiv \tfrac{1}{4}\lambda_2\nu_8^2\nu_1 - \tfrac{1}{4}\lambda_1\nu_3^2\nu_6 + \tfrac{1}{4}\lambda_2\nu_7^2\nu_4 - \tfrac{1}{4}\lambda_1\nu_2^2\nu_9 - \tfrac{1}{2}\lambda_2\nu_7\nu_8\nu_5$$
$$+ \tfrac{1}{2}\lambda_1\nu_2\nu_3\nu_{10} - \lambda_2\nu_7\nu_9\nu_2 + \lambda_1\nu_2\nu_4\nu_7 - \lambda_2\nu_6\nu_8\nu_3$$
$$+ \lambda_1\nu_1\nu_3\nu_8 - 2\lambda_2\nu_6\nu_9\nu_5 + 2\lambda_1\nu_1\nu_4\nu_{10},$$
$$\alpha_8 \equiv \tfrac{1}{4}\lambda_2\nu_8^2\nu_2 - \tfrac{1}{4}\lambda_1\nu_3^2\nu_7 - \tfrac{1}{2}\lambda_2\nu_7\nu_8\nu_3 + \tfrac{1}{2}\lambda_1\nu_2\nu_3\nu_8 + \lambda_2\nu_8\nu_9\nu_1$$
$$- \lambda_1\nu_3\nu_4\nu_6 - \lambda_2\nu_6\nu_8\nu_4 + \lambda_1\nu_1\nu_3\nu_9 - \lambda_2\nu_7\nu_9\nu_5$$
$$+ \lambda_1\nu_2\nu_4\nu_{10} - 2\lambda_2\nu_6\nu_9\nu_3 + 2\lambda_1\nu_1\nu_4\nu_8,$$
$$\alpha_9 \equiv \tfrac{1}{4}\lambda_2\nu_8^2\nu_5 - \tfrac{1}{4}\lambda_1\nu_3^2\nu_{10} + \lambda_2\nu_9^2\nu_1 - \lambda_1\nu_4^2\nu_6 - \tfrac{1}{2}\lambda_2\nu_7\nu_8\nu_4$$
$$+ \tfrac{1}{2}\lambda_1\nu_2\nu_3\nu_9 + \lambda_2\nu_8\nu_9\nu_2 - \lambda_1\nu_3\nu_4\nu_7 - \lambda_2\nu_7\nu_9\nu_3$$
$$+ \lambda_1\nu_2\nu_4\nu_8 - 2\lambda_2\nu_6\nu_9\nu_4 + 2\lambda_1\nu_1\nu_4\nu_9,$$
$$\alpha_{10} \equiv \tfrac{1}{4}\lambda_2\nu_8^2\nu_3 - \tfrac{1}{4}\lambda_1\nu_3^2\nu_8 + \lambda_2\nu_9^2\nu_2 - \lambda_1\nu_4^2\nu_7 - \lambda_2\nu_7\nu_9\nu_4 + \lambda_1\nu_2\nu_4\nu_9$$
$$+ \lambda_2\nu_8\nu_9\nu_5 - \lambda_1\nu_3\nu_4\nu_{10},$$
$$\alpha_{11} \equiv \tfrac{1}{4}\lambda_2\nu_8^2\nu_4 - \tfrac{1}{4}\lambda_1\nu_3^2\nu_9 + \lambda_2\nu_9^2\nu_5 - \lambda_1\nu_4^2\nu_{10} + \lambda_2\nu_8\nu_9\nu_3 - \lambda_1\nu_3\nu_4\nu_8,$$
$$\alpha_{12} \equiv \lambda_2\nu_9^2\nu_3 - \lambda_1\nu_4^2\nu_8 + \lambda_2\nu_8\nu_9\nu_4 - \lambda_1\nu_3\nu_4\nu_9,$$
$$\alpha_{13} \equiv \lambda_2\nu_9^2\nu_4 - \lambda_1\nu_4^2\nu_9.$$

3.6.3 Polynominal Resolvent of the Twelfth Degree

Equation 3.132 is of the twelfth degree in the semiparameter p of the transfer orbit. However, under the transformation

$$p = s^2$$

a substantial amount of algebra can be avoided to obtain

$$F^{**} = \alpha_1 s^{12} + \alpha_2 s^{11} + \alpha_3 s^{10} + \alpha_4 s^9 + \alpha_5 s^8$$
$$+ \alpha_6 s^7 + \alpha_7 s^6 + \alpha_8 s^5 + \alpha_9 s^4 + \alpha_{10} s^3$$
$$+ \alpha_{11} s^2 + \alpha_{12} s + \alpha_{13}. \tag{3.133}$$

The twelfth-order resolvent F^{**} possesses three parameters, λ_1, λ_2, and s. The λ's are multipliers whose range is by definition constrained to lie in the interval

$$0 \leq (\lambda_1, \lambda_2) \leq 1. \tag{3.134}$$

Physically, $\sqrt{\lambda_1}$ represents the total fraction of impulse applied at the first terminal. Similarly, $\sqrt{\lambda_2}$ represents the total fraction of impulse applied at the second terminal. Hence, for $\lambda_2 = 0$, the twelfth-order polynomial collapses to a lower-order equation and represents a resolvent for minimization of impulse at only the first terminal,[1] $\mathbf{r}_1$. Likewise, for $\lambda_1 = 0$, the twelfth-order resolvent collapses to a lower order equation which represents a resolvent for minimization of impulse only at the second terminal, $\mathbf{r}_2$. If $\lambda_1 = \lambda_2 = 1$, F^{**} also collapses to a lower-order equation [10] and represents a resolvent for minimization of equally weighted velocity increments at the first and second terminals. This resolvent is of degree eight with $\alpha_1 = \alpha_2 = \alpha_{12} = \alpha_{13} \equiv 0$. Finally, choosing the λ's according to inequality 3.134 results in the general two-impulse twelfth-order resolvent of the two-impulse problem.

The spurious roots obtained from the stationary condition $F^{**} = 0$ can be eliminated by direct substitution into (3.101) and (3.121) with the understanding that all roots where Δ is not minimum are rejected. Once the semi-parameter p has been determined, it follows that (3.101), that is,

$$\dot{\mathbf{r}}_1 = (\omega_1 p^{-1/2} + \omega_2 p^{1/2})\mathbf{U}_1 + (\omega_3 p^{1/2})\mathbf{U}_2, \tag{3.135}$$

yields the velocity vector at the first terminal. Hence, since $\mathbf{r}_1$ and $\dot{\mathbf{r}}_1$ are known, the transfer orbit is considered to be determined. Any auxiliary elements which are desired can be computed from the set $(\mathbf{r}_1, \dot{\mathbf{r}}_1)$.

3.6.4 Minimum Time Two-Impulse Transfer

In many instances the problem of minimum time trajectories is of interest. For example, in cases of life support equipment failure, it is imperative to return the astronaut to a final trajectory state as soon as possible.

The problem to be solved in this section is the determination of minimum time trajectories between two specified terminals, that is, between the two position vectors $\mathbf{r}_1$ and $\mathbf{r}_2$ illustrated in Figure 3.10.

As previously discussed for the case of fuel critical transfers, if the absolute minimum one- or two-impulse time transfer is desired, that is, if it is desired to find the best departure and arrival points on the initial and final orbits, then a stepping procedure, letting $\mathbf{r}_1$ and $\mathbf{r}_2$ vary in true anomaly from 0 to 2π, should be undertaken. For each step the suboptimum is determined by the analysis developed herein; the absolute global minimum is then the minimum of all the local minima. The problem of determining the global minimum time of transfer trajectory is a very difficult one. The procedure described in this section has the advantage of determining both local and global minima, even though the global minimum must be sought by a direct search procedure. It is known, however, that the determination breaks

[1] See Section 3.5.2 for the explicit reduction to a quartic equation.

down to the selection of the minimum of a set of local minima. A direct search for the global minimum would involve a great number of unknowns and a very complicated iterative procedure which may possess convergence difficulties. Furthermore, with the analysis given here it is possible to locate the true global minimum, but with a direct iterative procedure to determine the global minimum it is not possible to obtain the desired results for the very important two-impulse transfer case that is of primary importance in present-day studies [9].

3.6.5 Time Minimization Process

The problem to be solved in this section is the minimization of the time interval required to travel between position vectors $\mathbf{r}_1$ and $\mathbf{r}_2$ where, as illustrated in Figure 3.13, $\mathbf{r}_1$ is contained in orbit plane 1 and $\mathbf{r}_2$ is contained in orbit plane 2. For convenience the time interval is denoted by τ, the modified time variable, which is defined by

$$\tau \equiv k(t_2 - t_1), \tag{3.136}$$

where k = the central planet gravitational constant, t_1 = universal departure time from orbit 1, and t_2 = universal arrival time at orbit 2. The minimum time problem, however, is not physically meaningful unless an important constraint is imposed on the transfer process; namely, the total velocity increment available for a given transfer is limited by

$$|\dot{\mathbf{r}}_1 - \dot{\mathbf{r}}_1^*| + |\dot{\mathbf{r}}_2 - \dot{\mathbf{r}}_2^*| \leq \Delta V_{\max}, \tag{3.137}$$

where $\dot{\mathbf{r}}_1^*$ and $\dot{\mathbf{r}}_2^*$ are the known velocity vectors at the departure and arrival orbits and $\Delta V_{\max}$ is the maximum available velocity impulse which is available from the propulsion system. Equivalently,

$$\sqrt{\lambda_1}[(\dot{\mathbf{r}}_1 - \dot{\mathbf{r}}_1^*)\cdot(\dot{\mathbf{r}}_1 - \dot{\mathbf{r}}_1^*)]^{1/2} + \sqrt{\lambda_2}[(\dot{\mathbf{r}}_2 - \dot{\mathbf{r}}_2^*)\cdot(\dot{\mathbf{r}}_2 - \dot{\mathbf{r}}_2^*)]^{1/2} \leq \Delta V_{\max} \tag{3.138}$$

with λ_1 and λ_2 introduced as arbitrary positive multipliers. By carrying the λ's it is possible to minimize transfer times using one, two, or a set ratio of the impulses at each of the terminals. Inequality 3.138 is important for physical considerations because, if it did not control, a minimum time orbit would be characterized by a straight line trajectory between $\mathbf{r}_1$ and $\mathbf{r}_2$ with an infinite amount of energy; that is, inequality 3.138 is a propulsion system constraint.

Inequality 3.138 can be written more compactly as

$$\sqrt{\lambda_1}[\Delta V_1^2(p)]^{1/2} + \sqrt{\lambda_2}[\Delta V_2^2(p)]^{1/2} \leq \Delta V_{\max}, \tag{3.139}$$

where p is semiparameter of the transfer orbit and where, from (3.125) and (3.126),

$$\Delta V_1^2(p) = \nu_1 p + \nu_2 p^{1/2} + \nu_3 p^{-1/2} + \nu_4 p^{-1} + \nu_5, \tag{3.140}$$

$$\Delta V_2^2(p) = \nu_6 p + \nu_7 p^{1/2} + \nu_8 p^{-1/2} + \nu_9 p^{-1} + \nu_{10}. \tag{3.141}$$

As discussed in Section 3.6.1, it should be noted that the ν coefficients are known from the initial data of the problem, that is, from $\mathbf{r}_1$ and $\mathbf{r}_2$.

Inequality 3.139 can be expressed in equation form (Section 2.3) as

$$\delta - \{\sqrt{\lambda_1}[\Delta V_1^2(p)]^{1/2} + \sqrt{\lambda_2}[\Delta V_2^2(p)]^{1/2}\} = \epsilon^2, \tag{3.142}$$

where ϵ is a new independent parameter and where, for convenience, $\delta = \Delta V_{\text{max}}$.

By applying Lagrange's rule it is possible to form the minimization function M as

$$M(p) = \tau + \mu(\delta - \sqrt{\lambda_1}[\Delta V_1^2]^{1/2} - \sqrt{\lambda_2}[\Delta V_2^2]^{1/2} - \epsilon^2). \tag{3.143}$$

In (3.143) μ is an unknown Lagrange multiplier. In order to minimize τ the standard procedure would demand that the partial derivatives of M with respect to the independent variables vanish. If two position vectors are specified, however, six constants of the orbit are known and only one degree of freedom is available. This one degree of freedom is absorbed by (3.142). Hence the condition of optimality becomes

$$\frac{\partial M}{\partial \epsilon} = 0, \tag{3.144}$$

or

$$-2\mu\epsilon = 0. \tag{3.145}$$

If $\mu = 0$, no constraint would be imposed. Therefore, for the case $\epsilon = 0$, the controlling equation of the minimum time orbit determination problem becomes

$$\sqrt{\lambda_1}[\Delta V_1^2]^{1/2} + \sqrt{\lambda_2}[\Delta V_2^2]^{1/2} = \delta. \tag{3.146}$$

3.6.6 Polynomial Resolvent of the Eighth Degree

Consider (3.146) written as

$$\sqrt{\lambda_1}[\Delta V_1^2]^{1/2} = \delta - \sqrt{\lambda_2}[\Delta V_2^2]^{1/2} \tag{3.147}$$

which, on squaring, becomes

$$\lambda_1 \, \Delta V_1^2 - \lambda_2 \, \Delta V_2^2 - \delta^2 = -2\delta\sqrt{\lambda_2}[\Delta V_2^2]^{1/2}. \tag{3.148}$$

On squaring again, the minimum time condition can be written more explicitly as

$$\lambda_1^2\,\Delta V_1^4 + \lambda_2^2\,\Delta V_2^4 + \delta^4 - 2\lambda_1\lambda_2\,\Delta V_1^2\,\Delta V_2^2 - 2\lambda_1\,\delta^2\,\Delta V_1^2 - 2\lambda_2\,\delta^2\,\Delta V_2^2 = 0. \quad (3.149)$$

Equations 3.140 and 3.141 provide the necessary reduction of condition 3.149 to a function of p. Hence, by computing ΔV_1^4, ΔV_2^4, and $\Delta V_1^2\,\Delta V_2^2$, it is possible to obtain the form

$$F^* = \alpha_1 p^2 + \alpha_2 p^{3/2} + \alpha_3 p + \alpha_4 p^{1/2} + \alpha_5 + \alpha_6 p^{-1/2} + \alpha_7 p^{-1} + \alpha_8 p^{-3/2} + \alpha_9 p^{-2}, \quad (3.150)$$

where

$$\begin{aligned}
\alpha_1 &\equiv \lambda_1^2\nu_1^2 + \lambda_2^2\nu_6^2 - 2\lambda_1\lambda_2\nu_1\nu_6,\\
\alpha_2 &\equiv 2\lambda_1^2\nu_1\nu_2 + 2\lambda_2^2\nu_6\nu_7 - 2\lambda_1\lambda_2(\nu_2\nu_6 + \nu_1\nu_7),\\
\alpha_3 &\equiv \lambda_1^2(\nu_2^2 + 2\nu_1\nu_5) + \lambda_2^2(\nu_7^2 + 2\nu_6\nu_{10}) - 2\lambda_1\lambda_2(\nu_5\nu_6 + \nu_2\nu_7 + \nu_1\nu_{10})\\
&\quad - 2\lambda_1\,\delta^2\nu_1 - 2\lambda_2\,\delta^2\nu_6,\\
\alpha_4 &\equiv \lambda_1^2(2\nu_1\nu_3 + 2\nu_2\nu_5) + \lambda_2^2(2\nu_6\nu_8 + 2\nu_7\nu_{10})\\
&\quad - 2\lambda_1\lambda_2(\nu_3\nu_6 + \nu_5\nu_7 + \nu_1\nu_8 + \nu_2\nu_{10})\\
&\quad - 2\lambda_1\,\delta^2\nu_2 - 2\lambda_2\,\delta^2\nu_7,\\
\alpha_5 &\equiv \lambda_1^2(\nu_5^2 + 2\nu_1\nu_4 + 2\nu_2\nu_3) + \lambda_2^2(\nu_{10}^2 + 2\nu_6\nu_9 + 2\nu_7\nu_8)\\
&\quad - 2\lambda_1\lambda_2(\nu_4\nu_6 + \nu_3\nu_7 + \nu_2\nu_8 + \nu_1\nu_9 + \nu_5\nu_{10})\\
&\quad - 2\lambda_1\,\delta^2 v_5 - 2\lambda_2\,\delta^2\nu_{10} + \delta^4,\\
\alpha_6 &\equiv \lambda_1^2(2\nu_2\nu_4 + 2\nu_3\nu_5) + \lambda_2^2(2\nu_7\nu_9 + 2\nu_8\nu_{10})\\
&\quad - 2\lambda_1\lambda_2(\nu_2\nu_9 + \nu_4\nu_7 + \nu_5\nu_8 + \nu_3\nu_{10})\\
&\quad - 2\lambda_1\,\delta^2\nu_3 - 2\lambda_2\,\delta^2\nu_8,\\
\alpha_7 &\equiv \lambda_1^2(\nu_3^2 + 2\nu_4\nu_5) + \lambda_2^2(\nu_8^2 + 2\nu_9\nu_{10}) - 2\lambda_1\lambda_2(\nu_3\nu_8 + \nu_5\nu_9 + \nu_4\nu_{10})\\
&\quad - 2\lambda_1\,\delta^2\nu_4 - 2\lambda_2\,\delta^2\nu_9,\\
\alpha_8 &\equiv 2\lambda_1^2\nu_3\nu_4 + 2\lambda_2^2\nu_8\nu_9 - 2\lambda_1\lambda_2(\nu_4\nu_8 + \nu_3\nu_9),\\
\alpha_9 &\equiv \lambda_1^2\nu_4^2 + \lambda_2^2\nu_9^2 - 2\lambda_1\lambda_2\nu_4\nu_9.
\end{aligned}$$

The algebraic form (3.150) can be transformed on removal of radicals to the eighth-degree polynomial

$$F^{**} = \alpha_1 s^8 + \alpha_2 s^7 + \alpha_3 s^6 + \alpha_4 s^5 + \alpha_5 s^4 + \alpha_6 s^3 + \alpha_7 s^2 + \alpha_8 s + \alpha_9, \quad (3.151)$$

where for convenience $p = s^2$.

The eighth-order resolvent F^{**} possesses three parameters, λ_1, λ_2, and s. The λ's are multipliers whose range by definition is constrained to lie in the interval

$$0 \leq (\lambda_1, \lambda_2) \leq 1. \tag{3.152}$$

Physically, $\sqrt{\lambda_1}$ represents the total fraction of impulse applied at the first terminal. Similarly, $\sqrt{\lambda_2}$ represents the total fraction of impulse applied at the second terminal. Hence, for $\lambda_2 = 0$, the eighth-order resolvent collapses to a lower order equation[2] for the determination of minimum time trajectories using a single impulse at the first terminal, $\mathbf{r}_1$. If $\lambda_1 = \lambda_2 = 1$, F^{**} represents a resolvent for transfer time minimization with equally weighted velocity increments at the first and second terminals. Finally, choosing the λ's according to inequality 3.152 results in the general eighth-order resolvent of the two-impulse minimum time problem.

The spurious roots obtained from the stationary condition $F^{**} = 0$ can be eliminated by directly computing the time required to go from $\mathbf{r}_1$ to $\mathbf{r}_2$ and choosing the p corresponding to the smallest time interval τ. The determination of the orbit, that is, the determination of $\dot{\mathbf{r}}_1$, the velocity vector at the first terminal, can be accomplished by use of (3.101). Since, $\mathbf{r}_1$ and $\dot{\mathbf{r}}_1$ are known, the orbit and any additional elements can be computed.

3.6.7 Fixed Time of Flight Transfer

If rendezvous must be accomplished with a fixed time of flight, the problem is reduced to the deterministic two position vectors and time interval problem discussed in ([1], Chapter 6); any of the six methods discussed there suffices to determine the orbit. The velocity requirements to perform the maneuver are given directly by

$$\Delta V = |\dot{\mathbf{r}}_1 - \dot{\mathbf{r}}_1^*| + |\dot{\mathbf{r}}_2 - \dot{\mathbf{r}}_2^*|,$$

where $\dot{\mathbf{r}}_1^*$ and $\dot{\mathbf{r}}_2^*$ are the known velocity vectors on the departure and arrival orbits.

3.7 SUMMARY

This chapter introduced the fundamental concepts of impulsive velocity additions and the one-to-one correspondence of velocity increment with fuel expenditure. With these approximations it was possible to arrive at analytic solutions to various transfer processes.

[2] See Section 3.5.3 for the explicit reduction to a quartic equation.

Hohmann transfers or periapsis to apoapsis transfers and apoapsis to periapsis transfers were discussed. A relatively new transfer process, involving an intermediate impulse, called the bielliptic transfer was also introduced. It was demonstrated that for several criteria, depending on the radius ratio of the initial and final orbits, the bielliptic maneuver is superior to the Hohmann maneuver.

A general velocity transfer function was developed for bielliptic transfers with plane changes. It was shown how this function can be used to obtain other types of transfer modes with plane changes.

For transfers between elliptic orbits the cotangential transfer procedure was introduced. This mode of transfer represents a near optimum fuel critical transfer maneuver.

The three-dimensional intercept maneuver was introduced next. This mode of transfer consists in leaving a coasting orbit by application of an instantaneous velocity increment which places the transfer vehicle on a collision course with a specified target. Both fuel critical and time critical intercept maneuvers were introduced. It was shown that both of these maneuvers correspond to the solution of quartic equations in the square root of the semi-parameter of the transfer orbit. Fixed time of flight transfers must be solved by iterative procedures.

Two-impulse transfers, in which not only target position but also velocity are matched, were developed. In the two-impulse maneuver the fuel critical transfer for arbitrarily weighted impulses was shown to depend on the solution of a twelfth-degree algebraic equation in the square root of the semiparameter of the transfer conic. The time critical two-impulse maneuver was shown to be reducible to an eighth-degree algebraic equation. Fixed time of flight rendezvous operations must be solved by means of iterative techniques.

It was pointed out that the general three-dimensional problem of best overall points of departure and arrival for two orbits can be handled by repeated application of the optimum analytical formulations developed in this chapter. This optimization can be performed by systematically varying the initial and final arrival terminals.

EXERCISES

1. A spacecraft in a geocentric circular orbit with a semimajor axis of 1.5 e.r. is to descend to another coplanar circular orbit via a Hohmann transfer. What velocity impulse is required to perform the maneuver if the final orbit has a semimajor axis of 1.1 e.r.?
2. It is desired to have a reconnaissance flight to a Syncom satellite, orbiting in a circular orbit with semimajor axis 6 e.r., from a circular near Earth orbit of 1.2 e.r. The observation probe is to return to the original orbit

after a Hohmann transfer with no apogee injection. What is the coplanar phase angle between the reconnaissance probe and Syncom immediately before initiation of the mission?

3. An orbital transfer is to be performed between an inner geocentric circular orbit with semimajor axis 1.5 e.r. to an outer circular orbit with semimajor axis 17 e.r. Perform the most optimum fuel saving maneuver. Assume that the orbits are coplanar.

4. Two circular geocentric orbits are inclined to each other by an angle of 45°. Perform a bielliptic transfer from the departure orbit, $a = 2$ e.r., to the arrival orbit, $a = 8$ e.r., assuming an initial plane change of 5° and a secondary plane change of 15° applied at 6 e.r. Is the change of plane distribution optimum from the point of view of fuel expenditure? If not, what is the optimum distribution? What is the ΔV expenditure?

5. Space station Omicron-C is to receive a shipment of mercury for its heating system. The shipment of metal is in a low Earth circular orbit with a semimajor axis of 1.07 e.r. If the circular orbit of Omicron-C is inclined to the supply orbit by an angle of 15° and an initial inplane Hohmann transfer to the line of intersection of both orbits is made, at which point all the plane change is performed, what velocity increment is required to perform the maneuver?

6. Verify the general inward bielliptic formula corresponding to (3.80).

7. Compute the minimum one-impulse transfer from geocentric space station Omega 3 with state vector

$$x_1 = 1.1, \qquad y_1 = 1.05, \qquad z_1 = 1.02,$$

$$\dot{x}_1 = 0.01, \qquad \dot{y}_1 = 0.50, \qquad \dot{z}_1 = 0.80,$$

where the position components are in e.r. and velocity components are in c.s.u., to deliver a probe to geocentric position

$$x = 10.0, \qquad y = 8.0, \qquad z = 5.0.$$

8. For Problem 7, what is the minimum time solution? Assume a velocity budget of 0.001 c.s.u.

REFERENCES

[1] P. R. Escobal, *Methods of Orbit Determination*, John Wiley and Sons, New York 1965.

[2] W. Hohmann, *Die Erreichbarkeit der Himmelskörper*, Oldenbourg, Munich, 1925.

[3] R. F. Hoelker and R. Silber, *The Bi-elliptic Transfer between Circular Coplanar Orbits*, Army Ballistic Missile Agency, Redstone Arsenal, Alabama, DA-TM-2-59, January 1959.

[4] H. L. Roth, *Preliminary Investigation Relative to Multiple Rendezvous between Circular Orbits*, Aerospace Corporation, SSD-TDR-63-179, January 1964.

[5] D. F. Lawden, *Optimal Trajectories for Space Navigation*, Butterworth and Co. London, 1963.
[6] D. F. Lawden, "Orbital Transfer Via Tangential Ellipses," *Journal of the Interplanetary Society*, November 1952.
[7] H. L. Roth, "Transfer from an Arbitrary Inertial Flight Condition to a Point Target," *Journal of the Aerospace Sciences*, September 1961.
[8] P. R. Escobal, *Generalized Two-Impulse Minimum Velocity Rendezvous and Transfer Analysis*, TRW Systems, 3400-6023-RU-000, 1966.
[9] P. R. Escobal, *One and Two-Impulse Minimum Time Rendezvous and Transfer Analysis*, TRW Systems, 3400-6028-RU-000, 1966.
[10] G. Lee, "An Analysis of Two-Impulse Orbital Transfer," *AIAA Journal*, October 1964.
[11] G. A. McCue, "Optimum Two-Impulse Transfer and Rendezvous between Inclined Elliptical Orbits," *AIAA Journal*, August 1963.
[12] H. L. Roth, *Use of the Bi-Elliptic Transfer to Accomplish Single Rendezvous*, Aerospace Corporation Memo A63-1741.5-7, May 1963.
[13] R. H. Battin, *Astronautical Guidance*, McGraw-Hill Book Company, New York, 1964.
[14] R. Deutsch, *Orbital Dynamics of Space Vehicles*, Prentice-Hall, Englewood Cliffs, N.J., 1963.
[15] R. E. Moritz, *On Mathematics*, Dover Publications, New York, 1958.
[16] G. S. Stern, "Optimum Deorbit Positioning for Minimum Impulse Reentry," Paper presented at the AAS Spaceflight Mechanics Specialist Conference, July 1966.
[17] L. Ting, "Optimum Orbital Transfer by Impulses," *American Rocket Society Journal*, November, 1960.
[18] J. M. Horner, "Optimum Two-Impulse Transfer between Arbitrary Coplanar Terminals," *American Rocket Society Journal*, January 1962.
[19] H. Munick, "Optimum Orbital Transfer between Coplanar Orbits," *American Rocket Society Journal*, July 1962.
[20] H. G. Moyer, "Minimum Impulse Coplanar Circle-Ellipse Transfer," *AIAA Journal*, April 1965.
[21] J. M. Eggleston, "Optimum Time to Rendezvous," *American Rocket Society Journal*, November 1960.
[22] T. N. Edelbaum, "Minimum Impulse Transfers in the Near Vicinity of a Circular Orbit," *Journal of the Aerospace Sciences*, March 1967.

4 Optimum Interplanetary Transfers

We discover ourselves upon a planet, itself almost imperceptible in the vast extent of the solar system, which in turn is only an insensible point in the immensity of space. The sublime results to which this discovery has led should console us for our extreme littleness, and rank which it assigns to the Earth.

LAPLACE [10]

4.1 INTERPLANETARY TRAJECTORIES

Manned exploration of the different bodies of the solar system will be one of the most challenging chapters in man's history. As discussed in Chapter 1, the physical environment peculiar to each planet of the solar system presents an intriguing adventure which will shed new light on the many unsolved mysteries of various branches of science. Journeys to other worlds are very complicated from an astrodynamical point of view. The ballistics necessary to establish an interplanetary orbit with sufficient accuracy to ensure a successful mission can be implemented only through the techniques of special perturbations. These numerical procedures are discussed in Chapter 7. The main intent of this chapter is to develop rapid, but accurate, analytic solutions to the interplanetary orbit determination problem. Trajectory techniques developed by analytic means are the only efficient method of determining sufficiently valid starting conditions which can be input into the more accurate special perturbations procedures. Hence, without the rapid estimating methods to be discussed, the analyst would be lacking the initial conditions to start the numerical process of integration and would have to guess wildly at the necessary initial state variables required to target the objective planet.

In this chapter several analytic methods for determining interplanetary orbits are discussed. To introduce these techniques a discussion of the method of patched conics is undertaken and several relationships pertinent to more sophisticated transfer procedures are established. With these principles in mind it becomes possible to discuss optimum interplanetary transfer modes. These optimum transfer modes include minimum fuel, time of flight, and initial weight trajectories to planetary bodies. The special case of interplanetary flight to the Moon is not discussed in this chapter. By their specific nature lunar trajectories deserve more detailed treatment and are discussed in Chapter 5.

Some of the many contributors to this area of special interest have been Lawden [2], Fimple [3], Breakwell [4], and Battin [5]. Recently Escobal [6], Crichton and Escobal [7], and Ross [14] have investigated special aspects of the interplanetary process.

4.1.1 Escape from Primary Planet

A fundamental difference between planetocentric and interplanetary maneuvers is the extra added energy required by interplanetary vehicles to overcome or escape the gravitational attraction of the departure planet. Consider an interplanetary probe or spaceship escaping from terminal D as illustrated in Figure 4.1.

If the spaceship is to reach the arrival terminal A in specified transfer time Δt, it must be in a heliocentric orbit which passes through the terminals $\mathbf{r}_{dp}$ and $\mathbf{r}_{ap}$, where $\mathbf{r}_{dp}$ and $\mathbf{r}_{ap}$ are, respectively, the heliocentric position vectors of the departure planet (dp) and arrival planet (ap). As may be evident (see [1], Chapter 5), the two-position vectors and time interval completely determine the orbit of the interplanetary transfer vehicle. Therefore,

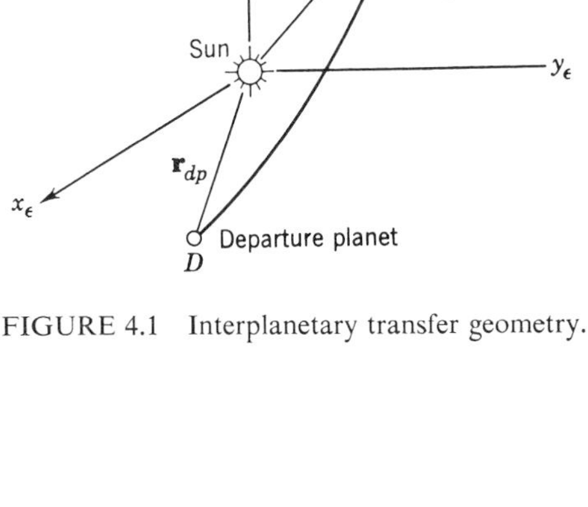

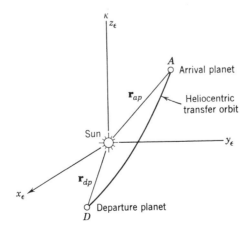

FIGURE 4.1 Interplanetary transfer geometry.

at terminal D it follows that the heliocentric transfer orbit has a very definite inertial velocity vector, call it $\dot{\mathbf{r}}_d$. The departure planet also has a specific velocity vector $\dot{\mathbf{r}}_{dp}$. It follows that $\dot{\mathbf{r}}_h$, that is,

$$\dot{\mathbf{r}}_h \equiv \dot{\mathbf{r}}_d - \dot{\mathbf{r}}_{dp}, \tag{4.1}$$

is the required velocity vector which must be added to the escape hyperbola velocity vector of a given vehicle to place it on the heliocentric orbit, at least kinematically. In fact, if the planet had no mass, this would be exactly the

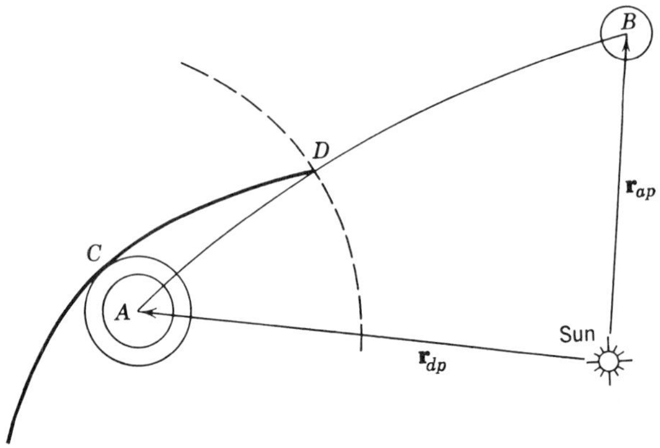

FIGURE 4.2 Escape maneuver.

additive velocity increment that would have to be provided by the vehicle propulsive system. However, the gravitational acceleration caused by the finite planetary mass must also be overcome. The kinematic transfer speed is of course given by

$$V_h^2 = \dot{\mathbf{r}}_h \cdot \dot{\mathbf{r}}_h = (\dot{\mathbf{r}}_d - \dot{\mathbf{r}}_{dp}) \cdot (\dot{\mathbf{r}}_d - \dot{\mathbf{r}}_{dp}). \tag{4.2}$$

Let it be assumed, as usual, that initially the interplanetary vehicle is coasting in a circular orbit with semimajor axis a about the departure planet. The vis-viva integral

$$V^2 = \mu\left(\frac{2}{r} - \frac{1}{a}\right), \tag{4.3}$$

where V is the planetocentric speed, μ is the sum of masses of vehicle and planet, and r is the magnitude of the radius vector from the planetary dynamical center to the orbiting space vehicle, can be used to compute the speed required to remove the spaceship to infinity or, more accurately, to a very removed point from the departure planet. Consider the escape maneuver illustrated in Figure 4.2.

For a specified trip time let it be assumed that a heliocentric orbital arc AB has been determined between the two known terminals $\mathbf{r}_{dp}$ and $\mathbf{r}_{ap}$. At point C a vehicle in a circular orbit is given an impulse ΔV which places the orbit on the hyperbolic escape arc CD. If it is assumed that at point D the effect of the departure planet becomes negligible and it is desired to be on orbital segment AB, then the kinematic velocity constraint described by (4.1) must hold true at this point. Since orbital segment CD is very short compared to the distance between the arrival and departure planets, let it be assumed that the velocity vector on arc AB at point D is identical with the velocity vector on arc AB at point A. This is by no means a serious approximation because an iteration can be used to eliminate the small discrepancy, if it is so desired. Since both $\dot{\mathbf{r}}_d$ and $\dot{\mathbf{r}}_{dp}$ are known, use of this approximation enables the magnitude of $\dot{\mathbf{r}}_h$ to be obtained directly from (4.2). The speed V_h at point D can now be used to determine the semimajor axis of the escape hyperbola from the vis-viva integral as

$$\frac{1}{a} = \frac{2}{r} - \frac{V_h^2}{\mu}. \tag{4.4}$$

For purposes of computation, point D, represents a very far removed point; more exactly, with respect to the departure planet this point is an infinite distance away. It is the point at which the departure planet's gravitational influence has become insignificant with respect to the heliocentric forces of attraction. Hence letting r tend to infinity in (4.4) yields

$$\frac{1}{a} = -\frac{V_h^2}{\mu}. \tag{4.5}$$

At point C, to leave the circular parking orbit, the following tangential velocity increment must be added over and above the coast orbit speed:

$$\Delta V = V - V_c,$$

where V is the hyperbolic periapsis speed and V_c is the circular orbit speed. The hyperbolic speed at periapsis can be obtained by a second application of the vis-viva equation so that

$$\Delta V = \left[\mu\left(\frac{2}{r} - \frac{1}{a}\right)\right]^{1/2} - V_c,$$

where $1/a$ is obtained from (4.5) and r is the radius of the circular parking orbit. Making the appropriate substitutions yields

$$\Delta V = [V_p^2 + V_h^2]^{1/2} - V_c, \tag{4.6}$$

where $V_p = \sqrt{2}V_c$. The parabolic speed V_p is a useful concept and reflects the fact that (4.6) is really an energy relationship; that is, V_p is the speed required to remove a vehicle to an infinitely distant location from the departure planet. Experience has shown that (4.6) provides the analyst with a single relationship which approximates the true physics of the escape maneuver to a good degree of accuracy and yet makes tractable the solution of many interplanetary problems. Extensive use of this equation is made in the following sections.

4.1.2 The Method of Patched Conics

A simple method which finds much use in interplanetary transfer techniques is called the method of patched conics. In this technique the trajectory from the starting terminal to the final terminal is segmented into various parts wherein each part of the trajectory is a two-body orbit. Hence, if the space vehicle is very close to a neighboring planet, a planetocentric Keplerian segment defines the motion; similarly, once the space vehicle has overcome the predominant part of the planetary attraction, a Keplerian heliocentric orbit is assumed and the trajectory is continued. The geometry of the patched conic technique is illustrated in Figure 4.3.

Switching of planetocentric to heliocentric orbital segments is accomplished at an arbitrary but physically reasonable boundary such as the sphere of equal or nearly equal gravitational attractions between the planet and the Sun. Hence, if a spaceship at point S of Figure 4.3 is about to enter a region or sphere of stronger gravitational influence than is presently controlling

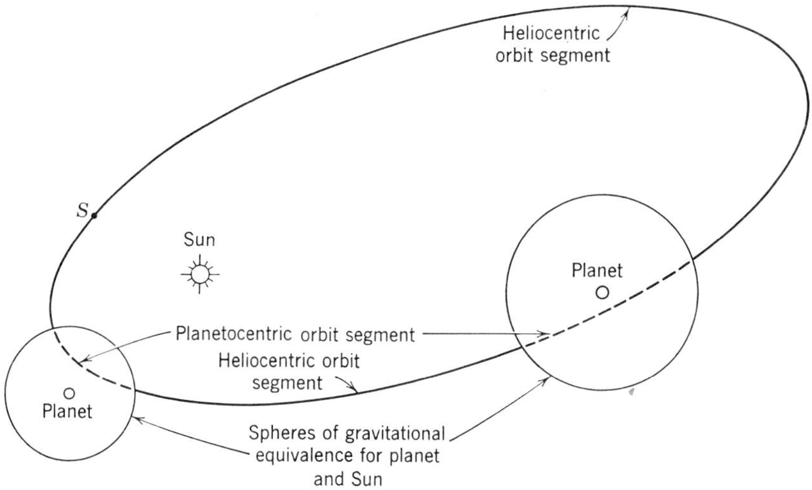

FIGURE 4.3 Segmented trajectory consisting of patched conic segments.

motion, the final heliocentric conditions form the initial conditions for the planetocentric orbit after crossing the dividing boundary. Similarly, on exit from the sphere the planetocentric conditions form the initial conditions for the next heliocentric leg of the journey. In practice the sphere or controlling boundary is referred to as the *mean sphere of influence* or in abbreviated form as the MSI.

The concept of the mean sphere of influence was originated by Laplace who stated that it is important always to refer the motion of a body to the mass or to the center of attraction which dominates the actual motion of the body under consideration. If three bodies are considered and the ratio of the perturbing to two-body acceleration is computed from the differential equations of motion relative to body 1, a partial index of the perturbative influence becomes available. Similarly, if the ratio of the perturbing to two-body acceleration is computed relative to body 3, another index of the perturbations induced on body 2, the space vehicle, becomes available. By equating these two ratios and making simplifying approximations it is possible to obtain the relationship [5]

$$r_I = r_D\left(\frac{m_1}{m_3}\right)^{2/5}, \tag{4.7}$$

where r_I = the radius of the sphere of influence, r_D = the distance between body 1 and body 3, m_1 = mass of body 1 about which sphere is placed, and m_3 = mass of body 3. This equation, though by no means exact, represents the envelope of influence, that is, a sphere with sufficient theoretical justification to form a working or usable model. In (4.7) the mass of body 2, the space vehicle has been assumed to be equal to zero. For the planets of the solar system a direct computation yields the radius of influence of the respective planets. These radii and the associated gravitational constants are collected in Table 4.1.

Table 4.1 Planetary Data

Planet	Radius of Influence (km)	Gravitational Constant (km^3/sec^2)
Sun	∞	$0.132715445 \times 10^{12}$
Mercury	0.111780×10^6	0.216855300×10^5
Venus	0.616960×10^6	0.324769500×10^6
Earth	0.924820×10^6	0.398603200×10^6
Mars	0.577630×10^6	0.429778000×10^5
Jupiter	0.48141×10^8	0.126710600×10^9
Saturn	0.54774×10^8	0.379187000×10^8
Uranus	0.51755×10^8	0.580329200×10^7
Neptune	0.86952×10^8	0.702607200×10^7
Pluto	0.35812×10^8	0.331788600×10^6

The actual patching process can be implemented in the following fashion. Suppose that at epoch time $t = t_0$ the state vector of a space vehicle is known through knowledge of the rectangular position and velocity elements

$$x_0,\ y_0,\ z_0,\ \dot{x}_0,\ \dot{y}_0,\ \dot{z}_0.$$

By the standard methods outlined in ([1], Chapter 3) it is possible to determine the classical elements

$$a,\ e,\ T,\ \mathbf{P},\ \mathbf{Q},$$

where, internal to the MSI, $a \equiv$ orbital semimajor axis, $e \equiv$ orbital eccentricity, $T \equiv$ universal time of periapsis passage, $\mathbf{P} \equiv$ unit vector from dynamical center pointing at periapsis, and $\mathbf{Q} \equiv$ unit vector advanced to $\mathbf{P}$ by a right angle in the plane and direction of motion. If the space vehicle is to cross the MSI, it follows that $e \neq 0$, and from the equation of a conic, that is,

$$r = \frac{p}{1 + e\cos v}, \tag{4.8}$$

the following relationships are obtained

$$\cos v_I = \frac{1}{e}\left(\frac{p}{r_I} - 1\right),$$

$$\sin v_I = \pm[1 - \cos^2 v_I]^{1/2}. \tag{4.9}$$

The duality of signs in (4.9) yields the two puncture points of the orbital segment with the MSI. Therefore at the exit point, corresponding to the latest time, the orbital plane positions and velocities can be obtained from

$$x_\omega = r_I \cos v_I, \qquad y_\omega = r_I \sin v_I, \tag{4.10}$$

$$\dot{x}_\omega = -\left(\frac{\mu}{p}\right)^{1/2} \sin v_I, \qquad \dot{y}_\omega = \left(\frac{\mu}{p}\right)^{1/2} (\cos v_I + e). \tag{4.11}$$

The planetocentric positions and velocities can now be obtained by means of the standard mappings

$$\mathbf{r}_I = x_\omega \mathbf{P} + y_\omega \mathbf{Q},$$

$$\dot{\mathbf{r}}_I = \dot{x}_\omega \mathbf{P} + \dot{y}_\omega \mathbf{Q}. \tag{4.12}$$

Usually the fundamental plane of the adopted coordinate system, the equator, is not in the ecliptic. Therefore a rotation of coordinates as described in ([1], Appendix 1) must be performed. The following equations define the appropriate rotation for a geocentric system:

$$\begin{aligned} x_\epsilon &= x_I, & \dot{x}_\epsilon &= \dot{x}_I,\\ y_\epsilon &= y_I \cos\epsilon + z_I \sin\epsilon, & \dot{y}_\epsilon &= \dot{y}_I \cos\epsilon + \dot{z}_I \sin\epsilon,\\ z_\epsilon &= -y_I \sin\epsilon + z_I \cos\epsilon, & \dot{z}_I &= -\dot{y}_I \sin\epsilon + z_I \cos\epsilon, \end{aligned} \tag{4.13}$$

where ϵ, the obliquity of the ecliptic obtained from (1.22), is taken as constant with very little error. The geocentric positions and velocities have been rotated into the ecliptic coordinate system, that is, into the celestial latitude-longitude coordinate system. However, a translation of (4.13) must still be effected. As illustrated in Figure 4.4, let it be assumed that $\mathbf{r}_{vp\epsilon}$ and $\dot{\mathbf{r}}_{vp\epsilon}$ represent the position and velocity vectors of the vehicle or spaceship relative

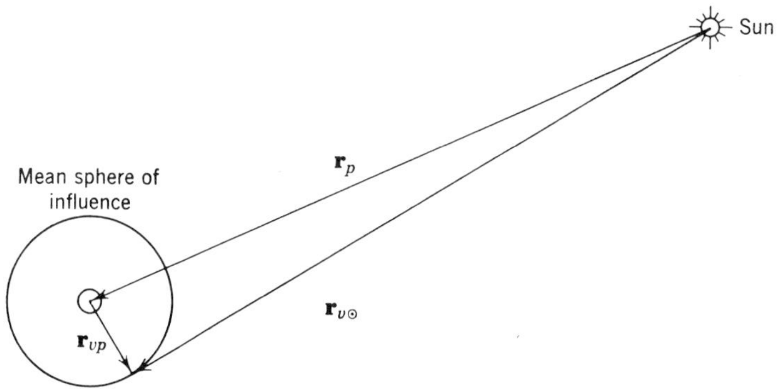

FIGURE 4.4 Orbital geometry.

to the planetocenter after rotation (4.13) has been performed. More explicitly,

$$\mathbf{r}_{vp} \equiv \mathbf{r}_{\epsilon},$$

$$\dot{\mathbf{r}}_{vp} \equiv \dot{\mathbf{r}}_{\epsilon}.$$

Furthermore, by the techniques of Section 1.3, at $t = t_I$ which is directly known from Kepler's equation via the known variable v_I, it becomes possible to obtain the position and velocities of the planet controlling the motion as

$$\mathbf{r}_p, \dot{\mathbf{r}}_p.$$

Hence the initial starting state at the boundary of the mean sphere of influence for the heliocentric leg of the journey is given by

$$\mathbf{r}_{v\odot} = \mathbf{r}_{vp} + \mathbf{r}_p, \tag{4.14}$$

$$\dot{\mathbf{r}}_{v\odot} = \dot{\mathbf{r}}_{vp} + \dot{\mathbf{r}}_p. \tag{4.15}$$

The generation of the Keplerian heliocentric orbit from $t = t_I$ until another planet is reached is carried on by the standard techniques of ([1], Chapter 3). Care must be exercised to ensure that another planetary sphere of influence does not interfere with the heliocentric trajectory. If this does happen, the procedure previously discussed should be reversed and the initial planetocentric conditions determined before continuing with the computation of the trajectory.

The patched conic technique has found frequent use in industry because of its inherent simplicity and also because a fairly good estimate of the true trajectory can be obtained by using this model.

4.2 MINIMUM FUEL INTERPLANETARY TRAJECTORIES

4.2.1 Preliminary Remarks

A rather interesting and important transfer mode from one planet to another is the minimum fuel transfer orbit. Perhaps, however, it will be beneficial

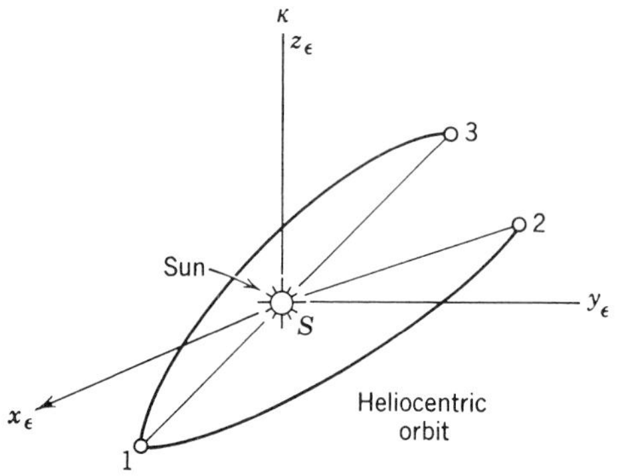

FIGURE 4.5 Ordinary interplanetary transfers.

to review briefly an ordinary interplanetary transfer between two planets. As illustrated in Figure 4.5, if it is desired to leave planet 1 and arrive at planet 2, the easiest way to accomplish the trip is to inject directly into the heliocentric transfer orbit which intersects the objective planet. Hence, as illustrated in Figure 4.5, for fixed time of transfer, Δt, the orbit which passes through terminals 1 and 2 can be determined by any of the six techniques discussed in ([1], Chapter 6). Following the discussion of Section 4.1.1, the velocity increment required to place the vehicle on the transfer trajectory would then be given by (4.6). Similarly, the retarding impulse at terminal 2 can be computed and the transfer can be considered determined. This type of interplanetary trajectory is called a planar transfer because the transfer mode is entirely carried out in plane $1S2$. It is a legal transfer and is used quite frequently. However, as the target planet begins to approach a condition of opposition, terminal 3, wherein the departure planet, Sun, and target planet are roughly lined up, the planar transfer problem becomes infeasible because the dynamics of a single plane transfer is forcing the orbit to go over the Sun. This situation is due to the relative inclination of the planets. To perform such a transfer the vehicle propulsive system must overcome tremendous velocity requirements because most of the planetary velocity at departure is not being utilized to aid in the transfer process. In this situation it becomes appealing to consider two-impulse transfers wherein a broke plane is used to accomplish the transfer. Hence, by applying an impulse somewhere between terminals 1 and 2 of Figure 4.5, a much cheaper fuel consumption maneuver becomes available to the astronaut. This is the transfer technique which is to be studied presently, and for which the equations for optimal transfer are developed. Furthermore, an automatic test that tells the analyst which mode to use, that is, direct or broken plane transfer, is also developed.

The analysis developed here follows the lines of development of Fimple [2]. However, the analysis is correct to the fifth order of the relative inclination i^*, between the velocity vectors in the departure and first transfer plane. The mechanics of the transfer as developed by Escobal [6] is more exact, at least within the limitations of patched conic theory.

4.2.2 Initial Coordinate System

The position and velocity vectors of a given planet are usually available in terms of mean elements or more accurate parameters relative to the ecliptic coordinate system, Section 1.3. Hence, for a given Julian date corresponding to initiation of the optimal transfer trajectory, it is assumed that the position and velocity vectors of the departure planet are known in the heliocentric ecliptic coordinate system denoted by the x_ϵ, y_ϵ, z_ϵ axes of Figure 4.6. In the analysis developed here it is advantageous to adopt a coordinate system that is fixed to the departure planet at time of initial launch into the heliocentric transfer maneuver. To this end the principal axis x_p is taken along the radius vector of the departure planet at time of launch. The y_p axis is advanced to x_p by a right angle and lies in the instantaneous plane of motion of the departure planet, thus defining the fundamental plane. Lastly z_p is picked such that x_p, y_p, and z_p form a right-handed system. If it is assumed that Ω, ω, and i, that is, the longitude of the ascending node, the argument of perihelion, and the orbital inclination of the departure planet, are not varying with time, then the position and velocity mappings from the ecliptic to the planetary coordinate system are given directly by

$$\begin{bmatrix} x_p \\ y_p \\ z_p \end{bmatrix} = [M] \begin{bmatrix} x_\epsilon \\ y_\epsilon \\ z_\epsilon \end{bmatrix}, \qquad \begin{bmatrix} \dot{x}_p \\ \dot{y}_p \\ \dot{z}_p \end{bmatrix} = [M] \begin{bmatrix} \dot{x}_\epsilon \\ \dot{y}_\epsilon \\ \dot{z}_\epsilon \end{bmatrix}, \tag{4.16}$$

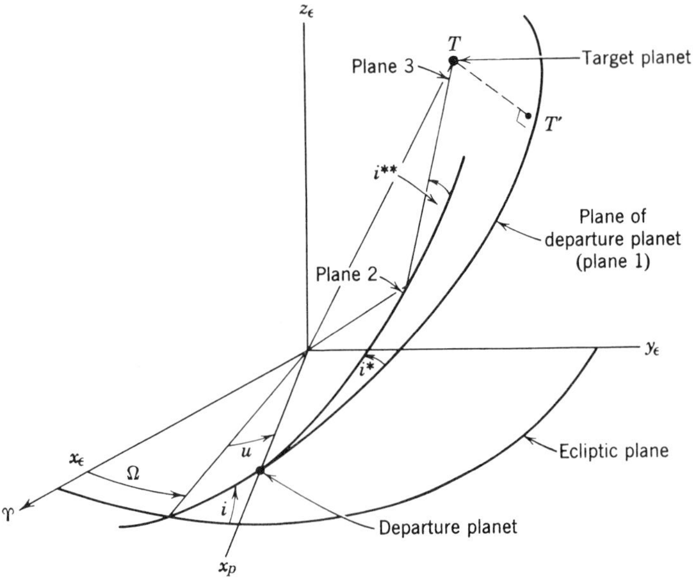

FIGURE 4.6 Planet orientation parameters.

where the epoch-dependent transformation matrix is

$$[M] = \begin{bmatrix} U_x & U_y & U_z \\ V_x & V_y & V_z \\ W_x & W_y & W_z \end{bmatrix} \tag{4.17}$$

with $u \equiv v + \omega$, and

$$\begin{aligned} U_x &= \cos u \cos \Omega - \sin u \sin \Omega \cos i, \\ U_y &= \cos u \sin \Omega + \sin u \cos \Omega \cos i, \\ U_z &= \sin u \sin i, \\ V_x &= -\sin u \cos \Omega - \cos u \sin \Omega \cos i, \\ V_y &= -\sin u \sin \Omega + \cos u \cos \Omega \cos i, \\ V_z &= \cos u \sin i, \\ W_x &= \sin \Omega \sin i, \\ W_y &= -\cos \Omega \sin i, \\ W_z &= \cos i. \end{aligned} \tag{4.18}$$

4.2.3 Transfer Process

For the present let it be assumed that a one-impulse transfer is utilized to satisfy the x_p and y_p coordinates of target planet T, call then x_{pT}, y_{pT}.

Hence for an assumed launch date and time differential any of the two-position vectors and time interval methods can be used to yield a fictitious transfer orbit constrained to lie in plane 1. As illustrated in Figure 4.6, this plane contains the departure point and the projection of T, that is, T'. Satisfaction of the true transfer now requires that impulses be applied normal to the departure plane in order to obtain the correct z_{pT} or final target point.

The immediate analysis developed here limits the transfer maneuver to two impulses, one applied at departure to attain a transfer in plane 2 at angle i_1 to plane 1, and a second impulse applied at point I, an intermediate point, to obtain an orbit in plane 3 inclined by an angle i_2 to plane 2. In the process of performing the maneuver the total out-of-plane velocity increment magnitude is minimized and point I is determined. Plane 3 is constrained to contain the final point T.

In closing this section it is well to note that, if the projection of the target planet on plane 1 is dropped to a point somewhere between T and T' (Figure 4.6), a slightly different transfer maneuver is realized. Actually, if T is not projected at all, then a true single-impulse transfer is determined between the respective terminals. As will be seen later, at times a single-impulse transfer is better than a double-impulse transfer. It may be necessary to repeat the analysis developed here for various projections of T' lying between point T and plane 1 to obtain the truly optimum two-impulse transfer. This does not pose any particular problems to the analysis. In this case, plane 1 should be redefined to be the instantaneous plane at the launch time defined by $\mathbf{r}_{dp}$ and $\dot{\mathbf{r}}_{dh}$, that is, the position vector of the first terminal or departure planet position and the required heliocentric velocity vector obtained from the two-position vectors and time interval method.

4.2.4 Position and Velocity Vectors at Secondary Impulse

Consider a vehicle moving in plane 1, as illustrated in Figure 4.7, with position and velocity vectors $\mathbf{r}_{dp}$ and $\dot{\mathbf{r}}_{dh}$. Note that $\dot{\mathbf{r}}_{dh}$ is the heliocentric transfer velocity in plane 1.

To effect a transfer maneuver from plane 1 into plane 2 a velocity increment must be added to $\dot{\mathbf{r}}_{dh}$ normal to plane 1, that is, $\Delta V_{dh}\mathbf{W}_1$. The result of this vector addition can be expressed as $\dot{\mathbf{r}}_{dpi}$, the new velocity vector, where

$$\dot{\mathbf{r}}_{dpi} = \dot{\mathbf{r}}_{dh} + \Delta V_{dh}\mathbf{W}_1 \tag{4.19}$$

or, from Figure 4.7,

$$\dot{\mathbf{r}}_{dpi} = \dot{\mathbf{r}}_{dh} + V_{dh} \tan i^*\mathbf{W}_1. \tag{4.20}$$

The subscript dp, as usual, stands for departure planet and the subscript, dpi for departure planet variables after rotation through angle i^*.

An auxiliary set of unit vectors, $\mathbf{U}_{dp}$ and $\mathbf{V}_{dpi}$, is introduced now and

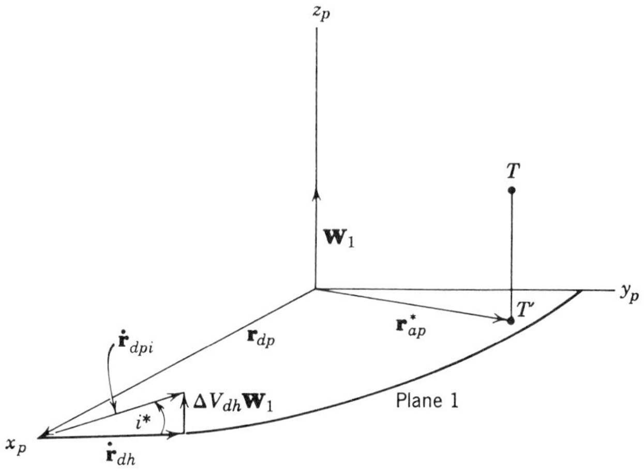

FIGURE 4.7 Plane change geometry.

defined by

$$\mathbf{U}_{dp} \equiv \frac{\mathbf{r}_{dp}}{r_{dp}}, \tag{4.21}$$

$$\mathbf{V}_{dpi} \equiv \frac{r_{dp}\dot{\mathbf{r}}_{dpi} - \dot{r}_{dpi}\mathbf{r}_{dp}}{\sqrt{\mu_3 p_{dpi}}}, \tag{4.22}$$

where μ_3 is the sum of masses of the space vehicle and the Sun.

Geometrically $\mathbf{U}_{dp}$ is a unit vector emanating from the dynamical center which points toward the vehicle, and $\mathbf{V}_{dpi}$ likewise is a unit vector advanced to $\mathbf{U}_{dp}$ by a right angle in the plane and direction of motion. Both of these vectors are evaluated at the departure terminal. As might be evident from the notation, $\mathbf{U}_{dp}$ is not functionally dependent on i^*, whereas $\mathbf{V}_{dpi}$ is directly linked to i^* through (4.20):

$$\mathbf{V}_{dpi} = \frac{r_{dp}(\dot{\mathbf{r}}_{dh} + V_{dh}\tan i^*\mathbf{W}_1) - (r_{dp}\dot{r}_{dpi})\mathbf{U}_{dp}}{\sqrt{\mu_3 p_{dpi}}}. \tag{4.23}$$

A simplification of (4.23) occurs if the standard parameter D_{dpi} is introduced as follows:

$$D_{dpi} \equiv \frac{\mathbf{r}_{dpi}\cdot\dot{\mathbf{r}}_{dpi}}{\sqrt{\mu_3}} = \frac{\mathbf{r}_{dp}\cdot\dot{\mathbf{r}}_{dpi}}{\sqrt{\mu_3}} = \frac{\mathbf{r}_{dp}\cdot(\dot{\mathbf{r}}_{dh} + V_{dh}\tan i^*\mathbf{W}_1)}{\sqrt{\mu_3}}. \tag{4.24}$$

Since $\mathbf{r}_{dp}$ and $\mathbf{W}_1$ are at right angles,

$$D_{dpi} = D_{dp} = \frac{\mathbf{r}_{dp}\cdot\dot{\mathbf{r}}_{dh}}{\sqrt{\mu_3}} = \frac{r_{dp}\dot{r}_{dh}}{\sqrt{\mu_3}}.$$

and (4.23) reduces to

$$\mathbf{V}_{dpi} = \frac{r_{dp}(\dot{\mathbf{r}}_{dh} + V_{dh} \tan i^* \mathbf{W}_1) - \sqrt{\mu_3}\, D_{dp} \mathbf{U}_{dp}}{\sqrt{\mu_3 p_{dpi}}}. \tag{4.25}$$

In terms of the unit vectors $\mathbf{U}_{dp}$ and $\mathbf{V}_{dpi}$ evaluated at time t_0, the position and velocity vectors at any future time of impulse application as denoted by the secondary subscript I, may be computed from ([1], Chapter 3):

$$\mathbf{r}_{2I} = x_{vI} \mathbf{U}_{dp} + y_{vI} \mathbf{V}_{dpi}, \tag{4.26}$$

$$\dot{\mathbf{r}}_{2I} = \dot{x}_{vI} \mathbf{U}_{dp} + \dot{y}_{vI} \mathbf{V}_{dpi}, \tag{4.27}$$

where

$$x_{vI} \equiv r_I \cos v^*, \qquad y_{vI} \equiv r_I \sin v^*,$$

$$\dot{x}_{vI} \equiv \left(\frac{\mu}{p_{dpi}}\right)^{1/2} [S_{vi} - \sin v^*], \qquad \dot{y}_{vI} \equiv \left(\frac{\mu}{p_{dpi}}\right)^{1/2} [C_{vi} + \cos v^*], \tag{4.28}$$

with

$$S_{vi} \equiv \frac{D_{dp}}{r_{dp}} (p_{dpi})^{1/2}, \qquad C_{vi} \equiv \frac{p_{dpi}}{r_{dp}} - 1, \qquad v^* \equiv v_I - v_1.$$

In order to effect the second plane change a velocity increment is added normal to plane 1 to force the incremented velocity vector to lie in the final transfer plane, as illustrated in Figure 4.8. The reason for not adding the

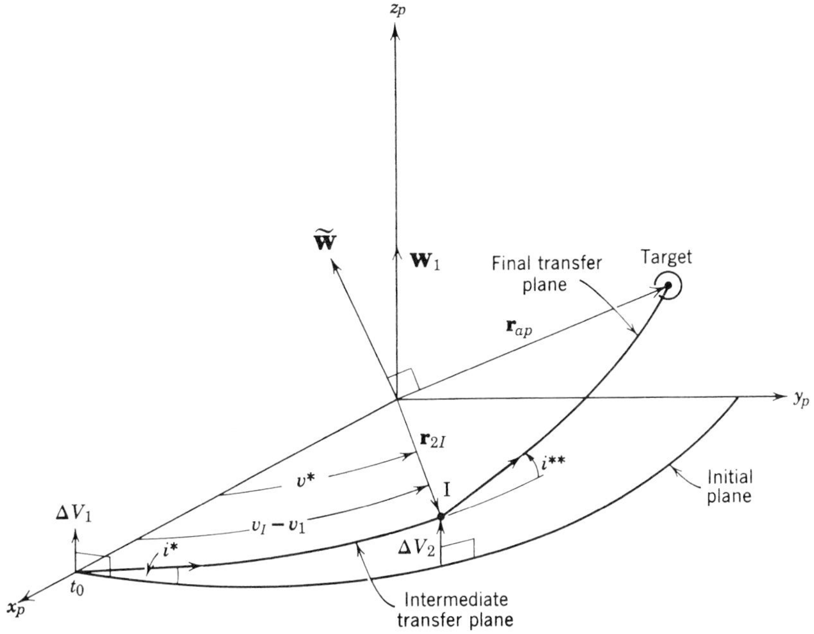

FIGURE 4.8 Impulse application points.

second velocity increment, call it ΔV_2, normal to plane 2 is that this maneuver would cause a side velocity increment,

$$\Delta V_2 \sin i^*, \tag{4.29}$$

to occur parallel to plane 1. This situation, under the construction employed here, would cause an error in the interception coordinates x_{pT} and y_{pT}. Actually, adding the velocity increment normal to plane 2 causes the error

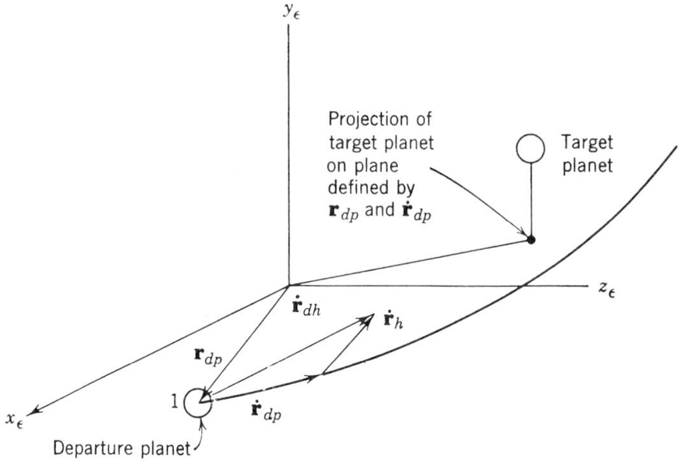

FIGURE 4.9 Velocity and position vectors at takeoff.

$\Delta V_2 \sin i^*$, which must be corrected. Firing normal to plane 1 also causes a loss of velocity increment approximately equal to $\Delta V_2 \sin i^*$. Apparently no true loss is incurred by adding ΔV_2 normal to plane 1; thus it is possible to form the new velocity vector after application of the secondary impulse as follows:

$$\dot{\mathbf{r}}_{3I} = \dot{\mathbf{r}}_{2I} + \Delta V_2 \mathbf{W}_1. \tag{4.30}$$

The magnitude of ΔV_2 is still to be determined.

4.2.5 Expressions for the Primary and Secondary Impulses

From the preceding sections it is assumed that a heliocentric transfer orbit has been obtained which passes through the planet located at terminal 1 of the transfer orbit and the projection of T on plane 1. Hence $\mathbf{r}_{dp}$ and $\dot{\mathbf{r}}_{dh}$, the initial heliocentric position and velocity vectors, are known.

Figure 4.9 shows that addition of $\dot{\mathbf{r}}_h$ to the planetary velocity at $\mathbf{r}_{dp}$ results in the required transfer trajectory. It should be noticed that $\dot{\mathbf{r}}_h$ is obtained by the vector subtraction of the heliocentric and planetary velocity vectors at the first terminal, that is, from (4.1).

With this understanding it is possible to employ (4.6) and explicitly write

$$\Delta V_1 = [V_p^{\,2} + (\dot{x}_{dp} - \dot{x}_{dh})^2 + (\dot{y}_{dp} - \dot{y}_{dh})^2 + (\dot{z}_{dp} - \dot{z}_{dh})^2 + V_{dh}^{\,2} \tan^2 i^*]^{1/2} - \left[\frac{\mu_2}{r_0}\right]^{1/2}, \quad (4.31)$$

where μ_2 is the sum of masses of the vehicle and departure planet and r_0 is the initial radius of the adopted circular parking orbit before application of the primary velocity impulse. The contribution of $\dot{z}_{dp}$ and $\dot{z}_{dh}$ when T is projected to T' is zero because of the coordinate system employed. For purposes of generality and future use, however, the $\dot{z}$ components are carried in this analysis. The contribution $V_{dh} \tan i^*$ is the velocity magnitude required for the out-of-plane maneuver. With this understanding the final expression for the primary impulse can be written in the following fashion:

$$\Delta V_1 = [\beta_1 + \beta_2 \tan^2 i^*]^{1/2} - \beta_3, \quad (4.32)$$

where

$$\beta_1 \equiv V_p^{\,2} + (\dot{\mathbf{r}}_{dp} - \dot{\mathbf{r}}_{dh}) \cdot (\dot{\mathbf{r}}_{dp} - \dot{\mathbf{r}}_{dh}),$$

$$\beta_2 \equiv V_{dh}^{\,2},$$

$$\beta_3 \equiv \left[\frac{\mu_2}{r_0}\right]^{1/2}.$$

Now it is necessary to obtain an analytic expression for the secondary impulse. Consider the determination of a third normal vector $\tilde{\mathbf{W}}$ perpendicular to the target planet radius of motion. The normal vector $\tilde{\mathbf{W}}$ is easily found at the desired intercept time from the position vectors $\mathbf{r}_{2I}$, $\mathbf{r}_{ap}$, and Figure 4.8 as

$$\tilde{\mathbf{W}} = (\mathbf{r}_{2I} \times \mathbf{r}_{ap}), \quad (4.33)$$

where the subscript *ap* denotes arrival planet. It should be noticed that $\tilde{\mathbf{W}}$ is dependent on i^* and v^*. In order to constrain $\dot{\mathbf{r}}_{3I}$ or the velocity vector to lie in plane 3 after application of the secondary impulse, the following condition is imposed:

$$\dot{\mathbf{r}}_{3I} \cdot \tilde{\mathbf{W}} = 0 \quad (4.34)$$

or, from (4.30),

$$(\dot{\mathbf{r}}_{2I} + \Delta V_2 \mathbf{W}_1) \cdot \tilde{\mathbf{W}} = 0, \quad (4.35)$$

which yields

$$\Delta V_2 = -\frac{\dot{\mathbf{r}}_{2I} \cdot \tilde{\mathbf{W}}}{\mathbf{W}_1 \cdot \tilde{\mathbf{W}}}. \quad (4.36)$$

Using (4.27) and (4.25), by direct substitution, the following reduction is possible:

$$\Delta V_2 = \left| -\frac{\zeta_1 \sin v^* + \zeta_4 \cos v^* + {}_2}{\gamma_4 \cos v^* + \gamma_5 \sin v^*} \right|, \tag{4.37}$$

where

$$\zeta_1 \equiv \left[\frac{\mu_3}{p_{dpi}}\right]^{1/2} \gamma_1 S_{vi},$$

$$\zeta_2 \equiv -\left[\frac{\mu_3}{p_{dpi}}\right]^{1/2} \gamma_1,$$

$$\zeta_4 \equiv \left[\frac{\mu_3}{p_{dpi}}\right]^{1/2} \gamma_2 C_{vi},$$

$$\gamma_1 \equiv [\mu_3 p_{dpi}]^{1/2}(d_1 + d_2 \tan i^*),$$

$$\gamma_2 \equiv [\mu_3 p_{dpi}]^{1/2}(d_3 + d_4 \tan i^*),$$

$$\gamma_4 \equiv \mu_3 p_{dpi} \boldsymbol{\epsilon}_1 \cdot \mathbf{W}_1,$$

$$\gamma_5 \equiv \mu_3 (p_{dpi})^{1/2} \boldsymbol{\epsilon}_2 \cdot \mathbf{W}_1,$$

$$d_1 \equiv \boldsymbol{\epsilon}_2 \cdot \mathbf{U}_{dp},$$

$$d_2 \equiv \boldsymbol{\epsilon}_3 \cdot \mathbf{U}_{dp},$$

$$d_3 \equiv r_{dp} \boldsymbol{\epsilon}_1 \cdot \dot{\mathbf{r}}_{dh},$$

$$d_4 \equiv r_{dp} V_{dh} \boldsymbol{\epsilon}_1 \cdot \mathbf{W}_1,$$

and

$$\boldsymbol{\epsilon}_1 \equiv \mathbf{U}_{dp} \times \mathbf{r}_{ap},$$

$$\boldsymbol{\epsilon}_2 \equiv r_{dp} \dot{\mathbf{r}}_{dh} \times \mathbf{r}_{ap} - (\mathbf{r}_{dp} \cdot \dot{\mathbf{r}}_{dh})(\mathbf{U}_{dp} \times \mathbf{r}_{ap}),$$

$$\boldsymbol{\epsilon}_3 \equiv r_{dp} V_{dh} \mathbf{W}_1 \times \mathbf{r}_{ap}.$$

The reduction of (4.37) is quite lengthy and in performing the actual computations quite a number of factors vanish identically. The complete reduction was checked by Crichton [11] and documentation of the operations performed.

4.2.6 Total Velocity Increment Minimization

If the assumption is made that the planetary vehicle or probe is captured by the objective or target planet,[1] the total mission velocity increment requirement is given by

$$\Delta = \Delta V_1 + |\Delta V_2|. \tag{4.38}$$

More explicitly, using (4.32) and (4.37),

$$\Delta = (\beta_1 + \beta_2 \tan^2 i^*)^{1/2} - \beta_3 + \left| -\frac{\zeta_1 \sin v^* + \zeta_4 \cos v^* + \zeta_2}{\gamma_4 \cos v^* + \gamma_5 \sin v^*} \right|. \tag{4.39}$$

[1] The analysis in this section is actually designed for flyby missions.

The parameter Δ is a function of i^* and v^*. Minimization of Δ requires that $\partial\Delta/\partial i^* = \partial\Delta/\partial v^* = 0$. Proceeding formally, $\partial\Delta/\partial v^*$ can be evaluated and equated to zero to yield F_1:

$$F_1 \equiv (\zeta_1\gamma_4 - \zeta_4\gamma_5) + \zeta_2\gamma_4 \sin v^* - \zeta_2\gamma_5 \cos v^* = 0 \qquad (4.40)$$

or, solving for $\sin v^*$,

$$\sin v^* = \frac{-\nu_5 \pm [\nu_5^2 - 4\nu_4\nu_6]^{1/2}}{2\nu_4}, \qquad (4.41)$$

where

$$\nu_4 \equiv \alpha_2^2 + \alpha_3^2, \qquad \nu_5 \equiv 2\alpha_1\alpha_2, \qquad \nu_6 \equiv \alpha_1^2 - \alpha_3^2,$$

and

$$\alpha_1 \equiv \gamma_4\zeta_1 - \gamma_5\zeta_4, \qquad \alpha_2 \equiv \gamma_4\zeta_2, \qquad \alpha_3 \equiv -\gamma_5\zeta_2.$$

Since the secondary impulse will be applied where $0 \leq v^* \leq \pi$, the spurious root of (4.41) can be rejected. The cosine of v^* is uniquely determined from

$$\cos v^* = -\frac{\alpha_2 \sin v^* + \alpha_1}{\alpha_3}. \qquad (4.42)$$

The second partial directive, $\partial\Delta/\partial i^*$, is exceedingly complicated because p, C_v, S_v, etc., are all functions of i^*. Expressing all elements and other parameters as power expansions to the fifth order of (i^*), it may be verified with patience that

$$F_2 \equiv \bar{\nu}_1 i^{*4} + \bar{\nu}_2 i^{*3} + \bar{\nu}_3 i^{*2} + \bar{\nu}_4 i^* + \bar{\nu}_5 = 0; \qquad (4.43)$$

the coefficients, $\bar{\nu} = \bar{\nu}(v^*)$, are tabulated in Section 4.2.9.

To solve (4.41) and (4.43) simultaneously the following rapidly converging method should be utilized. Since the coefficients of (4.41) are functionally dependent on i^* but are relatively insensitive to i^*, choose any value of i^*, for example $i^* = i_j^* = 0$, and compute the ν coefficients. This immediately permits the $\bar{\nu}$ coefficients to be computed via the formulas presented in Section 4.2.9. The closed form solution of the quartic by means of the algorithm in ([1], Appendix 3) immediately yields i_{j+1}^*, where the subscript implies that it is an improvement over the previously chosen value. Going through the procedure a second time usually yields no appreciable change in i^*.

For all realistic interplanetary transfers i^* is small because the relative inclination of the solar system planets likewise is small. In passing it is well to note that even for unrealistically high transfers, with relative inclinations between the departure plane and the first transfer plane of 0.25 radian $\approx$ 14 deg, $(i_1)^5 \simeq 0.00097$ radian $\simeq 0°.056$ and the elements p, C_v, etc., converge rapidly to their true values. The acceptable root of the quartic should possess the same sign as the z component of the target planet's position vector.

4.2.7 Examination of the Velocity Increment Function

Equation 4.39, for constant v^*, is an analytic function with certain properties. Certainly the minimum value of ΔV_1, from (4.39), is

$$(\Delta V_1)_{\min} = (\beta_1)^{1/2} - \beta_3 \tag{4.44}$$

for $i^* = 0$, and it increases positively for any other value of i^*. The absolute function $|\Delta V_2|$ is, of course, always positive and vanishes at $i^* = i_l^*$, where

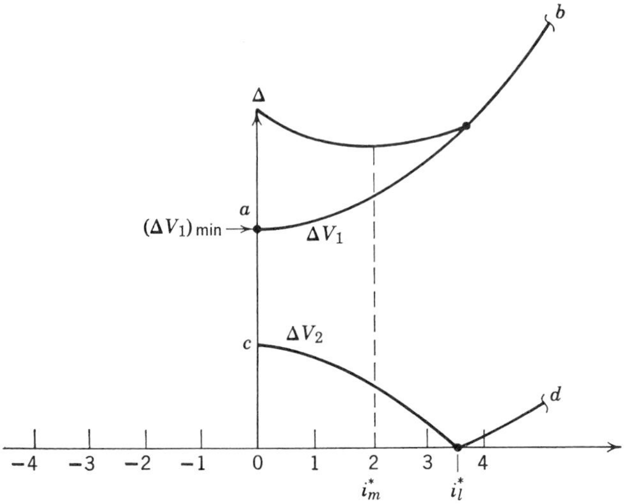

FIGURE 4.10 Graphical representation of velocity increment equation.

i_l^* is the inclination of the velocity vector of the one-impulse transfer trajectory, that is, the nonbroken plane transfer.

A little thought will reveal that ΔV_1 is a tangent function increasing to the one-half power, thus producing curve *ab* of Figure 4.10. The second impulse function must increase from both sides of i_l^* and yield curve $c\, i_l^* d$ of Figure 4.10. Addition of ΔV_1 and $|\Delta V_2|$ results in Δ, the total velocity increment function. Furthermore, since ΔV_1 increases in $(0, i_l^*)$ and $|\Delta V_2|$ decreases in the same interval, it is conceivable that a minimum exists, call it i_m^*. As may be evident from the graphical construction, the maximum value of i_m^*, the acceptable root of (4.43), for optimum transfer never exceeds i_l^*, as this would correspond to increased velocity requirements and, therefore, no minimum can exist for $i^* > i_l^*$.

The limiting i_l^* can always be found because it is the i^* of a one-impulse transfer. Therefore, by computing the transfer between terminals 1 and 2, with the desired travel time, i_l^* is determined. Then, if

$$i_m^* = i_l^*,$$

a single-impulse transfer is better than a double-impulse transfer. In opposite fashion, if

$$0 \leq i_m^* < i_l^*,$$

the double-impulse transfer is optimum.

4.2.8 Target Projection Equations

The preceding analysis tacitly assumed that the original relative departure velocity vector was contained in the plane of the departure planet. Thus a heliocentric orbit was placed between the radius vector at the departure planet and the projection of the target planet on the initial plane of motion of the departure planet. If T in Figure 4.6 had not been projected on plane 1, a single-impulse transfer would have resulted between the initial and final terminals. It appears feasible to expect that an optimum projection length of T' toward T or z_{p3}^* exists which, in combination with the preceding analysis, yields a true optimum. Hence the previous analysis should be performed for $z_{p3}^* = 0$, with no change in the equations developed, and for $z_{p3}^* \doteq n\, \Delta z_{p3}^*$, where n is a constant such that it divides the length $T'T$ into even increments. In essence, then, the initial one-impulse heliocentric transfer trace is taken slightly inclined to the instantaneous departure plane of the planet at terminal 1. The analysis outlined would be performed only as a second order refinement.

4.2.9 Algorithm for Various Transfer Modes

The objective of this section is to state a fairly extended algorithm for the computation of interplanetary trajectories. Four distinct transfer modes are discussed in the presentation of the algorithm. The first mode is an interplanetary transfer from a circular orbit via one impulse which will fly past the target planet. A second transfer mode is an interplanetary transfer from a circular orbit via one impulse and a capture maneuver into a circular orbit at the target planet. The third transfer mode is a transfer via one impulse from a circular orbit around the departure planet to an optimum midpoint where application of a second impulse causes the vehicle to fly past the target planet. Finally, the fourth transfer mode is a transfer via one impulse from a circular orbit around the departure planet to an optimum midpoint where application of a second impulse causes the vehicle to fly near the target planet, at which point a third impulse is applied to place the vehicle in a final circular orbit. These transfer modes are listed in Table 4.2. The last

Table 4.2 Interplanetary Transfer Modes

Mode	Impulses	Departure Impulse	Midcourse Impulse	Capture Impulse
β_1	1	ΔV_1	0	0
β_2	2	ΔV_1	0	ΔV_3
β_3	2	ΔV_1	ΔV_2	0
β_4	3	ΔV_1	ΔV_2	ΔV_3

two transfer modes minimize the total velocity increment applied to perform the maneuvers as discussed in the preceding sections.

The algorithm permits the firing of interplanetary probes to all the major planets, even for very difficult widespread transfers close to 180° where over the Sun trajectories cause severe fuel penalties. Hence, given any firing date, the transfer can be made for both flyby and capture missions. An automatic decision is made internal to the algorithm as to which transfer mode is best for a given firing date. The transfer geometry for the algorithm is illustrated in Figure 4.11. The optimum transfers, modes β_3 and β_4,

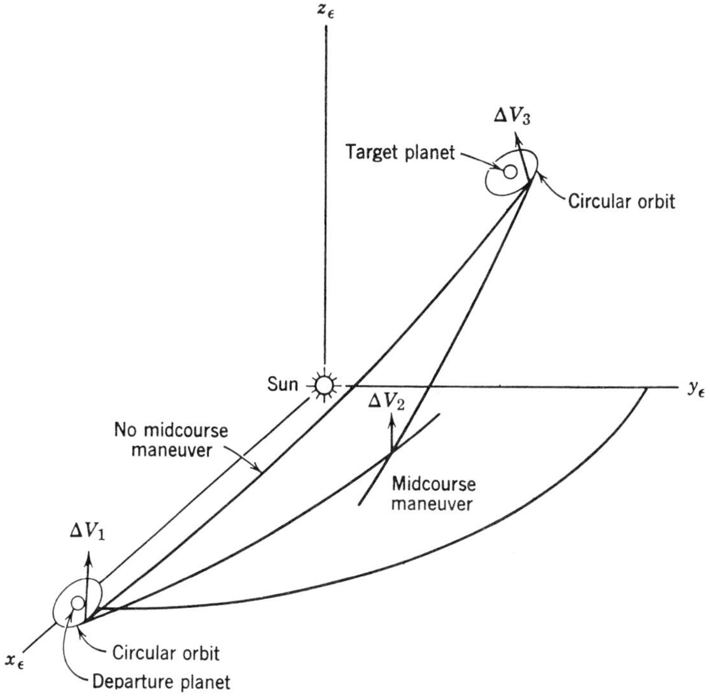

FIGURE 4.11 Transfer geometry.

discussed in Section 4.2.3 are valid to the order of i^{*5}. This is sufficient for most planetary missions. Analysis is based on patched conic theory with the departure and arrival capture modes making use of circular initial orbits.

Therefore, given the data

$T_l^{(1)}$ = year of launch,
$T_l^{(2)}$ = month of launch,
$T_l^{(3)}$ = day of launch,
$T_l^{(4)}$ = hour of launch,
$T_l^{(5)}$ = minute of launch,
$T_l^{(6)}$ = second of launch,
Δt = desired flight time,
μ_1 = sum of masses of Sun and departure planet,
μ_2 = sum of masses of vehicle and departure planet,
μ_3 = sum of masses of vehicle and Sun,
μ_4 = sum of masses of Sun and arrival planet,
μ_5 = sum of masses of vehicle and arrival planet,
$k_{\odot}$ = gravitational constant of the Sun,
k_{dp} = gravitational constant of departure planet,
k_{ap} = gravitational constant of arrival planet,
$\left[\frac{\text{a.u.}}{\text{e.r.}}\right]_d$ = conversion factor ratio from a.u. to e.r. of the departure planet ,
$\left[\frac{\text{a.u.}}{\text{e.r.}}\right]_a$ = conversion factor ratio from a.u. to e.r. of the arrival planet,
a_d = semimajor axis of circular departure orbit about launch planet,
a_a = semimajor axis of circular orbit about arrival planet,
Δ[J.D.] = Julian date increment,
m = Julian date increment divider,
N = number of impulses to be used ≤ 3,

proceed as follows. Compute the Julian date at launch;[2] that is, obtain

$$[\text{J.D.}]_d \tag{4.45}$$

and the Julian date at arrival from

$$[\text{J.D.}]_a = [\text{J.D.}]_d + \Delta t. \tag{4.46}$$

[2] See ([1], Section 1.3).

From the corresponding Julian dates obtain the ecliptic position and velocity vectors of the departure and arrival planets from Section 1.3, that is,

$$[\mathbf{r}_d, \dot{\mathbf{r}}_d]_\epsilon, \qquad [\mathbf{r}_a, \dot{\mathbf{r}}_a]_\epsilon \tag{4.47}$$

along with the ecliptic classical elements

$$[a_d, e_d, i_d, \Omega_d, u_{0d}]_\epsilon, \qquad [a_a, e_a, i_a, \Omega_a, u_{0a}]_\epsilon. \tag{4.48}$$

Compute the orientation vectors pointing toward the departure planet $\mathbf{U}$ and advanced to $\mathbf{U}$ in the direction and plane of motion, $\mathbf{V}$, from

$$\begin{bmatrix} U_x \\ U_y \\ U_z \end{bmatrix} = \begin{bmatrix} \cos u_{0d} \cos \Omega_d - \sin u_{0d} \sin \Omega_d \cos i_d \\ \cos u_{0d} \sin \Omega_d + \sin u_{0d} \cos \Omega_d \cos i_d \\ \sin u_{0d} \sin i_d \end{bmatrix}_\epsilon, \tag{4.49}$$

$$\begin{bmatrix} W_x \\ W_y \\ W_z \end{bmatrix} = \begin{bmatrix} \sin \Omega_d \sin i_d \\ -\cos \Omega_d \sin i_d \\ \cos i_d \end{bmatrix}_\epsilon, \tag{4.50}$$

$$\begin{bmatrix} V_x \\ V_y \\ V_z \end{bmatrix} = \begin{bmatrix} W_x \\ W_y \\ W_z \end{bmatrix} \times \begin{bmatrix} U_x \\ U_y \\ U_z \end{bmatrix}. \tag{4.51}$$

Rotate the departure and arrival planets from the ecliptic coordinate system to a coordinate system defined by the initial position of the departure planet; that is, defining

$$[M_p] \equiv \begin{bmatrix} U_x & U_y & U_z \\ V_x & V_y & V_z \\ W_x & W_y & W_z \end{bmatrix} \tag{4.52}$$

obtain

$$\begin{bmatrix} \mathbf{r}_{dp} \\ \dot{\mathbf{r}}_{dp} \end{bmatrix} = \begin{bmatrix} [M_p] & [0] \\ [0] & [M_p] \end{bmatrix} \begin{bmatrix} \mathbf{r}_d \\ \dot{\mathbf{r}}_d \end{bmatrix}, \tag{4.53}$$

$$\begin{bmatrix} \mathbf{r}_{ap} \\ \dot{\mathbf{r}}_{ap} \end{bmatrix} = \begin{bmatrix} [M_p] & [0] \\ [0] & [M_p] \end{bmatrix} \begin{bmatrix} \mathbf{r}_a \\ \dot{\mathbf{r}}_a \end{bmatrix}. \tag{4.54}$$

Determine a heliocentric orbit[3] between the radius vectors $\mathbf{r}_{dp}$ and $\mathbf{r}_{ap}$ for transfer time Δt; that is, obtain at the launch or first terminal

$$\mathbf{r}_{dl}, \quad \dot{\mathbf{r}}_{dl}, \qquad (\mathbf{r}_{dl} \equiv \mathbf{r}_{dp}). \tag{4.55}$$

[3] See ([1], Chapter 6) for any of the standard two-position vectors and time interval techniques.

The limiting value of the inclination i_l^* between the velocity vectors in the planes of motion of departure planet and vehicle can be obtained by determining a secondary heliocentric orbit as follows. As a first approximation set $\lambda = 0$, form the vector

$$\dot{\mathbf{r}}_{ap}^* = \begin{bmatrix} x_{ap} \\ y_{ap} \\ \lambda z_{ap} \end{bmatrix}, \tag{4.56}$$

and determine the heliocentric transfer orbit (see Figure 4.7) between $\mathbf{r}_{dp}$ and $\mathbf{r}_{ap}^*$ in order to obtain the positions and velocities at the first terminal as

$$\mathbf{r}_{dp}, \quad \dot{\mathbf{r}}_{dh}. \tag{4.57}$$

Now obtain i_l^* by forming the unit vectors

$$\mathbf{S}_{dl} \equiv \frac{\dot{\mathbf{r}}_{dl}}{[\dot{\mathbf{r}}_{dl} \cdot \dot{\mathbf{r}}_{dl}]^{1/2}},$$

$$\mathbf{S}_{dh} = \frac{\dot{\mathbf{r}}_{dh}}{[\dot{\mathbf{r}}_{dh} \cdot \dot{\mathbf{r}}_{dh}]^{1/2}}, \tag{4.58}$$

so that

$$i_l^* = \cos^{-1}(\mathbf{S}_{dl} \cdot \mathbf{S}_{dh}), \qquad -\frac{\pi}{2} \leq i_l^* \leq \frac{\pi}{2}. \tag{4.59}$$

Compute the parabolic escape speed from the departure planet:

$$V_p^2 = \frac{2\mu_2}{a_d}. \tag{4.60}$$

Compute the sum of the kinematic and parabolic speeds from

$$V_s^2 = V_p^2 + (\dot{\mathbf{r}}_{dp} - \dot{\mathbf{r}}_{dh}) \cdot (\dot{\mathbf{r}}_{dp} - \dot{\mathbf{r}}_{dh}). \tag{4.61}$$

Compute the circular speed at departure, the planetary departure speed, and the heliocentric departure speed from

$$V_c^2 = \frac{\mu_2}{a_d}, \tag{4.62}$$

$$V_{dp}^2 = \dot{\mathbf{r}}_{dp} \cdot \dot{\mathbf{r}}_{dp}, \tag{4.63}$$

$$V_{dh}^2 = \dot{\mathbf{r}}_{dh} \cdot \dot{\mathbf{r}}_{dh}. \tag{4.64}$$

Let

$$\begin{bmatrix} \beta_1 \\ \beta_2 \\ \beta_3 \end{bmatrix} = \begin{bmatrix} V_s^2 \\ V_{dh}^2 \\ V_c \end{bmatrix} \tag{4.65}$$

and obtain the auxiliary parameters

$$\mu = \mu_3,$$
$$\mathbf{W}_1 = \frac{\dot{\mathbf{r}}_{dp} \times \dot{\mathbf{r}}_{dh}}{[(\dot{\mathbf{r}}_{dp} \times \dot{\mathbf{r}}_{dh}) \cdot (\dot{\mathbf{r}}_{dp} \times \dot{\mathbf{r}}_{dh})]^{1/2}},$$
$$D_{dp} = \frac{\dot{\mathbf{r}}_{dp} \cdot \dot{\mathbf{r}}_{dh}}{\mu^{1/2}}. \tag{4.66}$$

Compute the δ coefficients (4.67):

$$\delta_{10} = \frac{r_{dp}^{\;2}\beta_2}{\mu} - D_{dp}^{\;2},$$
$$\delta_{11} = \frac{r_{dp}^{\;2}\beta_2}{\mu},$$
$$\delta_{12} = \frac{2}{3}\frac{r_{dp}^{\;2}\beta_2}{\mu},$$
$$\delta_{13} = \delta_{10}^{1/2},$$
$$\delta_{14} = \frac{1}{2}\left(\frac{\delta_{11}}{\delta_{10}^{1/2}}\right),$$
$$\delta_{15} = \frac{1}{2}\left(\frac{\delta_{12}}{\delta_{10}^{1/2}}\right) - \frac{1}{8}\left(\frac{\delta_{11}^2}{\delta_{10}^{3/2}}\right),$$
$$\delta_{16} = \frac{1}{\delta_{10}^{1/2}},$$
$$\delta_{17} = -\frac{1}{2}\left(\frac{\delta_{11}}{\delta_{10}^{3/2}}\right),$$
$$\delta_{18} = \frac{3}{8}\left(\frac{\delta_{11}^2}{\delta_{10}^{5/2}}\right) - \frac{1}{2}\left(\frac{\delta_{12}}{\delta_{10}^{3/2}}\right),$$
$$\delta_{19} = \frac{D_{dp}}{r_{dp}}\delta_{13},$$
$$\delta_{20} = \frac{D_{dp}}{r_{dp}}\delta_{14},$$
$$\delta_{21} = \frac{D_{dp}}{r_{dp}}\delta_{15},$$
$$\delta_{22} = \frac{\delta_{10}}{r_{dp}} - 1,$$
$$\delta_{23} = \frac{\delta_{11}}{r_{dp}},$$
$$\delta_{24} = \frac{\delta_{12}}{r_{dp}}. \tag{4.67}$$

As a first approximation let $i^* = i_n^* = 0$ and compute the following parameters (4.68)

$$
\begin{aligned}
V_{dpi}^2 &= \beta_2 \sec^2 i^*, \\
p_{dpi} &= \frac{{r_{dp}}^2}{\mu} V_{dpi}^2 - {D_{dp}}^2, \\
S_{vi} &= \frac{D_{dp}}{r_{dp}} p_{dpi}^{1/2}, \\
C_{vi} &= \frac{p_{dpi}}{r_{dp}} - 1, \\
\mathbf{U}_{dp} &= \frac{\mathbf{r}_{dp}}{r_{dp}}, \\
\boldsymbol{\epsilon}_1 &= \mathbf{U}_{dp} \times \mathbf{r}_{ap}, \\
\boldsymbol{\epsilon}_2 &= r_{dp}\dot{\mathbf{r}}_{dh} \times \mathbf{r}_{ap} - (\mathbf{r}_{dp} \cdot \dot{\mathbf{r}}_{dh})(\mathbf{U}_{dp} \times \mathbf{r}_{ap}), \\
\boldsymbol{\epsilon}_3 &= r_{dp}V_{dh}\mathbf{W}_1 \times \mathbf{r}_{ap}, \\
d_1 &= \boldsymbol{\epsilon}_2 \cdot \mathbf{U}_{dp}, \\
d_2 &= \boldsymbol{\epsilon}_3 \cdot \mathbf{U}_{dp}, \\
d_3 &= r_{dp}\boldsymbol{\epsilon}_1 \cdot \dot{\mathbf{r}}_{dh}, \\
d_4 &= r_{dp}V_{dh}\boldsymbol{\epsilon}_1 \cdot \mathbf{W}_1, \\
\gamma_1 &= (\mu p_{dpi})^{1/2}(d_1 + d_2 \tan i^*), \\
\gamma_2 &= (\mu p_{dpi})^{1/2}(d_3 + d_4 \tan i^*), \\
\gamma_4 &= \mu p_{dpi}(\boldsymbol{\epsilon}_1 \cdot \mathbf{W}_1), \\
\gamma_5 &= \mu(p_{dpi})^{1/2}(\boldsymbol{\epsilon}_2 \cdot \mathbf{W}_1), \\
\zeta_1 &= \left(\frac{\mu}{p_{dpi}}\right)^{1/2}(\gamma_1 S_{vi}), \\
\zeta_2 &= -\gamma_1\left(\frac{\mu}{p_{dpi}}\right)^{1/2}, \\
\zeta_4 &= \gamma_2\left(\frac{\mu}{p_{dpi}}\right)^{1/2} C_{vi}, \\
\alpha_1 &= \gamma_4\zeta_1 - \gamma_5\zeta_4, \\
\alpha_2 &= \gamma_4\zeta_2, \\
\alpha_3 &= -\gamma_5\zeta_2, \\
\nu_4 &= {\alpha_2}^2 + {\alpha_3}^2, \\
\nu_5 &= 2\alpha_1\alpha_2, \\
\nu_6 &= {\alpha_1}^2 - {\alpha_3}^2.
\end{aligned}
\tag{4.68}
$$

Compute the sine of the central transfer angle from

$$\sin v^* = \frac{-\nu_5 \pm [\nu_5^2 - 4\nu_4\nu_6]^{1/2}}{2\nu_4} \tag{4.69}$$

and reject the nonpositive root to obtain

$$\cos v^* = -\frac{\alpha_2 \sin v^* + \alpha_1}{\alpha_3}. \tag{4.70}$$

Determine the following auxiliary coefficients (4.71):

$$\phi_{11} = (\mu\delta_{19}d_1) \sin v^*,$$
$$\phi_{12} = (\mu\delta_{19}d_2) \sin v^*,$$
$$\phi_{13} = (\mu\delta_{20}d_1) \sin v^*,$$
$$\phi_{14} = \mu(\delta_{20}d_2 + \tfrac{1}{3}\delta_{19}d_2) \sin v^*,$$
$$\phi_{15} = (\mu\delta_{21}d_1) \sin v^*,$$
$$\phi_{16} = -\mu d_2(\tfrac{2}{15}\delta_{19} + \tfrac{1}{3}\delta_{20} + \delta_{21}) \sin v^*,$$
$$\phi_{21} = -\mu d_1 \sin^2 v^*,$$
$$\phi_{22} = -\mu d_2 \sin^2 v^*,$$
$$\phi_{23} = 0,$$
$$\phi_{24} = -\tfrac{1}{3}\mu d_2 \sin^2 v^*,$$
$$\phi_{25} = 0,$$
$$\phi_{26} = -\tfrac{2}{15}\mu d_2 \sin^2 v^*,$$
$$\phi_{31} = \mu d_3 \cos^2 v^*,$$
$$\phi_{32} = \mu d_4 \cos^2 v^*,$$
$$\phi_{33} = 0,$$
$$\phi_{34} = \tfrac{1}{3}\mu d_4 \cos^2 v^*,$$
$$\phi_{35} = 0,$$
$$\phi_{36} = \tfrac{2}{15}\mu d_4 \cos^2 v^*,$$
$$\phi_{41} = \mu\delta_{22}d_3 \cos v^*,$$
$$\phi_{42} = \mu\delta_{22}d_4 \cos v^*,$$
$$\phi_{43} = \mu\delta_{23}d_3 \cos v^*,$$
$$\phi_{44} = \mu(\delta_{23}d_4 + \tfrac{1}{3}\delta_{22}d_4) \cos v^*$$

$$
\begin{aligned}
\phi_{45} &= \mu\delta_{24}d_3 \cos v^*, \\
\phi_{46} &= \mu d_4(\tfrac{2}{15}\delta_{22} + \tfrac{1}{3}\delta_{23} + \delta_{24}) \cos v^*, \\
\phi_{61} &= \mu(\boldsymbol{\epsilon}_1 \cdot \mathbf{W}_1)\delta_{10} \cos v^*, \\
\phi_{63} &= \mu(\boldsymbol{\epsilon}_1 \cdot \mathbf{W}_1)\delta_{11} \cos v^*, \\
\phi_{65} &= \mu(\boldsymbol{\epsilon}_1 \cdot \mathbf{W}_1)\delta_{12} \cos v^*, \\
\phi_{71} &= \mu^{1/2}(\boldsymbol{\epsilon}_2 \cdot \mathbf{W}_1)\delta_{13} \sin v^*, \\
\phi_{73} &= \mu^{1/2}(\boldsymbol{\epsilon}_2 \cdot \mathbf{W}_1)\delta_{14} \sin v^*, \\
\phi_{75} &= \mu^{1/2}(\boldsymbol{\epsilon}_2 \cdot \mathbf{W}_1)\delta_{15} \sin v^*, \\
M_1 &= \phi_{61} + \phi_{71}, \\
M_3 &= \phi_{63} + \phi_{73}, \\
M_5 &= \phi_{65} + \phi_{75}, \\
N_1 &= -\sum_1^4 \phi_{j1}, \\
N_2 &= -\sum_1^4 \phi_{j2}, \\
N_3 &= -\sum_1^4 \phi_{j3}, \\
N_4 &= -\sum_1^4 \phi_{j4}, \\
N_5 &= -\sum_1^4 \phi_{j5}, \\
N_6 &= -\sum_1^4 \phi_{j6}, \\
\tilde{\omega}^2 &= \frac{\beta_2}{\beta_1}, \\
A_1 &= 0, \\
A_2 &= \tilde{\omega}^2 M_1{}^2 \beta_1^{1/2}, \\
A_3 &= 0, \\
A_4 &= (\tfrac{4}{3}\tilde{\omega}^2 - \tfrac{1}{2}\tilde{\omega}^4) M_1{}^2 \beta_1^{1/2} + 2M_1M_3\tilde{\omega}^2\beta_1^{1/2}, \\
A_5 &= 0.
\end{aligned}
\tag{4.71}
$$

If $0 \leq i^* \leq i_l^*$, let $s = -1$, otherwise let $s = 1$, and compute the $\bar{\nu}$ coefficients from

$$\begin{aligned}
\bar{\nu}_1 &= A_5 + s(5M_1N_6 + M_3N_4 - 3M_5N_2), \\
\bar{\nu}_2 &= A_4 + s(4M_1N_5 - 4M_5N_1), \\
\bar{\nu}_3 &= A_3 + s(3M_1N_4 - M_3N_2), \\
\bar{\nu}_4 &= A_2 + s(2M_1N_3 - 2M_3N_1), \\
\bar{\nu}_5 &= A_1 + s(M_1N_2).
\end{aligned} \tag{4.72}$$

Solve the quartic

$$\bar{\nu}_1 i^{*4} + \bar{\nu}_2 i^{*3} + \bar{\nu}_3 i^{*2} + \bar{\nu}_4 i^* + \bar{\nu}_5 = 0 \tag{4.73}$$

for all real i_i^* and choose $0 \leq i^* \leq i_l^*$. If

$$|i_n^* - i_{n+1}^*| < \delta^*, \tag{4.74}$$

where δ^* is a tolerance, continue with inequality 4.75; otherwise repeat equational loop (4.68) to (4.74) with the improved value of i^*. Determine if

$$0 \leq i^* \leq i_l^*. \tag{4.75}$$

If inequality 4.75 is satisfied, continue with (4.76); if not, the simple non-broken plane transfer is the best transfer.

Compute the separate velocity increments required to perform the maneuver from

$$\begin{aligned}
\Delta V_1 &= (\beta_1 + \beta_2 \tan^2 i^*)^{1/2} - \beta_3, \\
\Delta V_2 &= \left| \frac{\zeta_1 \sin v^* + \zeta_4 \cos v^* + \zeta_2}{\gamma_4 \cos v^* + \gamma_5 \sin v^*} \right|.
\end{aligned} \tag{4.76}$$

Determine the velocity and position vectors at the point of application of the secondary impulse as

$$\begin{aligned}
r_I &= \frac{p_{dpi}}{1 + C_{vi} \cos v^* - S_{vi} \sin v^*}, \\
\mathbf{V}_{dpi} &= \mathbf{W}_{dpi} \times \mathbf{U}_{dp}, \\
x_{vI} &= r_I \cos v^*, \\
y_{vI} &= r_I \sin v^*, \\
\dot{x}_{vI} &= \left(\frac{\mu}{p_{dpi}}\right)^{1/2} [S_{vi} - \sin v^*], \\
\dot{y}_{vI} &= \left(\frac{\mu}{p_{dpi}}\right)^{1/2} [C_{vi} + \cos v^*], \\
\mathbf{r}_{2I} &= x_{vI}\mathbf{U}_{dp} + y_{vI}\mathbf{V}_{dpi}, \\
\dot{\mathbf{r}}_{2I} &= \dot{x}_{vI}\mathbf{U}_{dp} + \dot{y}_{vI}\mathbf{V}_{dpi}.
\end{aligned} \tag{4.77}$$

Compute the time of flight between the departure terminal and the point of secondary impulse application,

$$\Delta t_1. \tag{4.78}$$

Compute the remaining flight time to the target planet as

$$\Delta t_2 = \Delta t - \Delta t_1. \tag{4.79}$$

Determine a heliocentric orbit between the radius vectors $\mathbf{r}_I$ and $\mathbf{r}_{ap}$ for transfer time Δt_2 to obtain the initial velocity vector at point I, that is,

$$\mathbf{r}_I, \quad \dot{\mathbf{r}}_I. \tag{4.80}$$

Obtain the position and velocity vectors of the space vehicle at the arrival terminal as an auxiliary computation:[4]

$$\mathbf{r}_{ap}, \quad \dot{\tilde{\mathbf{r}}}_{ap}.$$

Compute the kinematic arrival velocity

$$\Delta V_a^2 = (\dot{\mathbf{r}}_{ap} - \dot{\tilde{\mathbf{r}}}_{ap}) \cdot (\dot{\mathbf{r}}_{ap} - \dot{\tilde{\mathbf{r}}}_{ap}), \tag{4.81}$$

the parabolic speed

$$V_{pa}^2 = \frac{2\mu_5}{a_a}, \tag{4.82}$$

and determine the total velocity increment at the third terminal by subtracting the circular velocity:[5]

$$\Delta V_3 = [V_{pa}^2 + \Delta V_a^2]^{1/2} - \left[\frac{\mu_5}{a_a}\right]^{1/2}. \tag{4.83}$$

Compute the minimum two-impulse transfer velocity:

$$\Delta V_{M2} = \Delta V_1 + \Delta V_2. \tag{4.84}$$

If a refinement of velocity is required,[6] set $(\Delta V_{M2})_i = \Delta V_{M2}$, return to (4.56) and let $\lambda \simeq 5$ percent of $|z_p|$ and repeat equational loop (4.57) to (4.84). If $(\Delta V_{M2})_{i+1}$ is less than ΔV_{M2}, increment λ by 5 percent and continue repeating the analysis until the true minimum value of ΔV_{M2} is found.

If an optimal three-impulse mode ($N = 3$) is desired between launch times $[\text{J.D.}]_d \pm \Delta[\text{J.D.}]$, save ΔV_T, where

$$\Delta V_T = |\Delta V_1| + |\Delta V_2| + |\Delta V_3|, \tag{4.85}$$

[4] Note that this orbit is the same as that obtained immediately above except that different terminal conditions are produced. The final value for p obtained from state (4.80) should be used to avoid the iteration.

[5] The units of V_{pa}, V_a and the circular velocity must be consistent.

[6] This refinement is used to eliminate the slight errors caused by utilizing projection theory. See Section 4.2.8.

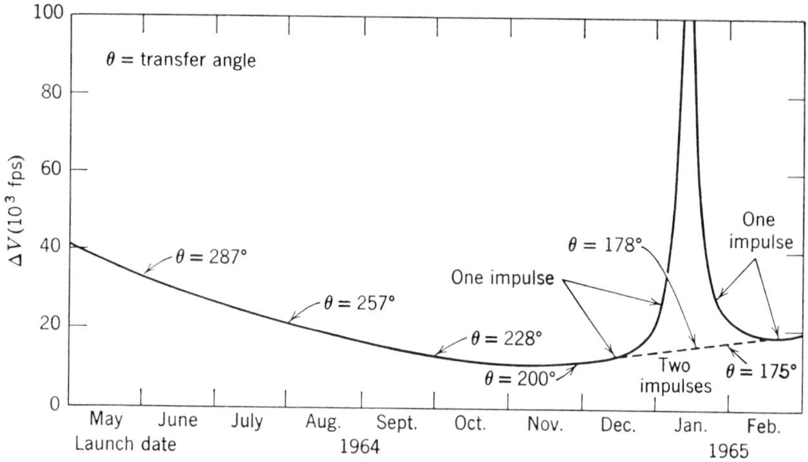

FIGURE 4.12 Velocity expenditure for Earth to Mars flyby.

and repeat the entire analysis starting at $[\text{J.D.}]_d - \Delta[\text{J.D.}]$ in increments of $\Delta[\text{J.D.}]/m$ to $[\text{J.D.}]_d + \Delta[\text{J.D.}]$ and select the corresponding minimum ΔV_T.

4.2.10 Numerical Results for Fuel Critical Orbits

The algorithm stated in Section 4.2.9 can be used to generate mission curves for various interplanetary transfer modes. Broken plane maneuvers, that is, the optimum application of intermediate impulses, and the one-impulse maneuver can be physically interpreted by an examination of Figures 4.12 and 4.13.

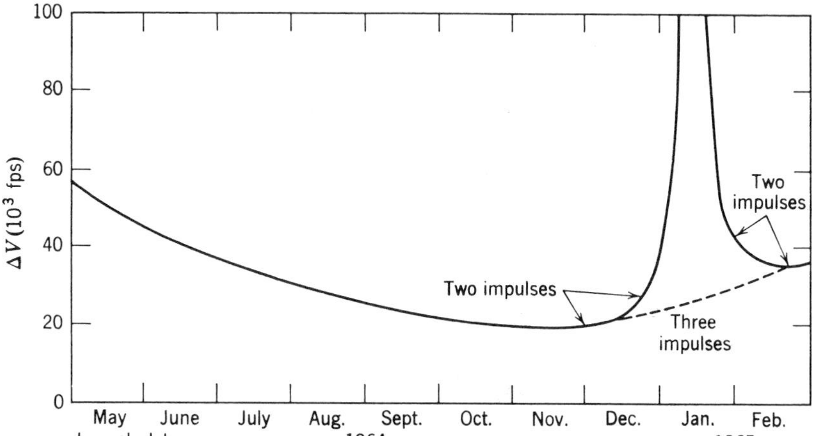

FIGURE 4.13 Earth to Mars rendezvous in circular orbit.

An Earth to Mars flyby is illustrated in Figure 4.12 from the point of view of propellant or, equivalently, velocity increment expenditure for various launch dates. The solid lines represent one-impulse flybys for 300-day missions. As these curves show, the optimum one-impulse transfer should have been initiated around November 1964. The budget for this transfer is approximately 11,000 fps over and above the 300-nautical mile coasting circular orbit speed. On either side of this optimum launch date the velocity budget increases rapidly. In fact, the size of the transfer velocity budget becomes prohibitive shortly after the first of the year. A launch possibility exists again around February. The central cusp of these curves corresponds to over the Sun transfers. To extend the launch window from November through March the two-impulse transfer must be used. The velocity budget for this curve is illustrated by the dashed line and rises from the true optimum of 11,000 fps to about 20,000 fps, which is still within feasible vehicle performance ranges.

Figure 4.13 illustrates the counterpart of Figure 4.12, that is, the rendezvous case in which the vehicle is put into a circular orbit about Mars at 300 nautical miles. In this case the optimum launch date is about mid-November for the one-impulse case and rapidly increases beyond present vehicle propulsive budgets shortly after. Initiation of the optimum mid-course maneuver permits a variable velocity increment budget ranging from about 20,000 to 30,000 fps for the November to February period.

The use of the broken plane transfer is very important in the rescue, abort, and ferrying of supplies for manned missions where the greatly increased launch window can mean the difference between life and death for future astronauts.

4.3 MINIMUM TIME INTERPLANETARY TRAJECTORIES

4.3.1 Preliminary Remarks

Interplanetary transfers to distant planets require fairly long transfer times; that is, the time difference from the departure terminal $\mathbf{r}_{dp}$ to the arrival terminal $\mathbf{r}_{ap}$ is usually large. It is perhaps not realistic to expect future astronauts to spend such long times within the small confines of interplanetary vehicles. In fact, real life conditions seem to indicate that usual transfer modes which stress conservation of fuel or, equivalently, velocity increment hardly seem appropriate in relation to the transit time imposed on the space passenger. Apparently, the ideal answer to a true optimum interplanetary transfer would consider a weighted combination of time and fuel conservation. The weighted mean or cost function could be represented by the linear relationship

$$w = A\,\Delta V + B\,\Delta t, \tag{4.86}$$

where ΔV is the total velocity expenditure of the transfer maneuver and Δt is the total transfer time between planets. The weights or parameters A and B are adopted constants which are functionally selected to satisfy real life constraints. For example, A and B would represent for a particular target planet a compromise situation reflecting the physiological resistance function of the astronaut, medical considerations, psychological strain factors, and finally cost of mission—which would most directly be reflected in the A coefficient.

Obviously, when B is taken to be identically zero, the case of minimum fuel transfer results. It can be treated analytically by the methods developed previously. In this section the case $A = 0$ that is, the pure minimum time situation, is investigated. This case has direct importance in rapid ferrying of supplies and in abort situations.

Perhaps it is well to mention that the transfer cases analyzed in this section are minimum time trajectories obtained from impulsive considerations. In the near future, ion and electrical propulsion systems will undoubtedly be added to space vehicles in order to have auxiliary thrust and thus reduce travel time as much as possible. In this light the transfer trajectories determined from the techniques developed here would reflect lower transfer time limits on trajectories to the distant planets. Application of additional thrust therefore results in further decreased times of transfer. The analysis in this section was developed by Escobal and Crichton [7] and assumes circular parking orbits.

4.3.2 Time Minimization Process

By use of patched conic theory it is possible to develop an expression for the impulse required at the first or departure terminal $\mathbf{r}_{dp}$. The geometry of the transfer process is illustrated in Figure 4.14. Specifically, the expression

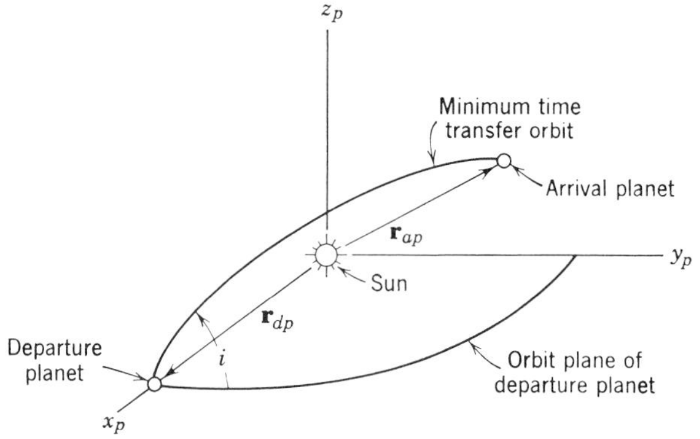

FIGURE 4.14 Minimum time interplanetary transfer.

for the velocity increment at the departure terminal is given by the familiar relationship derived in Section 4.1.1:

$$\Delta V_d = [V_p^{\,2} + V_h^{\,2}]^{1/2} - V_c, \tag{4.87}$$

where $V_p \equiv$ the parabolic escape speed, $V_h \equiv$ the heliocentric transfer speed increment, and $V_c \equiv$ the orbit speed on the circular departure orbit. It follows from Section 4.1.1 that the kinematic velocity increment that must be added to the transfer vehicle in order to place it on the transfer orbit which intersects the target planet can be written

$$V_h^{\,2} = [(\dot{\mathbf{r}}_{dp} - \dot{\mathbf{r}}_{dh}) \cdot (\dot{\mathbf{r}}_{dp} - \dot{\mathbf{r}}_{dh})]^{1/2}, \tag{4.88}$$

where the subscripts *dp* refer to the departure planet and *dh* to the transfer conic at departure.

In order to minimize the modified time of transfer, τ, between $\mathbf{r}_{dp}$ and $\mathbf{r}_{ap}$, where

$$\tau \equiv k_{\odot}(t_a - t_d) \tag{4.89}$$

with $k_{\odot} \equiv$ gravitational constant of the Sun, $t_a \equiv$ Julian date at arrival, and $t_d \equiv$ Julian date at departure, it suffices to form the Lagrange minimization function M, discussed in Section 2.3.1, as

$$M = \tau + \lambda(\Delta V_{\max}^2 - \Delta V^2 - \epsilon^2). \tag{4.90}$$

In (4.90), λ is an unknown constant Lagrange multiplier and

$$\Delta V_{\max}^2 - \Delta V^2 - \epsilon^2 = 0 \tag{4.91}$$

implies, as usual, the inequality

$$\Delta V^2 \leq \Delta V_{\max}^2, \tag{4.92}$$

or that the velocity increment magnitude of the transfer process is less than some available upper limit imposed by the vehicle propulsion system. The usual minimization procedure now requires that

$$\frac{\partial M}{\partial u_i} = 0, \qquad \frac{\partial M}{\partial \epsilon} = 0, \tag{4.93}$$

where u_i are the functional variables on which τ and ΔV depend.

Since the orbit must pass through $\mathbf{r}_{dp}$ and $\mathbf{r}_{ap}$, however, six conditions or degrees of freedom are now accounted for, and therefore only one more condition can be imposed. This condition is absorbed by (4.91). Hence

$$\frac{\partial M}{\partial \epsilon} = 0 \tag{4.94}$$

is the controlling condition. From (4.90) this implies

$$-\lambda\epsilon = 0 \tag{4.95}$$

or, for the case $\lambda \neq 0$, it follows that $\epsilon = 0$, and (4.91) yields

$$\Delta V_{\max}^2 = \Delta V^2 \tag{4.96}$$

as the only remaining relationship that can be imposed. As may be inferred from (4.87), the controlling equation of the orbit determination process is

$$[V_p^{\,2} + V_h^{\,2}]^{1/2} = \Delta V_{\max} + V_c. \tag{4.97}$$

Hence, by removing radicals and solving for $V_h^{\,2}$, it is possible to obtain

$$V_h^{\,2} = \alpha_0^{\,2}, \tag{4.98}$$

where

$$\alpha_0^{\,2} \equiv \Delta V_{\max}^2 + 2V_c\,\Delta V_{\max} - V_c^{\,2},$$

which is positive only if $\Delta V_{\max} > ([2]^{1/2} - 1)V_c$. More explicitly, by virtue of (4.88),

$$(\dot{\mathbf{r}}_{dp} - \dot{\mathbf{r}}_{dh}) \cdot (\dot{\mathbf{r}}_{dp} - \dot{\mathbf{r}}_{dh}) = \alpha_0^{\,2}. \tag{4.99}$$

To determine the transfer orbit it is necessary to eliminate $\dot{x}_{dh}, \dot{y}_{dh}, \dot{z}_{dh}$ from the previous symmetric form.

4.3.3 Polynominal Resolvent for Minimum Time One-Impulse Orbit

It has been shown in Section 3.5.1 that

$$\dot{\mathbf{r}}_{dh} = (\omega_1 p^{-1/2} + \omega_2 p^{1/2})\mathbf{U}_{dp} + (\omega_3 p^{1/2})\mathbf{U}_{ap}, \tag{4.100}$$

where

$$\omega_1 \equiv \frac{[\mu]^{1/2}[1 - \cos(v_{ap} - v_{dp})]}{\sin(v_{ap} - v_{dp})},$$

$$\omega_2 \equiv \frac{-[\mu]^{1/2}}{r_{ap}\sin(v_{ap} - v_{dp})},$$

$$\omega_3 \equiv \frac{[\mu]^{1/2}}{r_{dp}\sin(v_{ap} - v_{dp})},$$

with $\mu \equiv$ sum of masses of vehicle and Sun, $v_{ap} \equiv$ true anomaly of transfer orbit at arrival time, and $v_{dp} \equiv$ true anomaly of transfer orbit at departure time.

The unit vector $\mathbf{U}_{dp}$, pointing at the departure planet from the Sun, is directly known from

$$\mathbf{U}_{dp} = \frac{\mathbf{r}_{dp}}{r_{dp}}. \tag{4.101}$$

On the other hand, since the time of transit is unknown, $\mathbf{U}_{ap}$ is not known. At this point a suitable estimate of the optimum time must be adopted.

This estimate is refined as the minimum transfer time approaches optimality. Once the transfer time is estimated, it follows from an analytical ephemeris of the target planet, Section 1.3, that

$$\mathbf{U}_{ap} = \frac{\mathbf{r}_{ap}}{r_{ap}} \tag{4.102}$$

can be computed and the ω coefficients of (4.100) become numerically determined. To obtain the angular argument of the ω coefficients, use of the following equations extracted from ([1], Chapter 6) is required:

$$\cos(v_{ap} - v_{dp}) = \mathbf{U}_{dp} \cdot \mathbf{U}_{ap},$$

$$\sin(v_{ap} - v_{dp}) = \frac{x_{dp}y_{ap} - x_{ap}y_{dp}}{|x_{dp}y_{ap} - x_{ap}y_{dp}|}[1 - \cos^2(v_{ap} - v_{dp})]^{1/2}.$$

With this understanding and the use of (4.100), (4.99) can be transformed as follows:

$$\begin{aligned}[(\omega_1 p^{-1/2} + \omega_2 p^{1/2})U_{xdp} + (\omega_3 p^{1/2})U_{xap} - \dot{x}_{dp}]^2 \\ + [(\omega_1 p^{-1/2} + \omega_2 p^{1/2})U_{ydp} + (\omega_3 p^{1/2})U_{ydp} - \dot{y}_{dp}]^2 \\ + [(\omega_1 p^{-1/2} + \omega_2 p^{1/2})U_{zdp} + (\omega_3 p^{1/2})U_{zdp} - \dot{z}_{dp}]^2 = \alpha_0^2,\end{aligned} \tag{4.103}$$

where

$$\dot{\mathbf{r}}_{dp} \equiv \begin{bmatrix} \dot{x}_{dp} \\ \dot{y}_{dp} \\ \dot{z}_{dp} \end{bmatrix}.$$

Expanding (4.103) yields the compact expression

$$V_{hd}^2 = a_0 p + a_1 p^{1/2} + a_2 + a_3 p^{-1/2} + a_4 p^{-1} = \alpha_0^2, \tag{4.104}$$

where

$$a_0 \equiv \omega_2^2 + \omega_3^2 + 2\omega_2\omega_3\mathbf{U}_{dp} \cdot \mathbf{U}_{ap},$$

$$a_1 \equiv -2\omega_2\mathbf{U}_{dp} \cdot \dot{\mathbf{r}}_{dp} - 2\omega_3\mathbf{U}_{ap} \cdot \dot{\mathbf{r}}_{dp},$$

$$a_2 \equiv 2\omega_1\omega_2 + 2\omega_3\omega_1\mathbf{U}_{dp} \cdot \mathbf{U}_{ap} + \dot{\mathbf{r}}_{dp} \cdot \dot{\mathbf{r}}_{dp},$$

$$a_3 \equiv -2\omega_1\mathbf{U}_{dp} \cdot \dot{\mathbf{r}}_{dp},$$

$$a_4 \equiv \omega_1^2.$$

Equation 4.104 can be transformed under the substitution $p = s^2$ to the quartic

$$a_0 s^4 + a_1 s^3 + (a_2 - \alpha_0^2)s^2 + a_3 s + a_4 = 0. \tag{4.105}$$

The quartic can be solved in closed form [1] for the square root of the semiparameter of the orbit, $[p]^{1/2}$, that minimizes the transfer time from the departure to the arrival planet. The computation of the coefficients of the quartic required the position of the arrival planet to be known. Since the minimum time to reach the target planet, τ_m, was not known, an estimate of the transfer time had to be made. The optimum orbit must be solved by a search procedure in which each orbit is in itself a suboptimum.

As will be discussed presently, once the computation of p is performed the time interval τ_c between $\mathbf{r}_{dp}$ and $\mathbf{r}_{ap}$ can be computed in closed form. The function F can therefore be formed, that is,

$$F \equiv \tau - \tau_c, \tag{4.106}$$

where τ is the estimated transfer time. If $F = 0$, the true global minimum time transfer for a specified launch date has been obtained; if $F \neq 0$, the estimated time can be varied and the analysis repeated. In essence, this procedure yields the zero of F which determines the optimum orbit. Either a Newton procedure or a direct search can be utilized to determine the zero of F.

4.3.4 Polynomial Resolvent for Minimum Time Two-Impulse Orbits

The preceding analysis does not consider the case of entering into an orbit about the target planet. Realistically, however, the total amount of fuel or velocity increment available for the interplanetary maneuver must be limited by the inequality

$$\Delta V_T \equiv \Delta V_d + \Delta V_a \leq \Delta V_{\max}, \tag{4.107}$$

where ΔV_d is the departure impulse and ΔV_a is the arrival impulse. If it is assumed that the departure and arrival orbits are circular, then from energy considerations it is possible to write, in analogy with (4.87), that

$$\Delta V_T = (V_{pd}^2 + V_{hd}^2)^{1/2} - V_{cd} + (V_{pa}^2 + V_{ha}^2)^{1/2} - V_{ca} \tag{4.108}$$

with the understanding that the secondary subscripts d and a, respectively, denote departure and arrival conditions. Furthermore it is possible to show, as in Chapter 3, that the condition for minimum time is given by

$$[\lambda_1]^{1/2}[(V_{pd}^2 + V_{hd}^2)^{1/2} - V_{cd}] + [\lambda_2]^{1/2}[(V_{pa}^2 + V_{ha}^2)^{1/2} - V_{ca}] = \Delta V_{\max}, \tag{4.109}$$

where $[\lambda_1]^{1/2}$ and $[\lambda_2]^{1/2}$ are constant multipliers carried along as weighting factors. Obviously, if $[\lambda_2]^{1/2} = 0$ and $[\lambda_1]^{1/2} = 1$, the previously discussed one-impulse transfer case results. Note that λ_1, λ_2 are chosen arbitrarily for a particular problem. In general, these weights would reflect utilization of engines of different specific impulses at the respective terminals.

Introducing the definition

$$\beta \equiv \Delta V_{\max} + [\lambda_1]^{1/2} V_{cd} + [\lambda_2]^{1/2} V_{ca} \tag{4.110}$$

and squaring (4.109) twice results in

$$\lambda_1^2 V_{hd}^4 + 2\lambda_1^2 V_{pd}^2 V_{hd}^2 + \lambda_2^2 V_{ha}^4 + 2\lambda_2^2 V_{pa}^2 V_{ha}^2 - 2\lambda_1 \beta^2 V_{hd}^2$$
$$- 2\lambda_2 \beta^2 V_{ha}^2 - 2\lambda_1\lambda_2 V_{pd}^2 V_{ha}^2 - 2\lambda_1\lambda_2 V_{pa}^2 V_{hd}^2 - 2\lambda_1\lambda_2 V_{hd}^2 V_{ha}^2 + \gamma = 0, \tag{4.111}$$

where the known constant γ is defined as

$$\gamma \equiv \beta^4 + \lambda_1^2 V_{pd}^4 + \lambda_2^2 V_{pa}^4 - 2\lambda_1\beta^2 V_{pd}^2 - 2\lambda_2 \beta^2 V_{pa}^2 - 2\lambda_1\lambda_2 V_{pd}^2 V_{pa}^2.$$

Equation 4.111 can be written more compactly as

$$(\lambda_1 V_{hd}^2 - \lambda_2 V_{hd}^2)^2 + (2\lambda_1^2 V_{pd}^2 - 2\lambda_1\lambda_2 V_{pa}^2 - 2\lambda_1\beta^2) V_{hd}^2$$
$$+ (2\lambda_2^2 V_{pa}^2 - 2\lambda_1\lambda_2 V_{pd}^2 - 2\lambda_2\beta^2) V_{ha}^2 + \gamma = 0. \tag{4.112}$$

By using an expression similar to (4.100), that is,

$$\dot{\mathbf{r}}_{ah} = (-\omega_1 p^{-1/2} + \omega_3 p^{1/2})\mathbf{U}_{ap} + (\omega_2 p^{1/2})\mathbf{U}_{dp}, \tag{4.113}$$

it is possible to obtain an expression for the arrival terminal velocity analogous to (4.104):

$$V_{ha}^2 = b_0 p + b_1 p^{1/2} + b_2 + b_3 p^{-1/2} + b_4 p^{-1} \tag{4.114}$$

with

$$b_0 \equiv \omega_3^2 + \omega_2^2 + 2\omega_3\omega_2 \mathbf{U}_{dp} \cdot \mathbf{U}_{ap},$$
$$b_1 \equiv -2\omega_3 \mathbf{U}_{ap} \cdot \dot{\mathbf{r}}_{ap} - 2\omega_2 \mathbf{U}_{dp} \cdot \dot{\mathbf{r}}_{ap},$$
$$b_2 \equiv -2\omega_1\omega_3 - 2\omega_2\omega_1 \mathbf{U}_{dp} \cdot \mathbf{U}_{ap} + \dot{\mathbf{r}}_{ap} \cdot \dot{\mathbf{r}}_{ap},$$
$$b_3 \equiv 2\omega_1 \mathbf{U}_{ap} \cdot \dot{\mathbf{r}}_{ap},$$
$$b_4 \equiv \omega_1^2.$$

For convenience the following auxiliary known constants can be introduced:

$$\lambda_3 \equiv 2\lambda_1^2 V_{pd}^2 - 2\lambda_1\lambda_2 V_{pa}^2 - 2\lambda_1\beta^2, \tag{4.115}$$
$$\lambda_4 \equiv 2\lambda_2^2 V_{pa}^2 - 2\lambda_1\lambda_2 V_{pd}^2 - 2\lambda_2\beta^2, \tag{4.116}$$

so that (4.112) can be written as

$$(\lambda_1 V_{hd}^2 - \lambda_2 V_{ha}^2)^2 + \lambda_3 V_{hd}^2 + \lambda_4 V_{ha}^2 + \gamma = 0. \tag{4.117}$$

Substitution of (4.104) and (4.114) into (4.117) results in the algebraic form

$$[(\lambda_1 a_4 - \lambda_2 b_4)p^{-1} + (\lambda_1 a_3 - \lambda_2 b_3)p^{-1/2} + (\lambda_1 a_1 - \lambda_2 b_1)p^{1/2}$$
$$+ (\lambda_1 a_0 - \lambda_2 b_0)p + (\lambda_1 a_2 - \lambda_2 b_2)]^2$$
$$+ (\lambda_3 a_4 + \lambda_4 b_4)p^{-1} + (\lambda_3 a_3 + \lambda_4 b_3)p^{-1/2}$$
$$+ (\lambda_3 a_1 + \lambda_4 b_1)p^{1/2} + (\lambda_3 a_0 + \lambda_4 b_0)p + (\lambda_3 a_2 + \lambda_4 b_2) + \gamma = 0. \tag{4.118}$$

Collecting coefficients of equal powers of p results in the expression

$$c_0p^2 + c_1p^{3/2} + c_2p + c_3p^{1/2} + c_4 + c_5p^{-1/2} + c_6p^{-1} + c_7p^{-3/2} + c_8p^{-2} = 0, \quad (4.119)$$

where

$$\begin{aligned}
c_0 &\equiv (\lambda_1a_0 - \lambda_2b_0)^2,\\
c_1 &\equiv 2(\lambda_1a_1 - \lambda_2b_1)(\lambda_1a_0 - \lambda_2b_0),\\
c_2 &\equiv (\lambda_1a_1 - \lambda_2b_1)^2 + 2(\lambda_1a_0 - \lambda_2b_0)(\lambda_1a_2 - \lambda_2b_2) + (\lambda_3a_0 + \lambda_4b_0),\\
c_3 &\equiv 2(\lambda_1a_3 - \lambda_2b_3)(\lambda_1a_0 - \lambda_2b_0) + 2(\lambda_1a_1 - \lambda_2b_1)(\lambda_1a_2 - \lambda_2b_2)\\
&\quad + (\lambda_3a_1 + \lambda_4b_1),\\
c_4 &\equiv (\lambda_1a_2 - \lambda_2b_2)^2 + 2(\lambda_1a_4 - \lambda_2b_4)(\lambda_1a_0 - \lambda_2b_0)\\
&\quad + 2(\lambda_1a_3 - \lambda_2b_3)(\lambda_1a_1 - \lambda_2b_1) + (\lambda_3a_2 + \lambda_4b_2) + \gamma,\\
c_5 &\equiv 2(\lambda_1a_4 - \lambda_2b_4)(\lambda_1a_1 - \lambda_2b_1) + 2(\lambda_1a_3 - \lambda_2b_3)(\lambda_1a_2 - \lambda_2b_2)\\
&\quad + (\lambda_3a_3 + \lambda_4b_3),\\
c_6 &\equiv (\lambda_1a_3 - \lambda_2b_3)^2 + 2(\lambda_1a_4 - \lambda_2b_4)(\lambda_1a_2 - \lambda_2b_2) + (\lambda_3a_4 + \lambda_4b_4),\\
c_7 &\equiv 2(\lambda_1a_4 - \lambda_2b_4)(\lambda_1a_3 - \lambda_2b_3),\\
c_8 &\equiv (\lambda_1a_4 - \lambda_1b_4)^2.
\end{aligned}$$

Equation 4.119 is of eighth degree in the square root of the semiparameter of the transfer orbit, and on removal of radicals it reduces to

$$c_0s^8 + c_1s^7 + c_2s^6 + c_3s^5 + c_4s^4 + c_5s^3 + c_6s^2 + c_7s + c_8 = 0, \quad (4.120)$$

where the transformation of algebraic forms is obtained via the relationship

$$p = s^2. \quad (4.121)$$

Solution of the resolvent permits the determination of the semiparameter of the orbit to be made. From this point on, the solution for the targeting of the planet proceeds as previously described for the one-impulse case. It should be noted that if $\lambda_1 = \lambda_2$, that is, for equally weighted impulses, $c_0 = c_1 = c_7 = c_8 \equiv 0$, and the octic defined by (4.120) collapses into a simple quartic with the spurious double root $s = 0$. In this case the quartic can be solved in closed form as discussed in ([1], Appendix 3).

4.3.5 Rejection of Spurious Roots

It is possible for the polynomial resolvents to produce spurious roots. Usually half as many real roots are produced as the order of the polynomial. However, the undesirable roots can be eliminated rapidly.

Consider that there exist s_i real roots to the particular polynomial that is to be solved. Since the ω coefficients, (4.100), are numerically known, it follows that $\dot{\mathbf{r}}_{dhi}$ can be evaluated very easily from

$$\dot{\mathbf{r}}_{dhi} = (\omega_1 s_i^{-1} + \omega_2 s_i)\mathbf{U}_{dp} + (\omega_3 s_i)\mathbf{U}_{ap}.$$

Furthermore, since $\mathbf{r}_{dp}$ is known, the transfer orbit eccentricity can be readily computed from the common calculations ([1], Appendix 2) as

$$r_{dp}^{\ 2} = \mathbf{r}_{dp} \cdot \mathbf{r}_{dp}, \tag{4.122}$$

$$r_{ap}^{\ 2} = \mathbf{r}_{ap} \cdot \mathbf{r}_{ap},$$

$$D_{dhi} = \frac{\mathbf{r}_{dp} \cdot \dot{\mathbf{r}}_{dhi}}{[\mu]^{1/2}},$$

$$\frac{V_{dhi}^2}{\mu} = \frac{\dot{\mathbf{r}}_{dhi} \cdot \dot{\mathbf{r}}_{dhi}}{\mu},$$

$$\frac{1}{a_i} = \frac{2}{r_{dp}} - \frac{V_{dhi}^2}{\mu},$$

$$e_i = \left[\left(1 - \frac{r_{dp}}{a_i}\right)^2 + \frac{1}{a_i} D_{dhi}^2\right]^{1/2}. \tag{4.123}$$

Hence for $e_i < 1$ it is possible to obtain τ_i from the chain calculation

$$C_{ei} = 1 - \frac{r_{dp}}{a_i},$$

$$S_{ei} = \frac{D_{dhi}}{[a_i]^{1/2}},$$

$$\sin(E_{ap} - E_{dp})_i = \frac{r_{ap}}{(a_i p_i)^{1/2}} \sin(v_{ap} - v_{dp}) - \frac{r_{ap}}{p_i}[1 - \cos(v_{ap} - v_{dp})]S_{ei},$$

$$\cos(E_{ap} - E_{dp})_i = 1 - \frac{r_{dp} r_{ap}}{a_i p_i}[1 - \cos(v_{ap} - v_{dp})],$$

$$(M_{ap} - M_{dp})_{ei} = (E_{ap} - E_{dp})_i + 2S_{ei}\sin^2\left[\frac{(E_{ap} - E_{dp})_i}{2}\right] - C_{ei}\sin(E_{ap} - E_{dp})_i,$$

$$n_i = k_{\odot}[\mu]^{1/2} a_i^{-3/2},$$

$$\tau_i = \frac{k_{\odot}(M_{ap} - M_{dp})_{ei}}{n_i}. \tag{4.124}$$

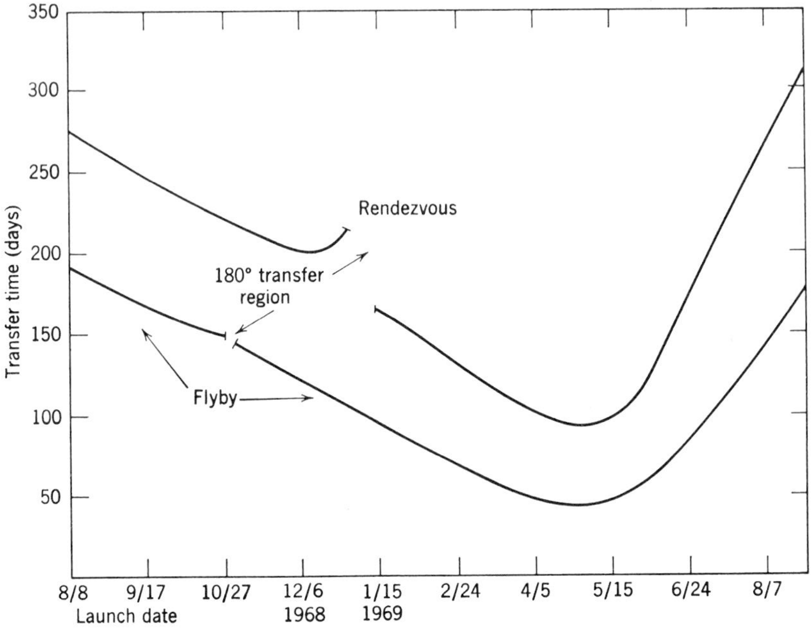

FIGURE 4.15 Earth to Mars time critical orbits.

Having obtained the matched pairs

$$p_i, \tau_i, \qquad i = 1, 2, \ldots, s,$$

where s is the number of real positive roots obtained from the polynomial resolvent, the elimination is obviously given by

$$\tau_{\min} = \min(\tau_1, \tau_2, \ldots, \tau_s).$$

The elimination logic permits very rapid calculations, and it has the advantage of ensuring that all possible candidates for the minimum time orbit are obtained. No other possibilities exist.

4.3.6 Numerical Results for Time Critical Orbits

Figures 4.15 and 4.16 represent direct minimum time transfers from Earth to Mars and from Earth to Jupiter, respectively, for both flyby and rendezvous missions. The abscissas represent the launch date from Earth, and the ordinates represent the total transfer time. These curves are for constant velocity budgets of 40,000 fps second (for Mars) and 90,000 fps (for Jupiter).

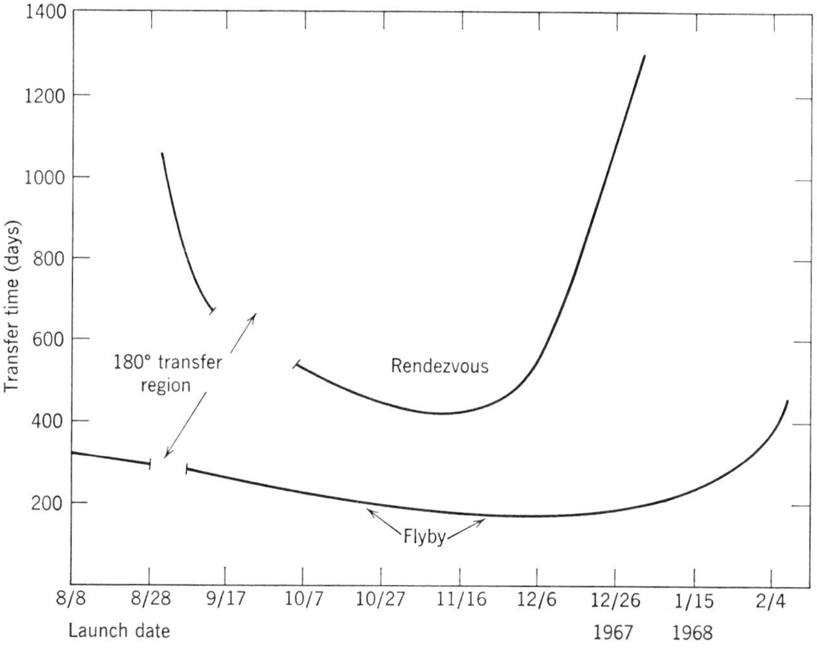

FIGURE 4.16 Earth to Jupiter time critical orbits.

The vehicle is launched from a geocentric 300-nautical mile parking orbit. For the rendezvous case, injection into a 300-nautical mile parking orbit at the target planet (for Mars) and a 6300-nautical mile parking orbit (for Jupiter) is assumed.

The particular launch dates chosen give favorable as well as unfavorable periods for interplanetary voyages owing to the unavoidable 180° transfer situations. In the case of Mars a flyby mission can be made in as few as 40 days with the assumed velocity budgets. This transfer time seems reasonable for future manned missions. The curves show a discontinuity when approaching the 180° transfer case; otherwise the curves are continuous. The curves display the minimum time envelope of the standard constant velocity impulse contours. However, the lengthy process of crossplotting many curves of launch date versus velocity impulse with constant trip time is circumvented by utilization of this method. The flyby and rendezvous missions to Mars and Jupiter are roughly similar in shape. The vertical displacement in trip time is caused by the gravitational field of the planets. The difference in the vehicle's velocity at Jupiter is more critical than in the Mars missions, and hence the Jupiter curves differ more noticeably.

4.4 MINIMUM INITIAL WEIGHT-IN-ORBIT INTERPLANETARY TRAJECTORIES

4.4.1 Preliminary Remarks

In many instances the problem of minimum weight in Earth orbit to accomplish an interplanetary mission is a controlling factor. For realistic trajectory transfers fuel or velocity expenditures for Earth departure, outbound midcourse correction, planet arrival, orbital injection, planet departure, etc., should be considered. Abrupt weight changes, that is, tank jettison losses and time-dependent weight changes (propellant vaporization and life support expendables), should also be noted. In this section a method of determining interplanetary trajectories based on minimum weight in Earth orbit is presented by developing a weight function which by the techniques of Chapter 2 can be minimized to yield the desired orbit. This weight function is developed in as simple a manner as possible in order to stress the basic principles involved. Any degree of sophistication can be included in the weight function once the basic technique is understood. The analysis in this section was developed by Sohn [18] and by Chovit and Klystra [19], [20].

4.4.2 Mission Payload Ratio Function

Consider a mission from Earth, or any other planet, to a planet during which a midcourse correction is executed, weights are jettisoned, time-dependent weight losses must be accounted for, and a final circularizing impulse is applied to enter orbit about the distant primary. The trajectory is illustrated in Figure 4.17, which depicts a trajectory from Earth to an objective planet. In the figure, w_{be} = the vehicle weight before leaving Earth,

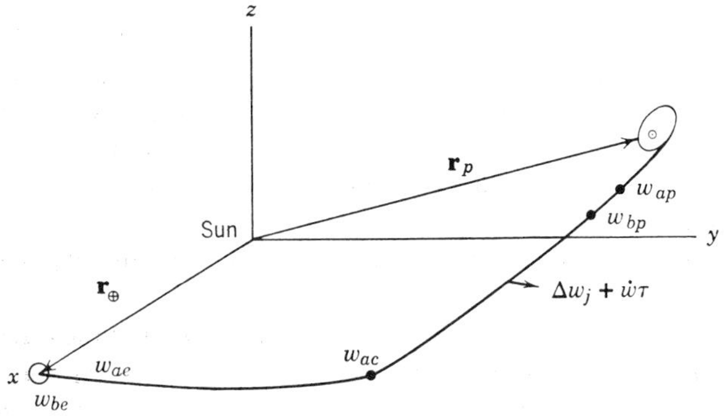

FIGURE 4.17 Interplanetary trajectory and associated weights.

w_{ae} = the vehicle weight after leaving Earth, w_{ac} = the vehicle weight after midcourse correction, Δw_j = the fixed weight jettisoned overboard, for example, probes or empty ranks, $\dot{w}$ = the rate of weight lost which depends on trip time, for example, life support expendables and vaporized propellant, w_{bp} = the vehicle weight before planetary arrival, w_{ap} = the vehicle weight after planetary arrival, and τ = the leg total trip time. To arrive at the planetary spacecraft payload ratio function, that is, the ratio of the total weight before leaving Earth to the final arrival weight,

$$P = \frac{w_{be}}{w_{ap}}, \tag{4.125}$$

it is possible to sum the weights consecutively starting with w_{ap} in terms of the individual stage payload ratios and discrete weight changes to obtain

$$P = \frac{1}{w_{ap}} \frac{w_{be}}{w_{ae}} \frac{w_{ae}}{w_{ac}} \left(\Delta w + \dot{w}\tau + \frac{w_{bp}}{w_{ap}} w_{ap} \right). \tag{4.126}$$

The function P can be written more compactly in terms of the individual stage payload ratios

$$R_e \equiv \frac{w_{be}}{w_{ae}}, \qquad R_c \equiv \frac{w_{ae}}{w_{ac}}, \qquad R_p \equiv \frac{w_{bp}}{w_{ap}}$$

as

$$P = \frac{1}{w_{ap}} R_e R_c (\Delta w + \dot{w}\tau + R_p w_{ap}), \tag{4.127}$$

where R_e is the leave Earth stage payload ratio, R_c is the outbound mid-course correction stage payload ratio, assumed independent of the trajectory parameters, and R_p is the arrive planet stage payload ratio. Equation 4.127 defines a typical payload ratio function. Many others are possible depending on the complexity of the mission, for example, in swingby or return missions. However, for purposes of illustration (4.127) suffices.

4.4.3 Optimization Equations for Minimum Weight Trajectories

The payload ratio function discussed in the preceding section depends on the total weight expended to perform the specified mission. However, the weight expended is related in part to the velocity increment expenditure for each maneuver, ΔV_i, through the well-known rocket equation (Section 2.5)

$$\Delta V_i = g I_{spi} \log M_i, \qquad i = 1, 2, \ldots, n, \tag{4.128}$$

where I_{sp} is the engine specific impulse, g is the appropriate units conversion factor, and M is propulsion mass ratio. The propulsion mass ratio is defined as the stage launch to burnout weight. Equation 4.128 is a fair approximation to the physics describing the staging of vehicles. With this understanding it follows functionally that

$$P = f(w_i) = \bar{f}(\Delta V_i). \tag{4.129}$$

More exactly, if the independent parameters of the interplanetary mission are specified as the planet arrive date t_{ap} and the total transfer time τ, the dynamical orbit is completely determined. This statement is true for fixed transfer time, because t_{ap} allows the direct computation of the planetary arrival radius vector $\mathbf{r}_p$, while $t_{ap} - \tau$ allows the direct computation of the Earth departure radius vector, $\mathbf{r}_\oplus$. Therefore, since $\mathbf{r}_p$, $\mathbf{r}_\oplus$, and τ completely determine the orbit (see [1], Chapter 6), the velocity change to leave Earth, ΔV_{le}, and the velocity change to arrive at the target planet, ΔV_{ap}, can be expressed functionally as

$$\begin{aligned} \Delta V_{le} &= \Delta V_{le}(\tau, t_{ap}), \\ \Delta V_{ap} &= \Delta V_{ap}(\tau, t_{ap}). \end{aligned} \tag{4.130}$$

With the previous understanding P can be functionally expressed as

$$P = P(\tau, t_{ap}). \tag{4.131}$$

By Section 2.2 an extreme value of P is obtained by imposing the conditions

$$\frac{\partial P}{\partial \tau} = 0, \qquad \frac{\partial P}{\partial t_{ap}} = 0. \tag{4.132}$$

Hence the explicit optimization equations are obtained from the differentiation of (4.127) as

$$\frac{\partial P}{\partial \tau} = \frac{R_c}{w_{ap}}\left[\left(\Delta w + \dot{w}\tau + R_p w_{ap}\right)\frac{\partial R_e}{\partial \Delta V_{le}}\frac{\partial \Delta V_{le}}{\partial \tau} + R_e\left(\dot{w} + w_{ap}\frac{\partial R_p}{\partial \Delta V_{ap}}\frac{\partial \Delta V_{ap}}{\partial \tau}\right)\right],$$

$$\frac{\partial P}{\partial t_{ap}} = \frac{R_c}{w_{ap}}\left[(\Delta w + \dot{w}\tau + R_p w_{ap})\frac{\partial R_e}{\partial \Delta V_{lc}}\frac{\partial \Delta V_{le}}{\partial t_{ap}} + R_e\left(w_{ap}\frac{\partial R_p}{\partial \Delta V_{ap}}\frac{\partial \Delta V_{ap}}{\partial t_{ap}}\right)\right],$$

or, since $\Delta w + \dot{w}\tau + R_p w_{ap} = w_{ac}$, by using the definitions

$$\omega_1 \equiv \frac{w_{ac}}{R_e}\frac{\partial R_e}{\partial \Delta V_{le}}, \qquad \omega_2 \equiv \frac{w_{bp}}{R_p}\frac{\partial R_p}{\partial \Delta V_{ap}} \tag{4.133}$$

and invoking conditions (4.132) it is possible to obtain the system

$$\begin{aligned} \omega_1\frac{\partial \Delta V_{le}}{\partial \tau} + \omega_2\frac{\partial \Delta V_{ap}}{\partial \tau} + \dot{w} &= 0, \\ \omega_1\frac{\partial \Delta V_{le}}{\partial t_{ap}} + \omega_2\frac{\partial \Delta V_{ap}}{\partial t_{ap}} &= 0. \end{aligned} \tag{4.134}$$

Before continuing with the solution of (4.134), some pertinent derivatives need to be determined.

4.4.4 Determination of the Weight Ratio Derivatives

It is possible to determine the appropriate weight derivatives by use of the identity

$$R \equiv \frac{w_b}{w_a} = \frac{M(1-\sigma)}{1 - M\sigma}, \tag{4.135}$$

where σ, the stage structure factor, is defined as

$$\sigma \equiv \frac{w_j}{w_j + w_p},$$

with w_j = stage jettison weight, w_p = stage propellant weight. Therefore, by direct differentiation, assuming that σ is a constant,

$$\frac{\partial R}{\partial \Delta V} = \left(\frac{M(1-\sigma)\sigma}{(1-M\sigma)^2} + \frac{1-\sigma}{1-M\sigma}\right)\frac{\partial M}{\partial \Delta V}. \tag{4.136}$$

However, since by (4.128)

$$\frac{\partial M}{\partial \Delta V} = \frac{\exp(\Delta V/gI_{sp})}{gI_{sp}}, \tag{4.137}$$

it is a simple matter to verify that

$$\frac{1}{R}\frac{\partial R}{\partial \Delta V} = [gI_{sp}(1 - M\sigma)]^{-1}. \tag{4.138}$$

The objection may be raised that σ is not actually constant. This in effect is true; however, since the process used to determine the actual orbit is iterative, the latest or most recent value of σ is always used in the successive solution of the equations. When the total iterative procedure, which is discussed presently, converges, the correct value of σ is automatically incorporated into (4.138). With this understanding,

$$\omega_1 = w_{ac}[I_{spe}(1 - M_e\sigma_e)]^{-1},$$
$$\omega_2 = w_{bp}[I_{spp}(1 - M_p\sigma_p)]^{-1}. \tag{4.139}$$

4.4.5 Obtaining Velocity Derivatives

The basic system to be solved, as noted previously, is given by (4.134). These equations contain the derivatives $\partial \Delta V_{le}/\partial\tau$, $\partial \Delta V_{le}/\partial t_{ap}$, $\partial \Delta V_{ap}/\partial\tau$, and $\partial \Delta V_{ap}/\partial t_{ap}$, which are discussed in this section from a numerical point of view. Geometrically, the required partial derivatives are illustrated in Figure 4.18. A plan view of trajectory data for constant ΔV is displayed in Figure 4.19. The display is called a *trajectory map*; it is employed to obtain the velocity partials. Consider choosing a specific Julian date, say $(\text{J.D.})_k$, $k = 1$, in Figure 4.19. Now, for various transfer times τ_i, the two planetary radius vectors $(\mathbf{r}_{\oplus i}, \mathbf{r}_{pi})$ can be obtained from Section 1.3 or Section 1.5, and the corresponding heliocentric transfer orbit determined. Knowing the heliocentric orbit in turn allows the appropriate ΔV_{ki} to be computed, that is, $(\Delta V_{le})_{ki}$ and $(\Delta V_{ap})_{ki}$. The ΔV_{ki} corresponding to $(\text{J.D.})_k$ can be stored and the process continued until the first row of points in Figure 4.19, where each point corresponds to the appropriate ΔV_{1i}, is generated. Finally, letting $k = 2, 3, 4$, etc., all the points in Figure 4.19 can be generated. It now follows from numerical experience that, over three discrete points, the velocity increments can be approximated by second-order curve fits as

$$
\begin{aligned}
\Delta V_{le} &= a_1\tau^2 + a_2\tau + a_3, \\
\Delta V_{ap} &= a_4\tau^2 + a_5\tau + a_6, \\
\Delta V_{le} &= a_7 t_{ap}^{\ 2} + a_8 t_{ap} + a_9, \\
\Delta V_{ap} &= a_{10} t_{ap}^{\ 2} + a_{11} t_{ap} + a_{12},
\end{aligned}
\tag{4.140}
$$

so that

$$
\begin{aligned}
\frac{\partial\, \Delta V_{le}}{\partial \tau} &= 2a_1\tau + a_2, \\
\frac{\partial\, \Delta V_{ap}}{\partial \tau} &= 2a_4\tau + a_5, \\
\frac{\partial\, \Delta V_{le}}{\partial t_{ap}} &= 2a_7 t_{ap} + a_8, \\
\frac{\partial\, \Delta V_{ap}}{\partial t_{ap}} &= 2a_{10} t_{ap} + a_{12}.
\end{aligned}
\tag{4.141}
$$

Equations 4.141 determine the partial derivatives required for the solution of system (4.134) explicitly in terms of the independent variables of the problem. The numerical curve-fitting procedure outlined in this section has been used with much success in actual optimization studies [19]. The evaluation of these partial derivatives requires the generation of a vast trajectory data map before the actual optimization process. The technique seems to be unavoidable especially when complicated payload ratio functions involving planetary powered swingbys are to be analyzed. A slight modification of this technique would be to determine directly the velocity increments simultaneously by using the heliocentric orbit determination process as frequently as required and not saving the data in the map.

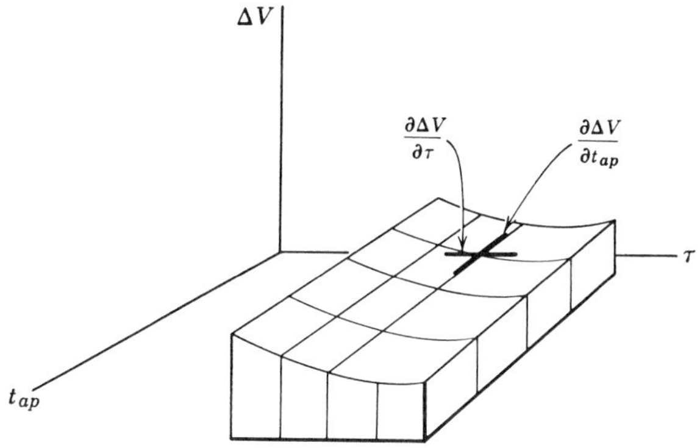

FIGURE 4.18 Velocity partial derivatives.

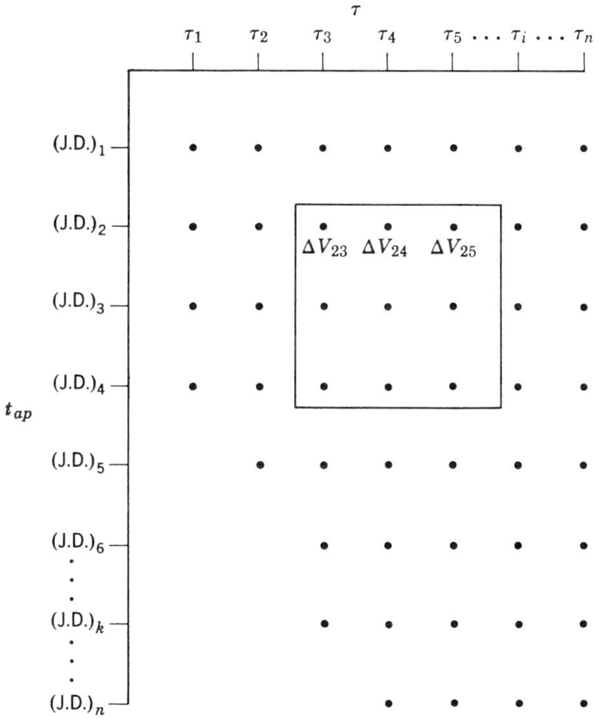

FIGURE 4.19 Julian date versus transfer time in days.

4.4.6 Iterative Procedure for the Orbit Determination Process

The computation of the minimum weight-in-orbit transfer trajectories must be done by means of iterative procedures. As discussed in Section 4.4.5, the first step of the process involves generating the appropriate trajectory maps and saving the triads (τ, t_{ap}, ΔV) for the curve-fitting procedure. This can be done apart from the actual optimization procedure. Once the trajectory data have been generated and stored, the second step requires that a nominal planetary arrive date $(t_{ap})_n$ and transfer time $(\tau)_n$ be chosen. About this nominal point a 3 × 3 matrix of surrounding points, as illustrated in Figure 4.19, is adopted, and the triads (τ, t_{ap}, ΔV) associated with these points are extracted from the trajectory map. These triads permit the curve fit coefficients a_i, $i = 1, 2, \ldots, 12$ to be determined. The Lagrange interpolating polynomial ([1], Chapter 7) can be used to obtain the actual coefficients with the least amount of difficulty. Third, a nominal vehicle or nominal vehicle weights are assumed for typical realistic state-of-the-art hardware. Finally, equations 4.134 are solved by means of relationships (4.139) and (4.141) to yield

$$\tau = -\frac{\omega_1 a_2 + \omega_2 a_5 + \dot{w}}{2(\omega_1 a_1 + \omega_2 a_4)}, \tag{4.142}$$

$$t_{ap} = -\frac{\omega_1 a_8 + \omega_2 a_{12}}{2(\omega_1 a_7 + \omega_2 a_{10})}. \tag{4.143}$$

Hence for the first approximation of the ω and a coefficients the corresponding times may be determined from (4.142) and (4.143). These equations provide the $(t_{ap})_{n+1}$ and $(\tau)_{n+1}$ estimates of the optimum transfer parameters. The process can now be repeated, making sure to update all weights and weight functions. Finally, when

$$|(t_{ap})_n - (t_{ap})_{n+1}| < \epsilon_1,$$

$$|(\tau)_n - (\tau)_{n+1}| < \epsilon_2,$$

where ϵ_1 and ϵ_2 are suitable tolerances, the procedure has converged and the optimum minimum weight orbit has been determined. In passing, it is well to note that a substantial increase in accuracy can be achieved by shrinking the increment between the elements in the 3 × 3 matrix of Figure 4.19 as the optimum orbit is approached. Numerical experience with this method has shown that the convergence is fairly stable and a typical optimization can be obtained quite rapidly. The interested reader is directed to [19] and [20] for a more detailed treatment of this type of analysis and the use of more complicated payload ratio functions.

4.5 SUMMARY

Interplanetary trajectories have been discussed in this chapter via the method of patched conics. This method considers that the total trajectory between two terminals in the solar system can be segmented into Keplerian arcs wherein each arc is subject to the action of the primary gravitational mass that is controlling motion. The arcs or orbital segments are joined at a fictitious boundary called the mean planetary sphere of influence in order to obtain the total interplanetary trajectory. It was pointed out that by means of this physically realistic model many complicated transfer computations are solvable analytically.

The problem of optimum secondary impulse trajectory transfers was discussed in great detail. It was shown that at times an in-plane transfer between two planets is not physically realistic because very high fuel requirements are likely to be incurred. In this case a broken plane transfer must be utilized to effect the transfer. The transfer orbit can be obtained by solution of a fourth-degree polynomial equation with very slowly varying coefficients. A rather lengthy algorithm which includes various transfer modes obtained as by-products from the broken plane transfer was developed.

Minimum time of transfer trajectories were shown to be solvable in terms of algebraic polynomials whose coefficients must be obtained by iterative methods. These transfer modes are physically realistic because astronauts cannot be confined to spacecraft cabins for unduly long periods of time. Hence with a given fuel supply it is desirable to reach the target planet as rapidly as possible. One- and two-impulse transfers of this type are discussed in detail and are shown to be solvable by iterative techniques.

The final portion of this chapter was devoted to the determination of minimum vehicle weight-in-orbit interplanetary trajectories. The method introduced included important systems constraints such as the jettisoning of weights and vaporization of the propellant. Since cost of interplanetary missions is correlated with minimum in-orbit departure weight, these techniques are of great practical importance from an economic point of view.

EXERCISES

1. The radius of influence of the Sun is not truly infinite. Assume that Alpha Centauri is twice as massive as our Sun and that it is $4\frac{1}{2}$ light years away. What is the magnitude of the Solar radius of influence?
2. A Voyager probe is leaving the planet Mars after transmitting photographs of the Martian surface. If the orbital elements of the probe relative to the Areocentric system are

$$a = -10.0 \text{ Mars radii}, \qquad e = 1.1,$$

what is the universal time of exit from the mean sphere of influence if the epoch universal time at which the osculating elements are evaluated, T, is 12:07 P.M. August 1973?

3. Realistic optimization procedures for manned interplanetary flight actually require a minimization of total vehicle weight. If it is assumed that the rocket equation,

$$\Delta V = gI_{sp} \log R,$$

where $I_{sp} \equiv$ rocket engine specific impulse, $M \equiv$ vehicle mass ratio, and $\Delta V \equiv$ velocity increment, represents the relationship between propulsion engine output and velocity increment, show that, for a one-impulse transfer, minimization of ΔV is equivalent to minimization of vehicle weight. Obtain the relationship between the λ multipliers used in Section 4.3 for optimization of two-impulse transfers in which different specific impulses are used for the departure and arrival impulses. Assume that the specific impulses of each engine are known constants, and recall from Chapter 2 that

$$R = \frac{\text{launch weight}}{\text{burnout weight}}.$$

4. A Jupiter flyby probe is planned which utilizes available booster hardware. The particular booster available has a payload capability of placing 10,000 lb into a 150-nautical mile geocentric circular parking orbit. Over and above the parking orbit speed the vehicle propulsive system has a 13,000 fps velocity increment available in order to perform the desired mission. Assume that both planets are on their respective line of nodes (in opposition) and perform a Hohmann maneuver. Is the mission feasible?
5. Why is the broken plane transfer of Section 4.2 of interest in manned missions? Discuss generally.
6. Consider the truncation of the quartic resolvent in i^* for the broken plane transfer discussed in Section 4.2. Show that, if only terms in i^{*2} are retained in both the quartic and in the expressions for $\sin v^*$ and $\cos v^*$, a closed form second-order solution to the broken plane transfer problem is possible.
7. A journey to the solar system of Alpha Temorious is under preliminary planning. Consider that all the mass in the solar system is concentrated at the Sun. Using the mass values in Table 1.5, find the velocity of escape from the solar system.
8. Verify the identity defined by (4.135), that is,

$$\frac{w_b}{w_a} = \frac{M(1 - \sigma)}{1 - M\sigma}.$$

9. Derive a payload ratio function for a complete round trip departing from Earth, arriving at Mars, leaving ΔW pounds on the Martian surface, and swinging by the planet Venus on the return leg of the journey. Assume that a propulsive correction is performed as the space vehicle passes Venus.

10. Derive an expression for the magnitude of the velocity increment required to escape from a planetocentric orbit which is an ellipse. Assume the impulse is applied at periapsis.

REFERENCES

[1] P. R. Escobal, *Methods of Orbit Determination*, John Wiley and Sons, New York, 1965.

[2] D. F. Lawden, *Optimal Trajectories for Space Navigation*, Butterworth and Co., London, 1963.

[3] W. R. Fimple, *Optimum Midcourse Plane Changes for Ballistic Interplanetary Trajectories*, United Aircraft Corporation Research Laboratories, Report A-110058-3, June 1962.

[4] J. V. Breakwell, R. W. Gillespie, and S. Ross, "Researches in Interplanetary Transfer," *American Rocket Society Journal*, August 1961.

[5] R. H. Battin, *Astronautical Guidance*, McGraw-Hill Book Company, New York, 1964.

[6] P. R. Escobal, *Analytic Resolvents for Minimum Secondary Impulses for Interplanetary Transfers*, TRW Systems, 9990-6810-RU000, November 1964.

[7] P. R. Escobal and G. A. Crichton, *Minimum Time Interplanetary Trajectories*, TRW Systems, 9990-7151-RU000, October 1965.

[8] R. W. Gillespie and S. Ross, *The Venus Swingby Mission Mode and Its Role in the Manned Explorations of Mars*, AIAA, Third Aerospace Sciences Meeting, AIAA Paper 66-37, New York, January 1966.

[9] D. N. Lascody, *Analytic Determination of Three-Dimensional Interplanetary Transfers*, Lockheed California Co., LR 16179, September 1962.

[10] P. S. Laplace, *Esposition du Système du Monde*, 1796.

[11] G. A. Crichton, *Analytic Computation of Formulas for Secondary Impulses*, TRW Systems, 3422.3-78, June 1966.

[12] G. A. Crichton, P. R. Escobal, T. J. Mucha, and H. L. Roth, *The Evolution and Application of the Interplanetary Optimization Trajectory Analyzer Concept*, TRW Systems 9863-6009-R0000, April 1966.

[13] T. J. Mucha, *The Method of Gradients for the Determination of Optimal Interplanetary Trajectories with Inequality Constraints*, TRW Systems, 9863-6001-R0000, July 1966.

[14] S. E. Ross, "A Systematic Approach to the Study of Non-Stop Interplanetary Round Trips," AAS Interplanetary Missions Conference, Los Angeles, California, January 1963.

[15] J. M. Deerwester, "Initial Mass Savings Associated with the Venus Swingby Mode of Mars Round Trips," AIAA Paper 65-89, AIAA Second Aerospace Sciences Meeting, New York, January 1966.

[16] R. R. Titus, "Powered Flybys of Mars," AIAA Paper 65-515, AIAA Second Annual Meeting, San Francisco, California, July 1965.

[17] R. L. Sohn, "Design of Spacecraft for Interplanetary Missions," International Astronautical Congress, Athens, Greece, September 1965.

18] R. L. Sohn, *Manned Mars Landing and Return Mission Study*, TRW Systems, 8572-6011-RU000, March 1964.

[19] A. R. Chovit, *Mission Oriented Advanced Nuclear System Parameters Study*, TRW Systems, 8423-6005-RU000, 8423-6013-RU000, March 1965.

[20] A. R. Chovit and C. D. Klystra, "Optimization of Manned Interplanetary Stopover Missions," AIAA Paper No. 65-513, AIAA Second Annual Meeting, San Francisco, California, July 1965.

[21] G. P. Sutton, *Rocket Propulsion Elements*, John Wiley and Sons, New York, 1958.

5 The Analytic Determination of Lunar Trajectories

The huge colossus which the works of Euler, Lagrange, and Laplace have raised demands the most prodigious force and exertion of thought if one is to penetrate into its inner nature and not merely rummage about on its surface. To dominate this colossus and not to fear being crushed by it demands a strain which permits neither rest nor peace till one stands on top of it and surveys the work in its entirety. Then only, when one has comprehended its spirit, is it possible to work justly and in peace at the completion of its details.

JACOBI [27]

5.1 LUNAR TRAJECTORIES

Man's first significant step beyond the near confines of our atmosphere will be the manned lunar mission. This step as presently envisioned in the Apollo project provides the process of celestial mechanics with a severe challenge. This challenge basically consists in developing both rapid and accurate methods that can be used to determine translunar trajectories. In this chapter the problem of concern is involved with the rapid computation of lunar trajectories by means of analytic techniques. These analytic techniques provide the necessary tools to generate the large amounts of data required to ensure successful missions and to provide approximate initial conditions to the more accurate techniques of special perturbations discussed in Chapter 7.

In the sections which follow immediately, attention is focused on the dynamics of three bodies, the Earth, the Moon, and the space vehicle. As an introduction the method of patched conics is discussed. This development is followed by a derivation and review of the characteristic equations of motion of the three-body system. Finally, some new developments for the practical computation of lunar trajectories are introduced.

The subject of this chapter has attracted the attention of a great number of scientists. In fact the actual problem of three bodies has been under investigation since the early days of celestial mechanics. To date no explicit solution has been found, however, certain approximations of profound significance have been obtained.

5.1.1 The Method of Patched Conics

A technique that has found much use in the rapid computation of lunar trajectories is called the method of patched conics. Its conception is due to Egorov [12]. This method basically assumes that Earth to Moon trajectories are composed of two Keplerian arcs. The two segments, namely the conic of a particle about the Earth without lunar perturbations and the conic of the same particle about the Moon without Earth perturbations, are joined at a point in space to produce the composite trajectory. The joining point in space is taken to lie on the surface of an approximate lunicentered sphere, called the *mean sphere of influence*, MSI, which was discussed in Chapter 4. Geometrically, the composite trajectory can be visualized in an inertial coordinate system as depicted in Figure 5.1.

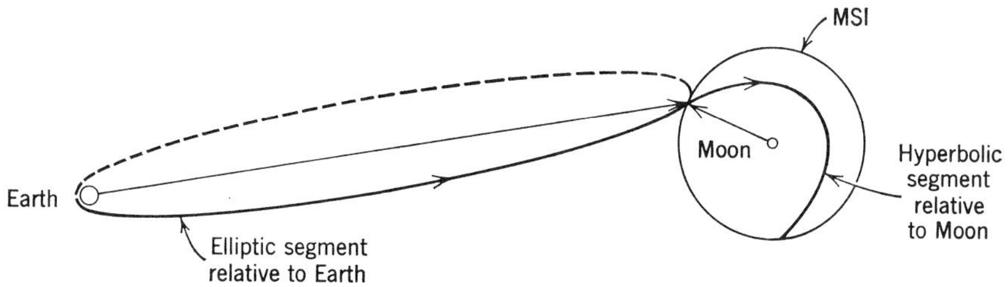

FIGURE 5.1 Patched conic geometry of Earth to Moon trajectories.

5.1.2 Basic Relationships

As may be evident from the geometric construction in Figure 5.1, position vectors can be matched exactly at the MSI. Likewise, velocities can be satisfied by assuming that the MSI moves or is carried along with the Moon so that, if the velocity outside the MSI boundary is known, the velocity inside the MSI is obtained by vectorial incrementation of the Earth-relative velocity by the Moon's velocity. Hence, at the boundary of the MSI defined in Chapter 4 as

$$r^* \equiv r_{☾}\left(\frac{m_{☾}}{m_{\oplus}}\right)^{2/5} \cong 10 \text{ e.r.}, \tag{5.1}$$

where $r^* \equiv$ radius of the MSI, $r_{☾} \equiv$ distance between Earth and Moon, $m_{☾} \equiv$ mass of the Moon, and $m_{\oplus} \equiv$ mass of the Earth, it follows at once that these conditions hold:

$$\mathbf{r}_{\oplus v} = \mathbf{r}_{☾} + \mathbf{r}_{☾ v}, \qquad \dot{\mathbf{r}}_{\oplus v} = \dot{\mathbf{r}}_{☾} + \dot{\mathbf{r}}_{☾ v}, \tag{5.2}$$

where the subscripts $\oplus v$, $☾ v$, and $☾$, respectively, are read vehicle with respect to the Earth, vehicle with respect to the Moon, and Moon with respect to the Earth. These particular constraints can be imposed with relative ease; however, in this model, accelerations are discontinuous across the MSI. In most cases this fault of the model does not represent a severe restriction.

Determinations of lunar trajectories can be segregated basically into two different types of problems. In the first type or class it is assumed that the initial state at the departure terminal is known. Hence, the initial conditions being known, the resulting trajectory is generated. The second type of problem has only a number of conditions (less than six) specified at the departure terminal, and the remaining conditions are imposed at the final or arrival terminal. These problems, commonly referred to as the initial and two-point boundary value problems, are discussed presently.

5.1.3 Determination of the Trajectory with Known Initial Conditions

A model for the computation of lunar trajectories which stresses rapidity of calculation can be obtained by the following approach. Consider the equation of the adopted sphere of influence in inertial coordinates relative to the right ascension-declination coordinate system discussed in ([1], Chapter 4), that is,

$$(x - x_☾)^2 + (y - y_☾)^2 + (z - z_☾)^2 = r^{*2}. \tag{5.3}$$

If a portion of the sphere of influence intersects the plane of motion of the Earth-centered conic, the equation of this intersection can be obtained via the standard mapping

$$\mathbf{r} = x_\omega \mathbf{P} + y_\omega \mathbf{Q}, \tag{5.4}$$

where, from ([1], Chapter 3), in terms of the classic orbital parameters a, e, i, Ω, and ω,

$$\begin{aligned}
x_\omega &= a(\cos E - e), \\
y_\omega &= a\sqrt{1 - e^2} \sin E, \\
P_x &= \cos\omega \cos\Omega - \sin\omega \sin\Omega \cos i, \\
P_y &= \cos\omega \sin\Omega + \sin\omega \cos\Omega \cos i, \\
P_z &= \sin\omega \sin i, \\
Q_x &= -\sin\omega \cos\Omega - \cos\omega \sin\Omega \cos i, \\
Q_y &= -\sin\omega \sin\Omega + \cos\omega \cos\Omega \cos i, \\
Q_z &= \cos\omega \sin i.
\end{aligned}$$

Substituting (5.4) into (5.3) yields the trace

$$x_\omega^2 + y_\omega^2 - 2\mathbf{r}_☾ \cdot \mathbf{P} x_\omega - 2\mathbf{r}_☾ \cdot \mathbf{Q} y_\omega = r^{*2} - r_☾^2. \tag{5.5}$$

The equation of this trace physically represents the intersection of the MSI with the orbit plane coordinate system [1] defined by the x_ω and y_ω axes. Furthermore, (5.4) yields

$$\frac{x_\omega + ae}{a} = \cos E,$$

$$\frac{y_\omega}{a\sqrt{1 - e^2}} = \sin E, \tag{5.6}$$

which, on elimination of the eccentric anomaly, provides the relationship

$$\frac{(x_\omega + ae)^2}{a^2} + \frac{{y_\omega}^2}{a^2(1 - e^2)} = 1, \tag{5.7}$$

or

$$y_\omega = \pm\sqrt{1 - e^2}[a^2(1 - e^2) - 2aex_\omega - {x_\omega}^2]^{1/2}. \tag{5.8}$$

Introducing the definitions

$$\beta = -2\mathbf{r}_☾ \cdot \mathbf{P}, \qquad \xi = -2\mathbf{r}_☾ \cdot \mathbf{Q}, \qquad \zeta = r^{*2} - r_☾^2 \tag{5.9}$$

and substituting (5.8) into (5.5) to eliminate y_ω yields the resolvent quartic

$$A_0{x_\omega}^4 + A_1{x_\omega}^3 + A_2{x_\omega}^2 + A_3x_\omega + A_4 = 0, \tag{5.10}$$

with

$$A_0 \equiv e^4,$$
$$A_1 \equiv 2\beta e^2 - 4ae^3[1 - e^2],$$
$$A_2 \equiv \xi^2[1 - e^2] + 2a^2e^2[1 - e^2]^2 - 2ae[1 - e^2] - 2e^2\zeta + \beta,$$
$$A_3 \equiv 2ae(\xi^2 + 2\zeta)[1 - e^2] + 2\beta a^2[1 - e^2]^2 - 4a^3e[1 - e^2]^3 - 2\beta\zeta,$$
$$A_4 \equiv (a^2[1 - e^2]^2 - \zeta)^2 - \xi^2a^2[1 - e^2]^2.$$

The quartic, by the methods discussed in ([1], Appendix 3), can be solved in closed form to yield the roots $x_{\omega i}$, $i = 1, 2, 3, 4$.

Several conditions can occur during an attempt to solve for the roots of the quartic. First, the quartic may be devoid of all real roots. In this case the Earth-centered conic does not intersect the MSI. Secondly, if the plane of the Earth-centered orbit is just tangent to the MSI, the quartic has two positive equal real roots; there is also a possibility of two positive equal roots if the Earth-centered conic is symmetric about the Earth-Moon line. Usually, the third case, the possession by the quartic of two complex and two real positive roots, is the most interesting one. It is also possible for the quartic to possess four real positive roots. This case can occur if the orbital eccentricity of the Earth-centered conic is very close to unity while the semi-major axis is greater than the Earth-Moon distance.

In passing, it should be emphasized that the parameters β, ξ, and ζ are not explicitly known before the initiation of computation; only a, e, T, i, Ω, and ω are assumed known. The parameters β, ξ, and ζ are not known because $\mathbf{r}_{☾}$ is not known at MSI passage. The problem is therefore an iterative one in which the time of transit from departure to MSI passage must be estimated and, from this estimate via the techniques of Section 1.4, $\mathbf{r}_{☾}$ determined. Once $\mathbf{r}_{☾}$ is known, the quartic is solved and the two possible MSI crossing values of x_ω are determined. The y_ω coordinate can be uniquely determined from (5.5) as

$$y_\omega = \frac{r^{*2} - x_\omega^{\ 2} - y_\omega^{\ 2} - \beta x_\omega - r_{☾}^{\ 2}}{\xi}, \tag{5.11}$$

where, from (5.8), the right side can be evaluated via

$$y_\omega^{\ 2} = (1 - e^2)[a^2(1 - e^2) - 2aex_\omega - x_\omega^{\ 2}]. \tag{5.12}$$

Equation 5.11 is important because it determines the sign of y_ω. The numerical determination of y_ω, however, is more accurately obtained via (5.12). Once the solution set $(x_\omega, y_\omega)_i$, $i = 1, 2$ is obtained, it is possible to determine the true anomaly at MSI passage from

$$r_i = [x_{\omega i}^{\ 2} + y_{\omega i}^{\ 2}]^{1/2}, \qquad i = 1, 2, \tag{5.13}$$

$$\cos v_i = \frac{x_{\omega i}}{r_i}, \tag{5.14}$$

$$\sin v_i = \frac{y_{\omega i}}{r_i}. \tag{5.15}$$

The true anomalies permit the eccentric, hyperbolic, and parabolic anomalies to be determined, depending on the magnitude of e, from

$$\cos E_i = \frac{\cos v_i + e}{1 + e\cos v_i}, \qquad \sin E_i = \frac{\sqrt{1 - e^2}\sin v_i}{1 + e\cos v_i}, \qquad e < 1, \tag{5.16}$$

$$\cosh F_i = \frac{\cos v_i + e}{1 + e\cos v_i}, \qquad \sinh F_i = \frac{\sqrt{e^2 - 1}\sin v_i}{1 + e\cos v_i}, \qquad e > 1, \tag{5.17}$$

$$D_i = \sqrt{2q}\tan \tfrac{1}{2}v_i, \qquad e = 1, \tag{5.18}$$

and the universal time at MSI passage, t^*, can be computed directly from the corresponding Keplerian forms ([1], Chapter 3):

$$k\sqrt{\mu}a^{-3/2}(t^* - T) = E_i - e\sin E_i, \tag{5.19}$$

$$k\sqrt{\mu}(-a)^{-3/2}(t^* - T) = e\sinh F_i - F_i, \tag{5.20}$$

$$k\sqrt{\mu}(t^* - T) = qD_i + \tfrac{1}{6}D_i^{\ 3}, \tag{5.21}$$

where T is the time of perifocal passage of the Earth-centered conic. Hence, having t^*, the position of the Moon, $\mathbf{r}_{☾}$, can be reevaluated, the parameters β, ξ, and ζ determined more accurately, and (5.10) resolved to yield an improved t^*. Once $t^*_{j+1} - t^*_j$, $j = 1, 2, \ldots, s$, does not vary, equations 5.2 provide the initial conditions required to determine the Moon-centered Keplerian orbit.

5.1.4 Auxiliary Initial Conditions

At times, polar parameters or, better, vehicle parameters, are more useful than the rectangular elements used in Section 5.1.2. A convenient set much used in mission analysis is

$$V, r, \gamma, A, \delta, \lambda_E, \tag{5.22}$$

where, relative to the Earth, $V \equiv$ vehicle velocity magnitude, $r \equiv$ vehicle position magnitude, $\gamma \equiv$ vehicle flight path angle, $A \equiv$ vehicle azimuth, $\delta \equiv$ vehicle declination, and $\lambda_E \equiv$ vehicle east longitude. These parameters are discussed in [1]. For convenience the transformation equations are listed here:

$$\begin{aligned}
x &= r \cos \delta \cos \theta, \\
y &= r \cos \delta \sin \theta, \\
z &= r \sin \delta, \\
V_S &= -V \cos \gamma \cos A, \\
V_E &= V \cos \gamma \sin A, \\
V_R &= V \sin \gamma, \\
\dot{x} &= V_S \sin \delta \cos \theta - V_E \sin \theta + V_R \cos \delta \cos \theta, \\
\dot{y} &= V_S \sin \delta \sin \theta + V_E \cos \theta + V_R \cos \delta \sin \theta, \\
\dot{z} &= -V_S \cos \delta + V_R \sin \delta,
\end{aligned} \tag{5.23}$$

where the local sidereal time θ is computed from

$$\theta = \theta_g + \dot{\theta}(t - t_g) + \lambda_E, \qquad 0 \le \theta < 2\pi. \tag{5.24}$$

The local sidereal time at universal time t is functionally dependent on λ_E through θ_g, the Greenwich sidereal time corresponding to universal time t_g, and $\dot{\theta}$, the constant sidereal rate of change ([1], Chapter 1). Hence an equivalence exists between x, y, z, $\dot{x}$, $\dot{y}$, $\dot{z}$, and V, r, γ, A, δ, λ_E.

In closing this section it is well to note that the resulting orbit inside the MSI defined by the MSI puncture position and velocity vectors, that is,

$$\mathbf{r}_l \equiv \mathbf{r}_{☾v} = \mathbf{r}_{\oplus v} - \mathbf{r}_{☾}, \qquad \dot{\mathbf{r}}_l \equiv \dot{\mathbf{r}}_{☾v} = \dot{\mathbf{r}}_{\oplus v} - \dot{\mathbf{r}}_{☾}, \tag{5.25}$$

may not satisfy desired constraints such as pericynthion distance. In this case the initial conditions must be varied until these constraints are satisfied.

Depending on the known parameters peculiar to a given problem, many computational variations of the two-point boundary value problem can be obtained.

5.2 THE PROBLEM OF THREE BODIES

A problem of much interest in the determination of lunar trajectories is the restricted problem of three bodies. In a classical sense this problem deals with the motion of the Earth, the Moon, and the space vehicle. It is assumed that the Moon moves in a circular orbit about the Earth and that the mass of the space vehicle is infinitesimal with respect to the total mass of the system. In this section the differential equations of motion of the three-body system are derived *without* imposing the assumption that the Moon moves in a circular orbit. These equations are fundamental to further developments that are treated in later sections.

5.2.1 Equations of Motion of the Three-Body Problem

An explicit derivation of the general barycentric equations of relative motion is given in ([1], Chapter 2). These equations of motion, with the coordinate frame fixed to the mass center of the system, are as follows:

$$\frac{d^2\mathbf{r}}{dt^2} = -k^2\left[\sum_{j=1}^{n} m_j\right]\frac{\mathbf{r}}{r^3} + k^2 \sum_{\substack{j=1 \\ j\neq 2}}^{n} m_j \mathbf{r}_{j2}\left(\frac{1}{r^3} - \frac{1}{r_{j2}}\right), \tag{5.26}$$

where, summarizing notation, the following associations are made.

- k The gravitational constant. Usually it is convenient to use the value of k for the Earth. Then it should be remembered that masses are the ratios of the masses of all bodies to the mass of the Earth.
- m_j With k chosen as above, this is the mass ratio of any mass to the mass of the Earth.
- n Number of bodies in the system.
- 2 Refers to the body under consideration (space vehicle).
- j Refers to the existing jth body in the system not counting body 2.
- $\mathbf{r}_{2j}$ The distance between the jth body and the body under consideration.
- $\mathbf{r}$ The distance between the barycenter and the body under consideration.
- t The ephemeris time associated with position $\mathbf{r}$.

If only three bodies are under consideration, it follows from (5.26) that

$$\frac{d^2\mathbf{r}}{dt^2} = -k^2[m_1 + m_2 + m_3]\frac{\mathbf{r}}{r^3} + k^2 m_1 \mathbf{r}_{12}\left[\frac{1}{r^3} - \frac{1}{{r_{12}}^3}\right] + k^2 m_3 \mathbf{r}_{32}\left(\frac{1}{r^3} - \frac{1}{{r_{32}}^3}\right). \tag{5.27}$$

The geometry related to (5.27) is depicted in Figure 5.2.

Equation 5.27 can be simplified greatly by the assumption that $m_2 = 0$. With the previous understanding, (5.27) can be written as

$$\frac{d^2\mathbf{r}}{dt^2} = -k^2(m_1 + m_3)\frac{\mathbf{r}}{r^3} + (k^2m_1\mathbf{r}_{12} + k^2m_3\mathbf{r}_{32})\frac{1}{r^3} - k^2m_1\frac{\mathbf{r}_{12}}{r_{12}^{\ 3}} - k^2m_3\frac{\mathbf{r}_{32}}{r_{32}^{\ 3}}. \tag{5.28}$$

A general property of a barycentric system, as discussed in ([1], Chapter 2) is that

$$k^2\sum_{j=1}^{n} m_j\frac{\mathbf{r}_{j2}}{r^3} - k^2\sum_{j=1}^{n} m_j\frac{\mathbf{r}}{r^3} = 0, \tag{5.29}$$

which implies that (5.28) reduces to

$$\frac{d^2\mathbf{r}}{dt^2} = -k^2m_1\frac{\mathbf{r}_{12}}{r_{12}^{\ 3}} - k^2m_3\frac{\mathbf{r}_{32}}{r_{32}^{\ 3}}. \tag{5.30}$$

Further simplification is afforded by introduction of the modified time variable τ, defined by

$$\tau \equiv k(t - t_0). \tag{5.31}$$

This change of independent variables, with the arbitrary choice of units such that $m_1 + m_3 = 1$, further reduces (5.30) to the compact expression

$$\frac{d^2\mathbf{r}}{d\tau^2} = -(1-\mu)\frac{\mathbf{r}_{12}}{r_{12}^{\ 3}} - \mu\frac{\mathbf{r}_{32}}{r_{32}^{\ 3}}, \tag{5.32}$$

where

$$\mu \equiv \frac{m_3}{m_1 + m_3} \simeq \frac{1}{81.3},$$

$$\mathbf{r}_{12} = \mathbf{r} - \mathbf{r}_1,$$

$$\mathbf{r}_{32} = \mathbf{r} - \mathbf{r}_3,$$

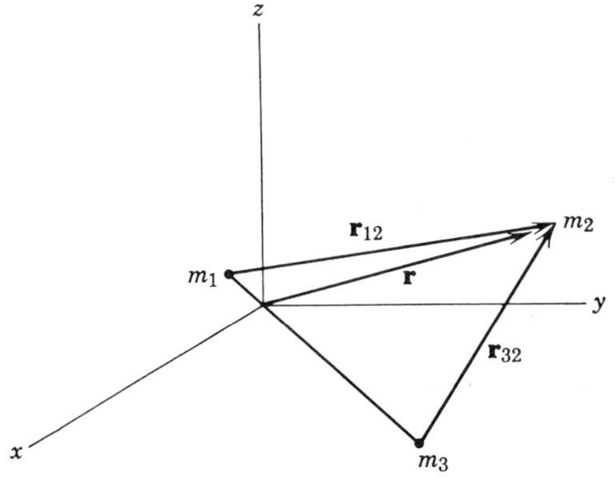

FIGURE 5.2 Geometry of the three-body problem.

with m_3 equal to the mass of the Moon and m_1 equal to the mass of the Earth. Equation 5.32 describes the motion of a body of infinitesimal mass subject to the attractions of two bodies of finite mass.

5.2.2 Transformation into a Rotating Coordinate System

As indicated by Moulton [2], if the unit of distance between body m_1 and body m_3 is taken to be unity, the mean motion of the finite bodies is

$$n = k[(1 - \mu) + \mu]a_{☾}^{-3/2} = 1. \tag{5.33}$$

The assumptions that $a_{☾}$ = constant and that body 3 moves in a circle about body 1 are implied in (5.33). Since the Earth and Moon, under the assumption

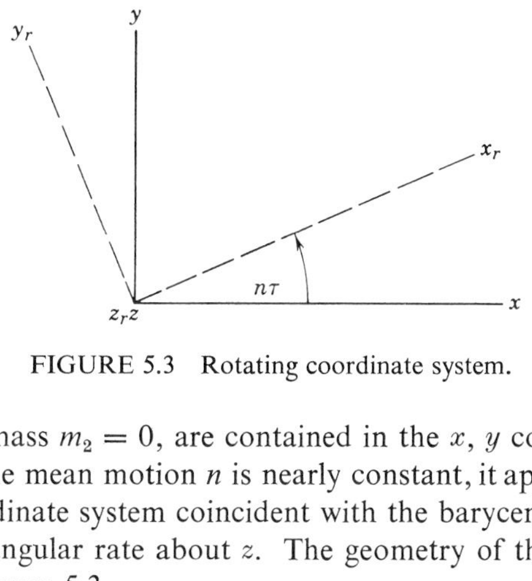

FIGURE 5.3 Rotating coordinate system.

that the vehicle mass $m_2 = 0$, are contained in the x, y coordinate plane of Figure 5.3 and the mean motion n is nearly constant, it appears appealing to introduce a coordinate system coincident with the barycentric system which is rotating with angular rate about z. The geometry of this transformation is described in Figure 5.3.

Consider the transformation

$$x = x_r \cos n\tau - y_r \sin n\tau,$$

$$y = x_r \sin n\tau + y_r \cos n\tau,$$

$$z = z_r, \tag{5.34}$$

where n is a function of τ, applied to (5.32). Differentiating (5.34) twice and inserting the respective derivatives for $\dot{\mathbf{r}}$ into the x and y components of (5.34) for all subscripts yields

$$\omega_1 \cos n\tau - \omega_2 \sin n\tau = 0,$$

$$\omega_1 \sin n\tau + \omega_2 \cos n\tau = 0, \tag{5.35}$$

where

$$\omega_1 \equiv \frac{d^2x_r}{d\tau^2} - 2n^* \frac{dy_r}{d\tau} - n^{*2}x_r - \dot{n}^* y_r + \frac{(1-\mu)}{r_{12}^{\ 3}}(x_r - x_{r1}) + \frac{\mu}{r_{32}^{\ 3}}(x_r - x_{r3})$$

$$\omega_2 \equiv \frac{d^2y_r}{d\tau^2} + 2n^* \frac{dx_r}{d\tau} + \dot{n}^* x_r - n^{*2}y_r + \frac{(1-\mu)}{r_{12}^{\ 3}}(y_r - y_{r1}) + \frac{\mu}{r_{32}^{\ 3}}(y_r - y_{r3})$$

and n^* and $\dot{n}^*$ are defined by the exact relationships

$$n^* \equiv n + \frac{dn}{d\tau}\tau,$$

$$\dot{n}^* \equiv 2\frac{dn}{d\tau} + \frac{d^2n}{d\tau^2}\tau. \tag{5.36}$$

The vectors $\mathbf{r}_{r1}$ and $\mathbf{r}_{r3}$ are the position vectors in the rotating barycentric coordinate system of the Earth and Moon, respectively.

Equations 5.35 can be solved to yield $\omega_1 = \omega_2 = 0$, and therefore the general equations of motion in a nonuniformly rotating system become

$$\frac{d^2x_r}{d\tau^2} - 2n^* \frac{dy_r}{d\tau} = n^{*2}x_r + \dot{n}^* y_r - (1-\mu)\frac{(x_r - x_{r1})}{r_{12}^{\ 3}} - \mu\frac{(x_r - x_{r3})}{r_{32}^{\ 3}}, \tag{5.37}$$

$$\frac{d^2y_r}{d\tau^2} + 2n^* \frac{dx_r}{d\tau} = n^{*2}y_r - \dot{n}^* x_r - (1-\mu)\frac{(y_r - y_{r1})}{r_{12}^{\ 3}} - \mu\frac{(y_r - y_{r3})}{r_{32}^{\ 3}}, \tag{5.38}$$

$$\frac{d^2z_r}{d\tau^2} = -(1-\mu)\frac{(z_r - z_{r1})}{r_{12}^{\ 3}} - \mu\frac{(z_r - z_{r3})}{r_{32}^{\ 3}}. \tag{5.39}$$

In (5.37), (5.38), and (5.39) all starred parameters are functions of τ, and therefore the complexity of the differential equations becomes very great. If (5.33) holds true, it follows that $n^* = n = 1$, $\dot{n}^* = 0$, and (5.37), (5.38), (5.39) reduce to

$$\frac{d^2x_r}{d\tau^2} - 2n \frac{dy_r}{d\tau} = n^2x_r - (1-\mu)\frac{(x_r - x_{r1})}{r_{12}^{\ 3}} - \mu\frac{(x_r - x_{r3})}{r_{32}^{\ 3}}, \tag{5.40}$$

$$\frac{d^2y_r}{d\tau^2} + 2n \frac{dx_r}{d\tau} = n^2y_r - (1-\mu)\frac{(y_r - y_{r1})}{r_{12}^{\ 3}} - \mu\frac{(y_r - y_{r3})}{r_{32}^{\ 3}}, \tag{5.41}$$

$$\frac{d^2z_r}{d\tau^2} = -(1-\mu)\frac{(z_r - z_{r1})}{r_{12}^{\ 3}} - \mu\frac{(z_r - z_{r3})}{r_{32}^{\ 3}}. \tag{5.42}$$

The basic difference between the two previous second-order differential systems is that the first system does not assume that the third body moves about the first body in a circle, whereas the second system, (5.40), (5.41), and (5.42), depends on this assumption.

For short periods of time such as the length of lunar trajectories, a compromise can be obtained by determining the starred parameters at initiation of the trajectory and holding these values constant. This modification allows for the correct phasing and positioning of the Moon.

5.2.3 The Jacobi Integral

An integral of motion for the restricted three-body problem can be obtained by introducing the potential U, defined by

$$U = \tfrac{1}{2}n^2(x_r^2 + y_r^2) + \frac{(1-\mu)}{r_{12}} + \frac{\mu}{r_{32}}, \tag{5.43}$$

so that (5.40), (5.41), and (5.42) can be written as

$$\frac{d^2x_r}{d\tau^2} - 2n\frac{dy_r}{d\tau} = \frac{\partial U}{\partial x_r}, \tag{5.44}$$

$$\frac{d^2y_r}{d\tau^2} + 2n\frac{dx_r}{d\tau} = \frac{\partial U}{\partial y_r}, \tag{5.45}$$

$$\frac{d^2z_r}{d\tau^2} = \frac{\partial U}{\partial z_r}. \tag{5.46}$$

Multiplying (5.44), (5.45), and (5.46) cyclically by $2\,dx_r/d\tau$, $2\,dy_r/d\tau$, $2\,dz_r/d\tau$ and adding yields

$$2\left(\frac{d^2x_r}{d\tau^2}\frac{dx_r}{d\tau} + \frac{d^2y_r}{d\tau^2}\frac{dy_r}{d\tau} + \frac{d^2z_r}{d\tau^2}\frac{dz_r}{d\tau}\right) = 2\left(\frac{\partial U}{\partial x_r}\frac{dx_r}{d\tau} + \frac{\partial U}{dy_r}\frac{dy_r}{d\tau} + \frac{\partial U}{\partial z_r}\frac{dz_r}{d\tau}\right). \tag{5.47}$$

Since U is explicitly free of time, it follows by direct integration that

$$\left(\frac{dx_r}{d\tau}\right)^2 + \left(\frac{dy_r}{d\tau}\right)^2 + \left(\frac{dz_r}{d\tau}\right)^2 = 2\int \frac{dU}{d\tau}\,d\tau$$

or

$$\left(\frac{dx_r}{d\tau}\right)^2 + \left(\frac{dy_r}{d\tau}\right)^2 + \left(\frac{dz_r}{d\tau}\right)^2 = V^2 = 2U - C, \tag{5.48}$$

where V is equal to the speed of the space vehicle and C is a constant of integration. Equation 5.48 is the celebrated Jacobi integral which relates the position and velocity of a particle in the three-body system. It is an analog of the well-known vis-viva integral of the two-body problem. The constant C can be determined from the initial conditions or known state, that is, from $\mathbf{r}_r$ and $\dot{\mathbf{r}}_r$.

5.2.4 Transformation of Variables

An interesting transformation of the three-body equations can be obtained in the following manner. Consider choosing x_r as the independent variable instead of modified time τ. Since

$$\frac{dx_r}{d\tau} = \frac{1}{d\tau/dx_r},$$
$$\frac{dy_r}{d\tau} = \frac{dy_r}{dx_r}\frac{dx_r}{d\tau},$$
$$\frac{dz_r}{d\tau} = \frac{dz_r}{dx_r}\frac{dx_r}{d\tau}, \tag{5.49}$$

it follows by direct differentiation that

$$\frac{d^2x_r}{d\tau^2} = \frac{1}{(d\tau/dx_r)^3}\cdot\frac{d^2\tau}{dx_r^{\,2}},$$
$$\frac{d^2y_r}{d\tau^2} = -\frac{1}{(d\tau/dx_r)^3}\cdot\frac{dy_r}{dx_r}\cdot\frac{d^2\tau}{dx_r^{\,2}} + \frac{1}{(d\tau/dx_r)^2}\frac{d^2y_r}{dx_r^{\,2}},$$
$$\frac{d^2z_r}{d\tau^2} = -\frac{1}{(d\tau/dx_r)^3}\cdot\frac{dz_r}{dx_r}\cdot\frac{d^2\tau}{dx_r^{\,2}} + \frac{1}{(d\tau/dx_r)^2}\frac{d^2z_r}{dx_r^{\,2}}. \tag{5.50}$$

By direct substitution of (5.50) into (5.37), (5.38), and (5.39) it is possible to obtain

$$\left(\frac{d\tau}{dx_r}\right)^{-3}\frac{d^2\tau}{dx_r^{\,2}} + 2n\left(\frac{d\tau}{dx_r}\right)^{-1}\frac{dy_r}{dx_r} = -S_x,$$
$$\left(\frac{d\tau}{dx_r}\right)^{-3}\frac{dz_r}{dx_r}\frac{d^2\tau}{dx_r^{\,2}} - \left(\frac{d\tau}{dx_r}\right)^{-2}\frac{d^2y_r}{dx_r^{\,2}} - 2n\left(\frac{d\tau}{dx_r}\right)^{-1} = -S_y,$$
$$\left(\frac{d\tau}{dx_r}\right)^{-3}\frac{dz_r}{dx_r}\frac{d^2\tau}{dx_r^{\,2}} - \left(\frac{d\tau}{dx_r}\right)^{-2}\frac{d^2z_r}{dx_r^{\,2}} = -S_z, \tag{5.51}$$

where

$$S_x \equiv n^2x_r - (1-\mu)\frac{(x_r - x_{r1})}{r_{12}^{\,3}} - \mu\frac{(x_r - x_{r3})}{r_{32}^{\,3}},$$
$$S_y \equiv n^2y_r - (1-\mu)\frac{(y_r - y_{r1})}{r_{12}^{\,3}} - \mu\frac{(y_r - y_{r3})}{r_{32}^{\,3}},$$
$$S_z \equiv -(1-\mu)\frac{(z_r - z_{r1})}{r_{12}^{\,3}} - \mu\frac{(z_r - z_{r3})}{r_{32}^{\,3}}.$$

The **S** vector can be simplified by arbitrarily choosing the initial coordinates $y_{r1} = y_{r3} = z_{r1} = z_{r3} = 0$. This implies that the x axis of the rotating coordinate system passes through the centers of bodies 1 and 3, the Earth and the Moon. Explicitly as functions of x_r, y_r, z_r, with the previous simplification, **S** can be written as

$$
\begin{aligned}
S_x &= n^2 x_r - (1-\mu)\frac{x_r - x_{r1}}{[(x_r - x_{r1})^2 + y_r^2 + z_r^2]^{3/2}} \\
&\quad - \mu\frac{x_r - x_{r3}}{[(x_r - x_{r3})^2 + y_r^2 + z_r^2]^{3/2}}, \\
S_y &= n^2 y_r - (1-\mu)\frac{y_r}{[(x_r - x_{r1})^2 + y_r^2 + z_r^2]^{3/2}} \\
&\quad - \mu\frac{y_r}{[(x_r - x_{r3})^2 + y_r^2 + z_r^2]^{3/2}}, \\
S_z &= -(1-\mu)\frac{z_r}{[(x_r - x_{r1})^2 + y_r^2 + z_r^2]^{3/2}} \\
&\quad - \mu\frac{z_r}{[(x_r - x_{r3})^2 + y_r^2 + z_r^2]^{3/2}},
\end{aligned}
\tag{5.52}
$$

where $\mathbf{r}_{r1}$ and $\mathbf{r}_{r3}$ are the initial position vectors of the Earth and the Moon in the rotating coordinate system.

5.2.5 Compact Form of the Transformed Equations

Consider (5.51) written in the compact notation

$$
\begin{aligned}
\frac{\tau''}{(\tau')^3} + 2n\frac{y'}{(\tau')} &= S_x, \\
\frac{\tau'' y'}{(\tau')^3} - \frac{y''}{(\tau')^2} - \frac{2n}{(\tau')} &= -S_y, \\
\frac{\tau'' z'}{(\tau')^3} - \frac{z''}{(\tau')^2} &= -S_z,
\end{aligned}
\tag{5.53}
$$

where primes denote differentiation with respect to x_r. If it is assumed that n is constant, then τ does not appear explicitly on the right side of (5.53), and it is possible to consider the transformation of variables

$$u = \tau', \tag{5.54}$$

so that

$$\frac{u'}{u^3} + 2n\frac{y'}{u} = -S_x, \tag{5.55}$$

$$\frac{u'y'}{u^3} - \frac{y''}{u^2} - \frac{2n}{u} = -S_y, \tag{5.56}$$

$$\frac{u'z'}{u^3} - \frac{z''}{u^2} = -S_z. \tag{5.57}$$

Equations 5.55, through 5.57 are useful in studies of three-body trajectories and have been used in recent small parameter perturbation studies.

5.2.6 Particular Solutions

In 1772 Lagrange was the first investigator to find particular solutions to the differential equations of motion of the three-body problem. These solutions are classified into three cases: the straight line, equilateral triangle, and conic section solutions. Among these cases, the straight line solutions wherein the bodies are positioned on a single line, and the equilateral triangle solutions wherein the bodies are located at the vertices of an equilateral triangle are of paramount importance for later developments in this chapter. The conic solutions, which show that the orbits are conic sections of arbitrary eccentricity but are constrained in that the ratios of the mutual distances of the bodies are constant, is of interest in a more general sense. The conic solutions are discussed at length in [2]. For our purposes here only, the first two cases are discussed. To develop these solutions some preliminary manipulation of the basis equations must be undertaken.

Consider the equations of relative motion developed in ([1], Chapter 2):

$$m_i\frac{d^2\mathbf{r}_i}{dt^2} = k^2\sum_{\substack{j=1\\ i\neq j}}^{n} m_i m_j \frac{\mathbf{r}_{ij}}{r_{ij}^{\,3}}, \tag{5.58}$$

where the following associations are made:

- k The gravitational constant of the body to which motion is referred.
- m_j With k chosen as above, this is the mass ratio of any mass to the mass of the body to which motion is referred, that is, normalized masses denoted by $m_j \doteq m_j/m_1$.
- n Number of bodies in the system.
- 1 Refers to the central body.
- 2 Refers to the body under consideration.
- j Refers to the existing jth body in the system not counting body 1 or 2.
- $\mathbf{r}_{2j}$ The distances between the jth body and the body under consideration.
- $\mathbf{r}_{1j}$ The distances between the jth body and the central body (origin).

Equation 5.58 can be written more compactly by introducing the potential for a system of finite particle masses:

$$U \equiv \tfrac{1}{2}k^2 \sum_{i=1}^{n} \sum_{j=1}^{n} \frac{m_i m_j}{r_{ij}}, \qquad i \neq j. \tag{5.59}$$

Since direct differentiation yields

$$\frac{\partial U}{\partial \mathbf{r}_i} = k^2 m_i \frac{\partial}{\partial \mathbf{r}_i} \sum_{j=1}^{n} \frac{m_j}{r_{ij}} = k^2 m_i \sum_{j=1}^{n} m_j \frac{\mathbf{r}_j - \mathbf{r}_i}{r_{ij}^3} = k^2 m_i \sum_{j=1}^{n} m_j \frac{\mathbf{r}_{ij}}{r_{ij}^3}, \tag{5.60}$$

equation 5.58 reduces to

$$\frac{d^2\mathbf{r}_i}{dt^2} = \frac{1}{m_i}\frac{\partial U}{\partial \mathbf{r}_i} \tag{5.61}$$

or, in component form for three bodies,

$$\frac{d^2x_i}{dt^2} = \frac{1}{m_i}\frac{\partial U}{\partial x_i}, \qquad \frac{d^2y_i}{dt^2} = \frac{1}{m_i}\frac{\partial U}{\partial y_i}, \qquad \frac{d^2z_i}{dt^2} = \frac{1}{m_i}\frac{\partial U}{\partial z_i}$$

with

$$U = k^2 \frac{m_1 m_2}{r_{12}} + k^2 \frac{m_2 m_3}{r_{23}} + k^2 \frac{m_3 m_1}{r_{31}}. \tag{5.62}$$

By choosing a set of uniformly rotating axes with origin at the mass center of the three-body system, (5.35), and introducing the modified time variable τ, (5.31), it is possible to write for $i = 1, 2, 3$

$$\frac{d^2x_{ri}}{d\tau^2} - 2n\frac{dy_{ri}}{d\tau} - n^2x_{ri} - \frac{1}{m_i}\frac{\partial U}{\partial x_{ri}} = 0, \tag{5.63}$$

$$\frac{d^2y_{ri}}{d\tau^2} + 2n\frac{dx_{ri}}{d\tau} - n^2y_{ri} - \frac{1}{m_i}\frac{\partial U}{\partial y_{ri}} = 0, \tag{5.64}$$

$$\frac{d^2z_{ri}}{d\tau^2} - \frac{1}{m_i}\frac{\partial U}{\partial z_{ri}} = 0. \tag{5.65}$$

If motion is constrained to be in the x_r, y_r plane and the bodies are in circular motion with respect to the origin, that is, fixed to the nonrotating system, it follows that

$$\frac{d^2\mathbf{r}_{ri}}{d\tau^2} = \frac{d\mathbf{r}_r}{d\tau} = 0 \tag{5.66}$$

so that (5.63) and (5.64), on evaluation of $\partial U/\partial x_{ri}$ and $\partial U/\partial y_{ri}$, can be written for the three bodies as

$$n^2 x_{r1} + m_2 \frac{x_{r2} - x_{r1}}{{r_{12}}^3} + m_3 \frac{x_{r3} - x_{r1}}{{r_{13}}^3} = 0, \tag{5.67}$$

$$n^2 x_{r2} + m_1 \frac{x_{r1} - x_{r2}}{{r_{12}}^3} + m_3 \frac{x_{r3} - x_{r2}}{{r_{23}}^3} = 0, \tag{5.68}$$

$$n^2 x_{r3} + m_1 \frac{x_{r1} - x_{r3}}{{r_{13}}^3} + m_2 \frac{x_{r2} - x_{r3}}{{r_{23}}^3} = 0, \tag{5.69}$$

$$n^2 y_{r1} + m_2 \frac{y_{r2} - y_{r1}}{{r_{12}}^3} + m_3 \frac{y_{r3} - y_{r1}}{{r_{13}}^3} = 0, \tag{5.70}$$

$$n^2 y_{r2} + m_1 \frac{y_{r1} - y_{r2}}{{r_{12}}^3} + m_3 \frac{y_{r3} - y_{r2}}{{r_{23}}^3} = 0, \tag{5.71}$$

$$n^2 y_{r3} + m_1 \frac{y_{r1} - y_{r3}}{{r_{13}}^3} + m_2 \frac{y_{r2} - y_{r3}}{{r_{23}}^3} = 0. \tag{5.72}$$

As indicated by Moulton [2], the fact that the origin is the center of mass can be utilized to simplify the previous system of equations because

$$m_1 x_{r1} + m_2 x_{r1} + m_3 x_{r1} = 0, \tag{5.73}$$

$$m_1 y_{r1} + m_2 y_{r2} + m_3 y_{r3} = 0. \tag{5.74}$$

Hence, arbitrarily choosing units such that $r_{12} = 1$, on substitution of (5.73) for (5.69) and (5.74) for (5.72), yields the equivalent system

$$n^2 x_{r1} + m_2(x_{r2} - x_{r1}) + \frac{m_3(x_{r3} - x_{r1})}{{r_{13}}^3} = 0, \tag{5.75}$$

$$n^2 x_{r2} + m_1(x_{r1} - x_{r2}) + \frac{m_3(x_{r3} - x_{r2})}{{r_{23}}^3} = 0, \tag{5.76}$$

$$m_1 x_{r1} + m_2 x_{r2} + m_3 x_{r3} = 0, \tag{5.77}$$

$$n^2 y_{r1} + m_2(y_{r2} - y_{r1}) + \frac{m_3(y_{r3} - y_{r1})}{{r_{13}}^3} = 0, \tag{5.78}$$

$$n^2 y_{r2} + m_1(y_{r1} - y_{r2}) + \frac{m_3(y_{r3} - y_{r2})}{{r_{23}}^3} = 0, \tag{5.79}$$

$$m_1 y_{r1} + m_2 y_{r2} + m_3 y_{r3} = 0. \tag{5.80}$$

This system is utilized in the next section to obtain two of the known solutions of the three-body problem.

5.2.7 The Straight Line and Equilateral Triangle Solutions

Imagine three masses, for example, the Earth, m_1, a satellite, m_2, and the Moon, m_3, all lying on the x_r axis such that $x_{r3} > x_{r2} > x_{r1}$; see Figure 5.4. For this configuration it follows that $y_{r1} = y_{r2} = y_{r3} = 0$ and (5.75) through (5.77), by virtue of the previously adopted units wherein $r_{12} = 1 = x_{r2} - x_{r1}$, become

$$n^2 x_{r1} + m_2 + \frac{m_3}{(x_{r3} - x_{r1})^2} = 0, \tag{5.81}$$

$$n^2(1 + x_{r1}) - m_1 + \frac{m_3}{(x_{r3} - x_{r1} - 1)^2} = 0, \tag{5.82}$$

$$m_1 x_{r1} + m_2(1 + x_{r1}) + m_3 x_{r3} = 0, \tag{5.83}$$

while (5.78) through (5.80) are satisfied identically. Solving (5.83) for x_{r3}, (5.81) for n^2, and substituting into (5.82) yields

$$\begin{aligned} [m_2 + (m_1 + m_2)x_{r1}][(m_1 + m_2 + m_3)x_{r1} + m_2]^2 \\ \times [(m_1 + m_2 + m_3)x_{r1} + m_2 + m_3]^2 \\ + [m_3^3(1 + x_{r1})][(m_1 + m_2 + m_3)x_{r1} + m_2 + m_3]^2 \\ - [m_3^3 x_{r1}][(m_1 + m_2 + m_3)x_{r1} + m_2]^2 = 0. \end{aligned} \tag{5.84}$$

To obtain Lagrange's quintic equation for the collinear solution the variable $x_{r3} - x_{r2}$, instead of x_{r1}, is used. Since $x_{r2} - x_{r1} \equiv 1$ and

$$m_1 x_{r1} + m_2 x_{r2} + m_3 x_{r3} = 0,$$

it follows that

$$x_{r1} = -\frac{m_2 + m_3 + m_3(x_{r3} - x_{r2})}{m_1 + m_2 + m_3}. \tag{5.85}$$

Substituting (5.85) into (5.84) and performing some algebraic manipulations yields the celebrated quintic

$$\begin{aligned} (m_1 + m_2)L^5 + (3m_1 + 2m_2)L^4 + (3m_1 + m_2)^3 - (m_2 + 3m_3)L^2 \\ - (2m_2 + 3m_2)L - (m_2 + m_3) = 0, \end{aligned} \tag{5.86}$$

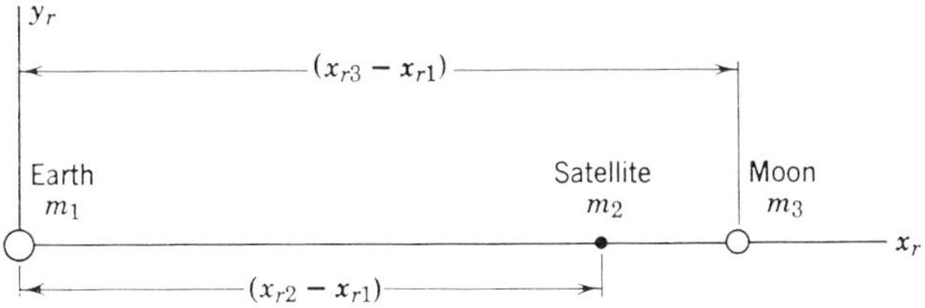

FIGURE 5.4 Straight line geometry.

where $L \equiv x_{r3} - x_{r2}$. From the theory of equations the quintic possesses only one real positive root because the coefficients change sign only one time. Neglecting m_2 and utilizing $m_1 = 1$, $m_2 = 1/81.3$, it is possible to obtain the root

$$x_{r3} - x_{r2} = 0.16308. \tag{5.87}$$

This solution of the three-body problem shows that a satellite placed at L_2 remains at this position. In Section 5.3 it is shown that this solution has very important ramifications in the theory of Earth to Moon trajectories. If the masses of the Earth, satellite, and Moon are cyclically interchanged, it follows that two other straight line solutions are obtained, call them L_1 and L_3. These solutions are discussed further in Section 5.3.

Another interesting solution to the system defined by (5.75) through (5.80) is obtained by letting $r_{12} = r_{23} = r_{13} = 1$, that is, letting the bodies lie at the vertices of an equilateral triangle so that

$$\begin{aligned}
(m_2 + m_3 - n^2)x_{r1} - m_2x_{r2} - m_3x_{r3} &= 0,\\
(m_1 + m_3 - n^2)x_{r2} - m_1x_{r1} - m_3x_{r3} &= 0,\\
m_1x_{r1} + m_2x_{r2} + m_3x_{r3} &= 0,\\
(m_2 + m_2 - n^2)y_{r1} - m_2y_{r2} - m_3y_{r3} &= 0,\\
(m_1 + m_3 - n^2)y_{r2} - m_1y_{r1} - m_3y_{r3} &= 0,\\
m_1y_{r1} + m_2y_{r2} + m_3y_{r3} &= 0.
\end{aligned} \tag{5.88}$$

In order that the homogeneous system of six equations in the six unknowns, x_{r1}, x_{r2}, x_{r3}, y_{r1}, y_{r2}, and y_{r3}, have a nontrivial solution, it is mandatory that the determinant of the system identically vanish. By direct computation

$$m_3^2(m_1 + m_2 + m_3 - n^2)^4 = 0,$$

and therefore

$$n^2 = m_1 + m_2 + m_3. \tag{5.89}$$

In closing this section it should be mentioned that equilateral triangle solutions of the three-body problem have actually been identified in nature. The theory in this section has shown that five individual solution points, or as they are better called, *libration points*, to the three-body problem exist. These points are illustrated in Figure 5.5, and physically they are locations at which bodies placed initially at rest at $L_1, \ldots, L_5$ remain captive at these points.

Actually, a satellite placed at L_1, L_2, or L_3 is not stable when compared to the L_4 and L_5 points. The stability of a satellite at the L_1, L_2, and L_3 points can be depicted by the analogy of a ball being balanced on the ridge of a smooth saddle. In contrast a satellite at the L_4 and L_5 points acts like a ball that has been placed in a shallow bowl.

The theory developed in the latter portion of this section, if applied to the Sun and the planet Jupiter, allows the computation of the L_4 and L_5 libration points for the Sun-Jupiter system. In 1906 Wolf discovered a small planet advanced to Jupiter by 60° in the region of the libration point. It

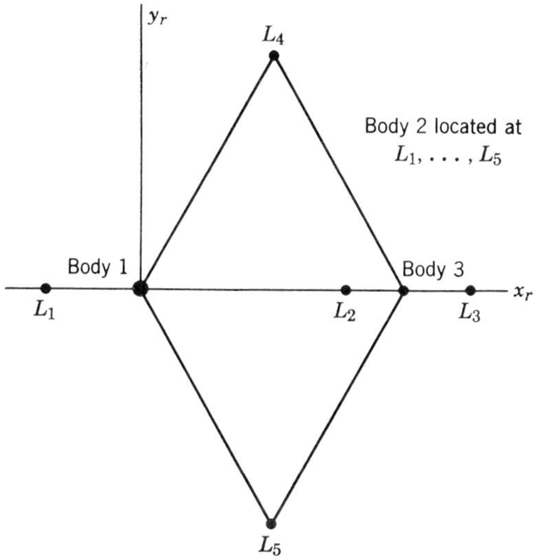

FIGURE 5.5 Libration points of the three-body problem.

was called Achilles and it became the vanguard of a collection of minor planets called the Trojan group. Soon Patroclus, Hector, Nestor, and others began to appear as further verification of the three-body theory.

5.3 LINEARIZATION FOR CERTAIN EARTH-MOON TRAJECTORIES IN THE RESTRICTED PROBLEM OF THREE BODIES

The basic difference between the patched conic method discussed in Section 5.1 and the method to be described presently lies in the introduction of a transitional regime between Earth and Moon. Hence, in contrast to regular patched conic theory, a finite region or domain is introduced at the crucial point wherein the accelerations of the Earth-Moon system are of equal importance. In patched conic theory the MSI forms an infinitely thin transitional region which on puncture varies Earth-relative to Moon-relative dynamical centers. Furthermore, in the method to be discussed, the concept of the MSI is abandoned and a new separation curve or surface is introduced between the two primary bodies of the three-body system. The new separation surface used between Earth and Moon was first proposed by

Forster [3], and in two dimensions it physically represents a curve[1] on one side of which a particle placed at rest with respect to the standard Earth-Moon rotating coordinate system, Figure 5.3, slowly commences to move toward the Earth and on the other side of which a particle placed at rest slowly commences to move toward the Moon. This curve is a mathematically well-defined curve which, in the theory of differential equations, is called the separatrix of the system of differential equations [4]. This type of curve, owing to the nature and definition of the libration points introduced in Section 5.2, that is, the points where a body placed at rest remains at rest,

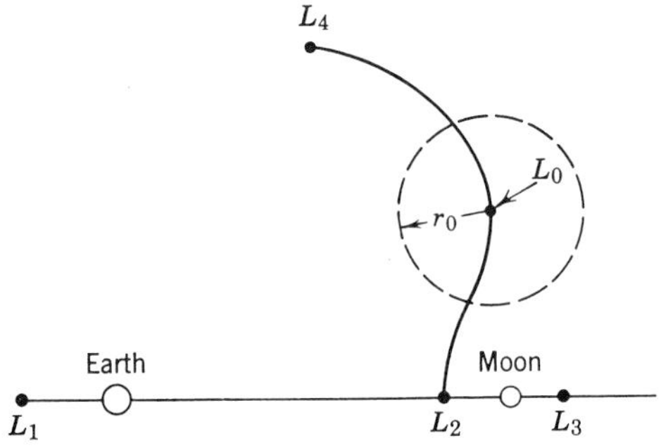

FIGURE 5.6 The separatrix of the Earth-Moon system.

must contain the libration points somewhere along its extension. The approximate geometry of the separatrix is illustrated in Figure 5.6.

The separatrix, of course, is symmetrical about the Earth-Moon line, but is not shown to conserve space. A second separatrix can also be generated between libration points L_3 and L_4. It follows from the previous heuristic development that curve L_2L_4 defines the middle of a region in which the importance of neither the Earth or the Moon is dominant.

Imagine a sphere of radius r_0 placed at any arbitrary point L_0 located on separatrix L_2L_4 as indicated by the dashed circle in Figure 5.6. If this circle is placed with its center lying on the separatrix of the system of differential equations, it appears intuitively acceptable to say that, if the equations of motion can be linearized about a point on curve L_2L_4, there is a neighborhood r_0 within which the accelerations of Earth and Moon are of the same order of magnitude. Perhaps therefore, though it is not known a priori, it may be possible to linearize the equations of motion for a vehicle passing through this region. If it is assumed that linearization of these equations is valid in this region, the computation of a lunar trajectory might proceed as follows.

[1] The same basic definition applies in three dimensions except that a surface replaces the curve. See Section 5.3.4.

A Keplerian orbit about the center of the Earth is determined from a set of initial conditions, the position and velocity vectors at some known epoch for example. The point of intersection of the Earth-centered Keplerian orbit with the separatrix is determined. This point forms the approximate center of the sphere of equal importance of Earth-Moon forces. The Keplerian orbit of a particle on this trajectory is now backed up to the boundary of the sphere, and the initial conditions before puncture are determined. These initial conditions form the starting values for the linearized equations of motion inside the sphere of radius r_0. The linearized equations can now be used to predict the puncture conditions at exit from the sphere. These exit values now can be used to yield the initial conditions for the remaining leg of the trajectory which is represented by a Moon-centered Keplerian orbit. Basically, this is Forster's model [5] for the computation of lunar orbits via analytic methods.

5.3.1 Generation of the Separatrix

For a rotating coordinate system whose origin is at the Earth-Moon barycenter and whose x axis traverses both the Earth and the Moon with the y axis in the Earth-Moon plane, the separatrix was defined as the curve on one side of which a particle placed at rest on the Earth side would start moving toward the Earth and on the other side of which a particle placed at rest would start moving towards the Moon. Only the initial movement is toward either of these bodies, and perhaps after sufficient time the particle could reverse direction and head for the opposite body. Consider the force diagram in Figure 5.7.

It follows that, if a particle placed infinitely close to curve L_2S_2 is to obey the definition implied by the separatrix, no component of force can exist along the unit normal **n** to S. If this were the case, a particle placed on S

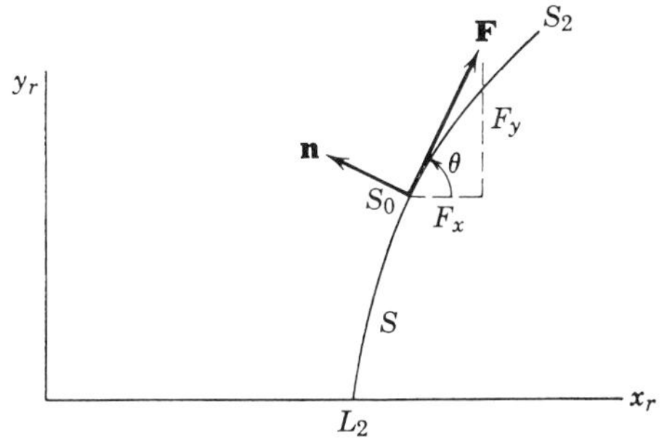

FIGURE 5.7 Separatrix force diagram.

would always have a definite tendency to move across S in a well-established direction and would thus violate the definition. Hence existing forces are tangent to S at any point on S; that is,

$$\mathbf{n} \cdot \mathbf{F} = 0, \tag{5.90}$$

where $\mathbf{F}$ is the local force at point S_0 as illustrated in Figure 5.7. It is well to note that, since the separatrix can be thought of as starting at the straight line solutions of the restricted three-body problem, curves satisfying (5.90) will emanate from the respective libration points. Actually the libration points are the trivial solutions of (5.90) for which $\mathbf{F} = \mathbf{0}$; all these points can be used as initial starting conditions for the integration of the differential equations defining the separatrix.

The differential equation defining any separatrix can be obtained at once, since

$$\frac{dy_r}{dx_r} = \tan\theta = \frac{F_y}{F_x} = \frac{\partial U/\partial y_r}{\partial U/\partial x_r}, \qquad \left.\frac{dy_r}{dx_r}\right|_{L_i} \to \pm\infty, \qquad i = 1, 2, 3, \tag{5.91}$$

where U is the previously introduced potential of the three-body problem defined in Section 5.2, namely,

$$U = \tfrac{1}{2}(x_r^{\,2} + y_r^{\,2}) + \frac{\mu}{r_{32}} + \frac{1-\mu}{r_{12}}, \tag{5.92}$$

with

$$r_{32} = (x_r - x_{r3})^2 + y_r^{\,2}, \qquad r_{12} = (x_r - x_{r1})^2 + y_r^{\,2}.$$

From Section 5.2.7, utilizing canonical units in which the distance between Earth and Moon is unity, it is possible to solve the appropriate quintics to obtain

$$L_2 \equiv 0.836916, \qquad L_3 \equiv 1.155682$$

as the location of the libration points of immediate interest. Hence, selecting the initial conditions

$$x_{r2} = L_2, \qquad x_{r3} = L_3,$$

$$y_{r2} = 0, \qquad y_{r3} = 0,$$

permits numerical integration of (5.91) to be performed; this in turn yields the sought-for separatrix. The curves from L_2 to L_4 and from L_3 to L_4 are defined in Table 5.1.

In passing it should be noted that in a rotating frame (x_r, y_r) the separatrix is a fixed curve, whereas in an inertial frame the separatrix is carried along with the Moon in the same manner as the MSI. An actual scale illustration of the separatrix curves is displayed in Figure 5.8. Near the Earth-Moon

Table 5.1 Shape of the Separatrix Curves

Mass of Moon = (0.01215045)

L_2L_4 and L_2L_5		L_3L_4 and L_3L_5	
x_r	y_r	x_r	y_r
0.836916	±0.000001	1.155682	±0.000001
0.847057	±0.079148	1.096641	±0.166765
0.901654	±0.218632	1.058019	±0.212638
0.927037	±0.304584	1.015824	±0.255293
0.927043	±0.334539	0.968110	±0.306466
0.900480	±0.420071	0.911087	±0.400230
0.853240	±0.508157	0.853240	±0.506442

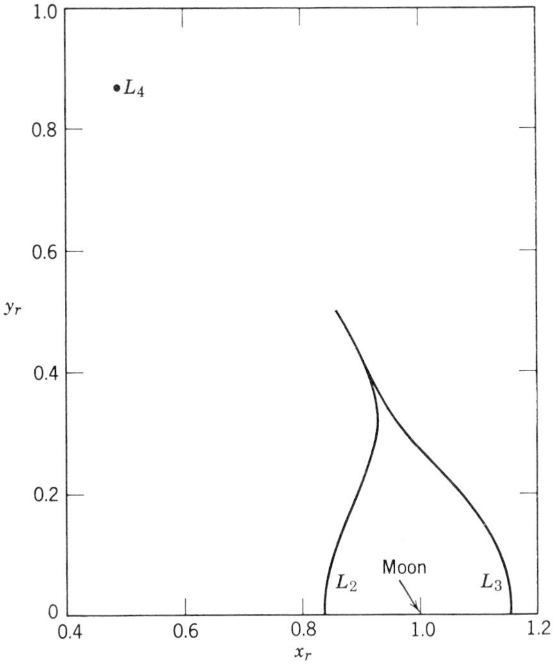

FIGURE 5.8 Separatrix curves.

line, direct analytic solutions for the separatrix can be obtained by series expansion. These expansions are convenient for the actual machine implementation of the model (see Exercise 4).

5.3.2 Transformation of the Equations of Motion

For simplicity let a convenient change of notation, which avoids repeated subscripts, be introduced. Following Forster's notation, the Jacobi potential in the rotating system, (5.92), can be expressed as

$$U(x_r, y_r) = \tfrac{1}{2}(x_r^2 + y_r^2) + \frac{\mu^*}{R} + \frac{\mu}{\rho}, \tag{5.93}$$

where $R \equiv$ distance between Earth and vehicle, $\rho \equiv$ distance between Moon and vehicle, and $\mu^* \equiv 1 - \mu$. Furthermore, it follows that by direct differentiation of U [5]

$$\begin{aligned}
\frac{\partial U}{\partial x_r} &= x_r - \frac{\mu^*}{R^3}(x_r + \mu) - \frac{\mu}{\rho^3}(x_r - \mu^*),\\
\frac{\partial U}{\partial y_r} &= y_r\left(1 - \frac{\mu^*}{R^3} - \frac{\mu}{\rho^3}\right),
\end{aligned} \tag{5.94}$$

$$\begin{aligned}
\frac{\partial^2 U}{\partial x_r^2} &= 1 + \frac{\mu^*}{R^3}\left(2 - \frac{3y_r^2}{R^2}\right) + \frac{\mu}{\rho^3}\left(2 - \frac{3y_r^2}{\rho^2}\right),\\
\frac{\partial^2 U}{\partial y_r^2} &= 1 - \frac{\mu^*}{R^3}\left(1 - \frac{3y_r^2}{R^2}\right) - \frac{\mu}{\rho^3}\left(1 - \frac{3y_r^2}{\rho^2}\right),\\
\frac{\partial^2 U}{\partial x_r\, \partial y_r} &= 3y_r\left[\frac{\mu^*}{R^5}(x + \mu) + \frac{\mu}{\rho^5}(x - \mu^*)\right].
\end{aligned} \tag{5.95}$$

Since the special objective of this method is to obtain a solution to the three-body equations of motion in the neighborhood of curve S, that is, of the separatrix, it is convenient to write the respective equations relative to a local point x_s, y_s on the separatrix. The geometry of this translation is illustrated in Figure 5.9.

Assuming that $\rho_s \equiv (x_s, y_s)$ are known coordinates which are constant, and letting

$$x_r = \xi + x_s, \qquad y_r = \eta + y_s \tag{5.96}$$

permits the system of equations defining the motion to be written as

$$\frac{d^2\xi}{d\tau^2} - 2\frac{d\eta}{d\tau} = \frac{\partial U}{\partial \xi}, \qquad \frac{d^2\eta}{d\tau^2} + 2\frac{d\xi}{d\tau} = \frac{\partial U}{\partial \eta}, \tag{5.97}$$

with respect to the η, ξ coordinate system [6]. Notice that planar motion is assumed by system (5.97).

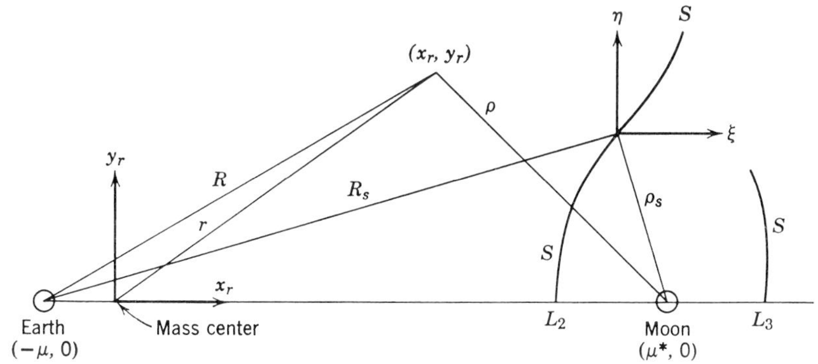

FIGURE 5.9 Translation geometry.

A two-dimensional Taylor expansion about x_s, y_s permits $\partial U/\partial \xi \equiv U_\xi$ and $\partial U/\partial \eta \equiv U_\eta$ to be written as

$$\frac{\partial U}{\partial \xi} = \frac{\partial U^s}{\partial x_r} + \xi\frac{\partial^2 U^s}{\partial x_r^{\,2}} + \eta\frac{\partial^2 U^s}{\partial x_r\,\partial y_r} + \cdots, \tag{5.98}$$

$$\frac{\partial U}{\partial \eta} = \frac{\partial U^s}{\partial y_r} + \xi\frac{\partial U^s}{\partial y_r\,\partial x_r} + \eta\frac{\partial U^s}{\partial y_r^{\,2}} + \cdots, \tag{5.99}$$

or more compactly as

$$\frac{\partial U}{\partial \xi} = U_{x_r}^{\ s} + \xi U^s_{x_r x_r} + \eta U^s_{x_r y_r} + \cdots, \tag{5.100}$$

$$\frac{\partial U}{\partial \eta} = U_{y_r}^{\ s} + \xi U^s_{y_r x_r} + \eta U^s_{y_r y_r} + \cdots, \tag{5.101}$$

where the superscript s on U indicates that the partial derivatives are evaluated at x_s, y_s. Hence, using (5.95), it is possible to obtain to the first order

$$\frac{\partial U}{\partial \xi} = F_1 + a_{11}\xi + a_{12}\eta, \tag{5.102}$$

$$\frac{\partial U}{\partial \eta} = F_2 + a_{12}\xi + a_{22}\eta, \tag{5.103}$$

where

$$F_1 \equiv x_s - \frac{\mu^*(x_s + \mu)}{R_s^{\,3}} - \frac{\mu(x_s - \mu^*)}{\rho_s^{\,3}},$$

$$F_2 \equiv y_s\left(1 - \frac{\mu^*}{R_s^{\,3}} - \frac{\mu}{\rho_s^{\,3}}\right),$$

$$a_{11} \equiv 1 + \frac{\mu^*}{R_s^{\,3}}\left(2 - \frac{3y_s^{\,2}}{R_s^{\,2}}\right) + \frac{\mu}{\rho_s^{\,3}}\left(2 - \frac{3y_s^{\,2}}{r_s^{\,2}}\right),$$

$$a_{22} \equiv 1 - \frac{\mu^*}{R_s^{\,3}}\left(1 - \frac{3y_s^{\,2}}{R_s^{\,2}}\right) - \frac{\mu}{\rho_s^{\,2}}\left(1 - \frac{3y_s^{\,2}}{r_s^{\,2}}\right),$$

$$a_{12} \equiv 3y_s\left[\frac{\mu^*(x_s + \mu)}{R_s^{\,5}} + \frac{\mu(x_s - \mu^*)}{\rho_s^{\,5}}\right].$$

With the previous first-order development the planar equations of motion may be written by virtue of (5.97) as

$$\ddot{\xi} - 2\dot{\eta} = F_1 + a_{11}\xi + a_{12}\eta, \qquad \ddot{\eta} + 2\dot{\xi} = F_2 + a_{12}\xi + a_{22}\eta, \tag{5.104}$$

with the initial conditions

$$\xi(0) \equiv \xi_0, \qquad \dot{\xi}(0) = \dot{\xi}_0,$$

$$\eta(0) \equiv \eta_0, \qquad \dot{\eta}(0) = \dot{\eta}_0.$$

Using the abbreviations

$$\alpha_0 \equiv \tfrac{1}{2}(a_{11} + a_{22} - 4),$$

$$\alpha \equiv [\alpha_0(1 + \gamma)]^{1/2},$$

$$\beta \equiv [\alpha_0(\gamma - 1)]^{1/2},$$

$$\gamma \equiv \left[1 - \frac{a_{11}a_{22} - a_{12}^{\,2}}{\alpha_0^{\,2}}\right]^{1/2}, \tag{5.105}$$

the solution of (5.104) can be found most easily by the Laplace transform method [7]; it is

$$\xi = (\alpha^2 + \beta^2)^{-1}\biggl\{\xi_0(\alpha^2 \cosh \alpha\tau + \beta^2 \cos \beta\tau) + \dot{\xi}_0(\alpha \sinh \alpha\tau + \beta \sin \beta\tau)$$

$$+ [F_1 + 2\dot{\eta}_0 + \eta_0 a_{12} + \xi_0(4 - a_{22})](\cosh \alpha\tau - \cos \beta\tau)$$

$$+ [2F_2 + (\dot{\eta}_0 + 2\xi_0)a_{12} - (\dot{\xi}_0 - 2\eta_0)a_{22}]\left(\frac{\sinh \alpha\tau}{\alpha} - \frac{\sin \beta\tau}{\beta}\right)$$

$$+ [F_2 a_{12} - F_1 a_{22}]\left(\frac{\cosh \alpha\tau - 1}{\alpha^2} + \frac{\cos \beta\tau - 1}{\beta^2}\right)\biggr\}, \quad (5.106)$$

$$\eta = (\alpha^2 + \beta^2)^{-1}\biggl\{\eta_0(\alpha^2 \cosh \alpha\tau + \beta^2 \cos \beta\tau) + \dot{\eta}_0(\alpha \sinh \alpha\tau + \beta \sin \beta\tau)$$

$$+ [F_2 - 2\dot{\xi}_0 + \xi_0 a_{12} + \eta_0(4 - a_{11})](\cosh \alpha\tau - \cos \beta\tau)$$

$$+ [-2F_1 + (\dot{\xi} - 2\eta_0)a_{12} - (\dot{\eta}_0 + 2\xi_0)a_{11}]\left(\frac{\sinh \alpha\tau}{\alpha} - \frac{\sin \beta\tau}{\beta}\right)$$

$$+ [F_1 a_{12} - F_2 a_{11}]\left(\frac{\cosh \alpha\tau - 1}{\alpha^2} + \frac{\cos \beta\tau - 1}{\beta^2}\right)\biggr\}. \quad (5.107)$$

These equations represent, to the first order in distance from the separatrix, the complete solution in the transition region. Once ξ and η have been determined, the positions in the x_r, y_r frame are given directly by (5.96).

5.3.3 Motion in the Transition Region

The developments in the preceding section describe the motion of the third body in the neighborhood of the separatrix, that is, in the region where the vehicle transits from the domain of preponderant influence of one primary to that of the other. Equations 5.106 and 5.107 permit analysis of the motion in the transition region similar to the analysis of motion near a libration point. As used here, analysis of the motion is concerned with the mathematical character of the solutions; for example, without regard to particular numerical coefficients, do the solutions have periodic or nonperiodic, polynomial or exponential dependence?

It is immediately apparent that the character of the motion is determined by α and β in (5.106) and (5.107), and more specifically by (5.105). From the mathematical point of view the motion is quite different depending on whether α and β are real, imaginary, or complex. This situation in turn is entirely determined by the a_{ik} in (5.102) and (5.103) which depend only on the point at which the trajectory approaches or intersects the separatrix.

At the Lagrangian points on the x_r axis, $y_s = 0$, $a_{12} = 0$, and both a_{22} and $c_0 \equiv a_{11}a_{22} - a_{12}^2$ are negative. It follows that both α and β are real and therefore the motion is given by hyperbolic and trigonometric time dependence, (5.106) and (5.107), in some neighborhood of the x_r axis. However, for trajectories intersecting the separatrix, c_0 eventually becomes positive and larger than α_0^2. As may be observed from (5.105), α and β then become complex. Therefore two classes of trajectories exist: Class I, for a vehicle near the separatrix where $c_0 < \alpha_0^2$, and Class II for a vehicle in a region of space where $c_0 > \alpha_0^2$.

It is shown in [9] that trajectories crossing the separatrix which emanates from L_2, at ordinates y_s and at distances from the small primary ρ_s limited by

$$\frac{y_s}{\rho_s} \ll \frac{\sqrt{3}}{3}, \tag{5.108}$$

are in Class I. For the Earth-Moon system, inequality 5.108 states that the motion of the vehicle is in Class I if the ordinate of intersection of the separatrix is about 5 e.r. or less.

For Class II trajectories the complex roots α and β of (5.105) are to be replaced by the real expressions

$$a = [\tfrac{1}{2}(\alpha_0 + c_0^{1/2})]^{1/2}, \qquad b = [\tfrac{1}{2}(-\alpha_0 + c_0^{1/2})]^{1/2}, \tag{5.109}$$

and the motion becomes expressible in terms of functions of the form (cosh at) (cos bt) or (cosh at) (sin bt), etc.

5.3.4 Extension of the Method into Three Dimensions

The technique discussed in this section is by no means constrained to only two dimensions. Equation 5.91 in three dimensions defines a surface of separation or three-dimensional separatrix surface. This separatrix can be generated numerically in much the same manner as the two-dimensional curve. Perhaps the curve of separation in the actual plane of motion or the intersection of the plane of the Keplerian Earth-centered conic with the separatrix surface can be generated to provide the critical boundary.

From this point on, all the preceding development can be carried out in three dimensions to yield the two previous coordinates, that is, η, ξ, and the new coordinate ζ. Hence, starting from (5.102) and (5.103), modified to read

$$\begin{aligned}
\frac{\partial U}{\partial \xi} &= F_1 + a_{11}\xi + a_{12}\eta + a_{13}\zeta, \\
\frac{\partial U}{\partial \eta} &= F_2 + a_{21}\xi + a_{22}\eta + a_{23}\zeta, \\
\frac{\partial U}{\partial \eta} &= F_3 + a_{31}\xi + a_{32}\eta + a_{33}\zeta,
\end{aligned} \tag{5.110}$$

the following system should be integrated analytically:

$$\ddot{\xi} - 2\dot{\eta} = F_1 + a_{11}\xi + a_{12}\eta + a_{13}\zeta,$$

$$\ddot{\eta} + 2\dot{\xi} = F_2 + a_{21}\xi + a_{22}\eta + a_{23}\zeta,$$

$$\ddot{\zeta} = F_3 + a_{31}\xi + a_{32}\eta + a_{33}\zeta, \tag{5.111}$$

where the new coefficients can be obtained by the methods of Section 5.3.2.

5.3.5 Numerical Results

It is interesting to compare numerical results obtained for motion within the transition region via linearization with exactly integrated data. Figures 5.10 through 5.12 illustrate such results for the Earth-Moon system [9].

Figure 5.12 displays a number of trajectories calculated by use of (5.106) and (5.107). The solid lines represent the analytic results while the dashed lines show the associated numerically integrated trajectory. A section of the separatrix as well as of the MSI has also been superimposed. The numbers along the lines are the corresponding flight times in hours.

The improvement gained by use of the linearized solution over use of a conic may be measured by the extent of change in the orbital elements a, the semimajor axis, and e, the eccentricity of a Keplerian orbit. Figures 5.10 and 5.11 illustrate the change in the Keplerian constants for trajectory number 4 of Figure 5.12. The graph illustrates that for this trajectory the method accounts for about 80 percent of the change in the dimensional elements, a and e.

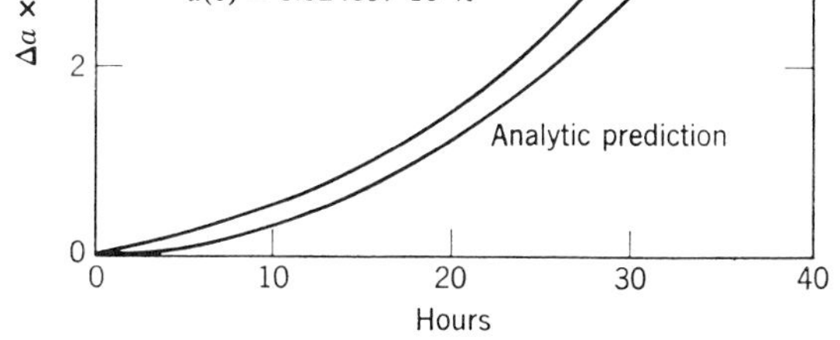

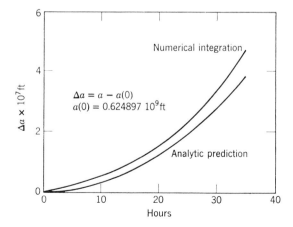

FIGURE 5.10 Change of semimajor axis as a function of time.

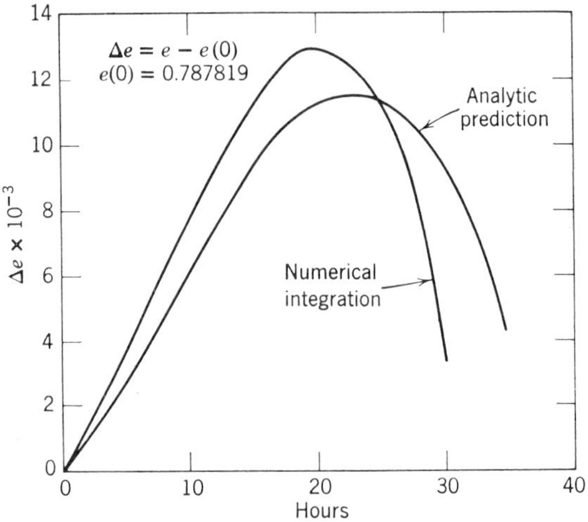

FIGURE 5.11 Change of eccentricity as a function of time.

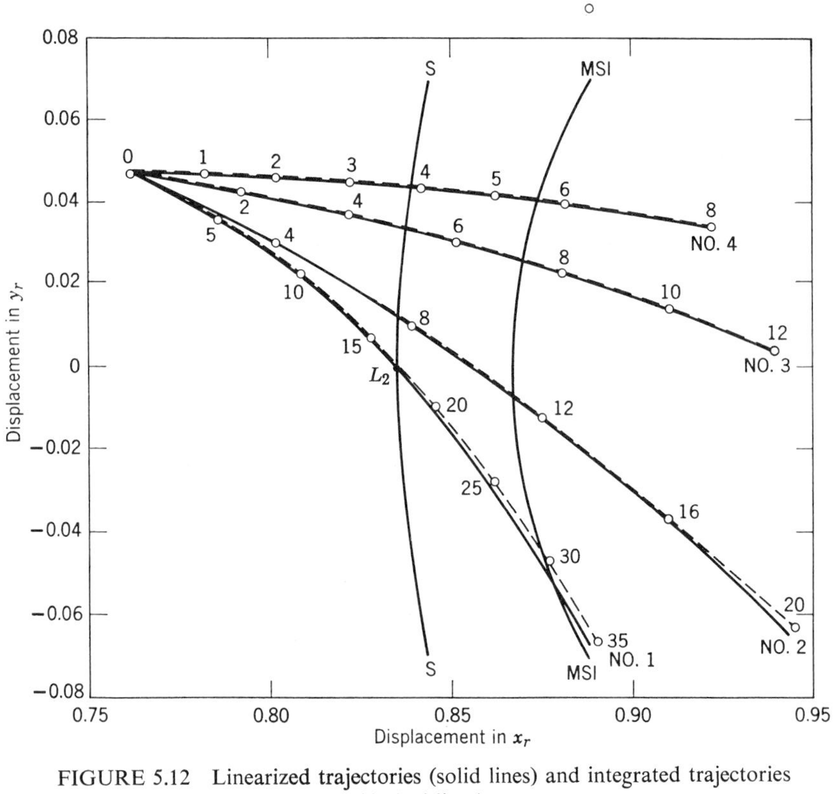

FIGURE 5.12 Linearized trajectories (solid lines) and integrated trajectories (dashed lines).

5.4 COMMENTS ON THE THREE-BODY PROBLEM

The problem of three bodies is concerned with the motion of three point masses in their mutual gravitational field. A numerical solution to this problem by means of the techniques of special and general perturbations is always possible; however, a general solution of the problem in the sense to be discussed presently has not yet been found. In fact, the problem has resisted the combined and directed efforts of the most eminent mathematicians. Such scientists as Lagrange, Levi-Civita, Birkhoff, and Poincaré have spent many long hours without obtaining a total solution of the problem.

Since the difficulties presented by the general problem of three bodies could not be overcome, three-body problems with reduced complexity were considered and in particular, the restricted three-body problem. In the restricted problem one mass is assumed infinitesimally small and hence does not disturb the other two masses which are assumed to revolve in stable elliptic orbits about each other. Application of further simplifications resulted in the circular and planar restricted problem in which the orbits of the two finite masses are assumed to be circles and the motion of the infinitesimal mass is taken to be in the plane of motion of the primaries. In this section the possibility of solution is discussed in general terms.

5.4.1 Possibility of Solution

The differential equations for the coordinates of the infinitesimal mass in the restricted problem were derived in Section 5.2. The system of three differential equations of the second order in the coordinates could be written as a system of six first-order differential equations in the three coordinates and the three velocity components; hence the system is said to be of the sixth order. Clearly, any relation which is valid for all time (be it algebraic, transcendental, etc.) between the coordinates and the velocity components constitutes a solution of one of these six equations and is, therefore, called an integral. Jacobi's integral, (5.48), is an example, and it reduces the restricted problem to the fifth order. No further integrals for this problem have been found as yet, and it is in this sense that the restricted three-body problem is one of the unsolved problems of dynamics. This, of course, does not preclude the possibility of finding approximate solutions which are valid for a limited time by special perturbations, etc.

If the restricted three-body analysis is further constrained to the planar case, the z coordinate and the $\dot{z}$ component of velocity are identically zero, the system becomes one of order four and, with Jacobi's integral, three further integrals would constitute a solution. Moulton [2] refers to a result of Jacobi's by which the last two integrals could be found. Thus, if only one more integral for the planar restricted problem were found, a complete solution of this problem would be known. Bruns has shown that new integrals do not exist in the form of *algebraic relations between coordinates and velocities in cartesian coordinates;* Poincaré has also shown that the new integral cannot be a transcendental relation with the elements as coordinates. In some other coordinates the solution may or may not depend on elementary functions—we do not know. In essence, the choice of variables and coordinate systems is infinite and no general rule has yet been stated that proves the impossibility of analytic solution. Not only does it appear possible, but it is certainly true that an integral in some form exists; the problem is to find its form.

This situation may be analogous to the attempts of many mathematicians, such as Gregory, to obtain the solution of polynomial equations of degree greater than four by the process of extraction of radicals. It was only when Abel showed the impossibility of such solutions that this matter appeared to be closed once and for all. Yet, even after Abel's celebrated proof, Klein [28] and others did manage to obtain the solution of quintic equations via the transformation of Tschirnhaus and the use of modular elliptic functions. Thus the problem of obtaining the roots of polynomials was solved analytically by the use of transformations and nonelementary functions. Therefore, that the restricted problem of three bodies is integrable by specialized transformations and the use of nonelementary functions.

5.4.2 Some Recent Developments

Recent investigators, such as Lagerstrom and Kevorkian [18], [19], [20], have made use of the method of matched asymptotic expansions in an attempt to derive analytic formulas valid over the entire Earth to Moon distance. This method assumes that a series expansion in the small parameter μ, where μ is the ratio of the mass of the Moon to the mass of the Earth, can be made in an outward direction from the Earth. In the same manner an inward expansion can also be made from the Moon toward the Earth. The two analytic expansions can then be joined at a limiting point between the Earth and the Moon and the coefficients of the series equated to yield composite formulas which define a set of analytic expressions for the elements of the Moon-centered hyperbola in terms of the initial conditions near Earth. Williams and Liu [29], [30] have extended the results of Lagerstrom and Kevorkian and documented the algebraic formulas. The method of matched asymptotic expansions requires more investigation. The main drawback of the technique is the overpowering algebraic manipulations that must be performed to yield suitable results.

From a different point of view Contopoulous [31] has obtained a new integral for the three-body problem in the form of a series expansion. The expansions are also lengthy and require much algebraic manipulation before meaningful results can be obtained.

Much work still remains to be performed with respect to the two previous techniques; however, the importance of these investigations stands on firm ground. The results to be obtained by such techniques will probably be more important from a theoretical than from a practical point of view.

Recently Escobal [33] and Escobal and Stern [34], [35] have introduced the concept of the hybrid patched conic technique. The technique uses a patched conic trajectory as a reference orbit and enables the resulting errors of the Keplerian orbits to be determined in a very straightforward manner at minimal expense in computing effort and algebraic manipulation. Even though analytic expressions of the corrections in the state variables are obtained explicitly, the method is not introduced in this chapter because it requires the reader to have knowledge of the Encke technique. The hybrid method is introduced in Chapter 7.

5.5 SUMMARY

The primary concern of this chapter was the development of analytic techniques for the computation of lunar trajectories. In the first portion of the chapter, patched conic techniques, in which Earth-centered and Moon-centered Keplerian conics are joined to provide a composite trajectory which approximates true lunar trajectories were discussed. These techniques are rapid and are therefore much used in industry for the determination of large numbers of lunar trajectories.

In order to obtain more exact and accurate techniques for the determination of lunar trajectories, an understanding of the three-body equations is required. Therefore considerable detail was given to the derivation of the pertinent equations of motion. Jacobi's integral, which relates the speed of an infinitesimal particle to its position, was derived. This integral was seen to be the three-body analog of the well-known vis-viva integral for the two-body problem.

Two of the classic known solutions of the restricted three-body problem, the straight line and the equilateral triangle solutions, were developed. It was shown that five points exist at which satellites placed at rest, with respect to a coordinate system rotating about the mass center of the system, have a tendency to remain. These points are called the libration points of the three-body system.

A new method for the computation of Earth to Moon trajectories in which a finite transition region is introduced between the Earth- and Moon-centered Keplerian joining points was introduced. By linearizing the three-body equations of motion in the vicinity of the force system separation surfaces, analytic solutions to the motion of a vehicle within that domain were obtained.

Finally, the integrability of the three-body problem was discussed and some recent developments in the area of three-body mechanics were mentioned.

EXERCISES

1. Why is it not possible, in the method of patched conics, to match exactly positions, velocities, and accelerations?

2. Assuming that the departure Keplerian orbit of the Earth to Moon trajectory is exactly parabolic, develop the resulting quartic required to match the Earth- and Moon-centered segments.

3. Show that, if a coordinate system u, v is introduced at L_2 with u aligned along x_r and v aligned along y_r, equations 5.94, that is,

$$\frac{\partial U}{\partial x_r} = x_r - \frac{\mu^*}{R^3}(x_r + \mu) - \frac{\mu}{\rho^3}(x_r - \mu^*),$$

$$\frac{\partial U}{\partial y_r} = y_r\left(1 - \frac{\mu^*}{R^3} - \frac{\mu}{\rho^3}\right),$$

can be developed to the second order in u and v as

$$\frac{\partial U}{\partial u} = (1 + 2\gamma_1)u + 3\gamma_2(u^2 - \tfrac{1}{2}v^2) + \cdots,$$

$$\frac{\partial U}{\partial v} = (1 - \gamma_1)v - 3\gamma_2 uv + \cdots,$$

where

$$\gamma_1 \equiv \frac{\mu^*}{(1 - L_2)^3} + \frac{\mu}{L_2^{\,3}}, \qquad \gamma_2 \equiv -\frac{\mu^*}{(1 - L_2)^4} + \frac{\mu}{L_2^{\,4}}.$$

4. From the results in Exercise 3 show that

$$\frac{dv}{du} = \frac{(1 - \gamma_1)v - 3\gamma_2 uv}{(1 + 2\gamma_1)u + 3\gamma_2(u^2 - \tfrac{1}{2}v^2)}$$

can be satisfied by the trial solution $v = Au^n + \cdots$, with $v(0) = 0$, hence $n = \tfrac{1}{2}$ and

$$A^2 = \frac{2(4\gamma_1 - 1)}{3\gamma_2}, \qquad \gamma_1 > 1.$$

This equation represents the analytic form of the separatrix in the vicinity of $u = 0$.

5. At what value of v does the analytic expression for the separatrix in Exercise 4 begin to deviate significantly from the true values displayed in Table 5.1?

6. Verify that the point of intersection of the Earth-centered conic with the parabolic separatrix curve in Exercise 4 is an algebraic equation of degree four with coefficients dependent on the time at intersection of the two respective curves. Obtain analytic expressions for the coefficients of the quartic.

7. Utilizing the diagram, show that the quintic equation defining the straight line solution of the three-body problem is the point of equilibrium of the centrifugal and gravitational forces. Use rotating coordinates and notice that the barycenter is at $x = \mu$.

8. A Moon probe that departed Earth on the Earth-Moon line at a distance of 1.4 e.r. with a speed of 28,000 ft/sec has crossed the Earth-Moon line for a second time at a distance of 40 e.r. Assuming the Moon is 60 e.r. from the Earth, what is the speed of the spacecraft at the crossing point?

REFERENCES

[1] P. R. Escobal, *Methods of Orbit Determination*, John Wiley and Sons, New York, 1965.
[2] F. R. Moulton, *An Introduction to Celestial Mechanics*, The Macmillan Company, New York, Chapter 8, 1914.
[3] K. Forster, *Linearization for Certain Earth-Moon Trajectories in the Restricted Problem of Three Bodies*, TRW Systems Independent Research Project, 9863-6005-R0-000, February 1966.
[4] H. T. Davis, *Introduction to Nonlinear Differential and Integral Equations*, Dover Publications, 1961, p. 270.
[5] K. Forster, "Differential Equations and Solutions for the Transition Region of the Three-Body Problem," TRW Systems, 3422.3-65, May 1966.
[6] E. Finlay-Fruendlich, *Celestial Mechanics*, Pergamon Press, London, 1958.
[7] R. V. Churchill, *Operational Mathematics*, McGraw-Hill Book Company, 1958.
[8] A. Wintner, *The Analytical Foundations of Celestial Mechanics*, Princeton University Press, 1941.
[9] K. Forster, P. R. Escobal, H. A. Lieske, "Motion of a Vehicle in the Transition Region of the Three-Body Problem," *Astronautica Acta*, to be published 1968.
[10] H. Poincaré, *Les Methodes Nouvelles de la Mécanique Céleste*, Volume I, Gauthier-Villars, Paris 1892.
[11] P. R. Escobal and F. H. Brinkmann, "Transformed Three-Body Equations of Motion," TRW Systems, 9882.3-167, October 1965.
[12] V. A. Egorov, "Certain Problems of Moon Flight Dynamics," *Russian Literature of Satellites*, Part I, International Physical Index, 1958.

[13] G. S. Stern, "Series Approximations of the Jacobi Integral," TRW Systems, 9882.3-188, December 1965.
[14] P. R. Escobal and F. H. Brinkmann, "The Lemniscate and the Restricted Problem of Three Bodies," TRW Systems Independent Research Project, 9863-6008-R0-000, March 1965.
[15] S. W. McCuskey, *Introduction to Celestial Mechanics*, Addison-Wesley Publishing Company, Reading, Mass., 1963, pp. 92–179.
[16] E. T. Whittaker, *A Treatise on the Analytical Dynamics of Particles and Rigid Bodies*, Cambridge University Press, Fourth Edition, 1959.
[17] Ali Hansen Nayfeh, "A Comparison of Three Perturbation Methods for Earth-Moon-Spaceship Problem," *AIAA Journal*, September 1965, pp. 1682–1687.
[18] P. A. Lagerstrom and J. Kevorkian, "Some Numerical Aspects of Earth-to-Moon Trajectories in the Restricted Three-Body Problem," AIAA Paper 63–389.
[19] P. A. Lagerstrom and J. Kevorkian, "Matched Conic Approximation to the Two Fixed Force-Center Problem," *The Astronomical Journal*, March 1963.
[20] P. A. Lagerstrom and J. Kevorkian, "Earth-to-Moon Trajectories with Minimal Energy," *Journal de Mécanique*, Vol. 2, No. 4, December 1963.
[21] V. Szebehely, *The Restricted Problem of Three Bodies*, Summer Institute in Dynamical Astronomy, General Electric Company, Philadelphia, Pa. 1961.
[22] J. Kevorkian, "Uniformly Valid Asymptotic Representation for All Times of the Motion of a Satellite in the Vicinity of the Smaller Body in the Restricted Three-Body Problem," *The Astronomical Journal*, Volume 67, No. 4, 1967, pp. 204–211.
[23] L. M. Perko, "Interplanetary Trajectories in the Restricted Three-Body Problem," *AIAA Journal*, December 1964, pp. 2187–2192.
[24] K. Stumpff, *On Lagrange's Theory of the Three-Body Problem*, NASA, TND-1417, January 1963.
[25] D. van Zelm Wadsworth, "A New Analytic Method for Rapid Computation of Earth-Moon Trajectories," AIAA Paper 65–514.
[26] P. Lanzano, "Periodic Solutions for the Restricted Three-Body Problem," Space Technology Laboratories, Inc., Report GM-TM-0165-00323, December 1958.
[27] E. T. Bell, *Men of Mathematics*, Simon & Schuster, New York, 1937.
[28] F. Klein, *The Icosahedron*, Dover Publications, New York, 1956.
[29] R. R. Williams and C. S. Liu, *The Matched Asymptotic Expansion Method for Lunar Trajectories*, TRW Systems, 9990-6078-R000, December 1966.
[30] R. R. Williams and C. S. Liu, *An Improved Asymptotic Expansion Method for Lunar Trajectories*, TRW Systems, 9990-6281-R000, July 1967.
[31] G. Contopoulous, "On the Existence of a Third Integral of Motion," *The Astronomical Journal*, February 1963.
[32] H. A. Lieske, *Lunar Instrument Carrier–Trajectory Studies*, RAND Corporation, Research Memorandum RM 1728, June 1956.
[33] P. R. Escobal, *Inclusion of Planetary Perturbations into Patched Conic Programs*, TRW Systems, 3431.2-19, March 1967.
[34] P. R. Escobal and G. S. Stern, *The Hybrid Patched Conic Technique*, TRW Systems, 05952-6170-R000, July 1967.
[35] P. R. Escobal and G. S. Sterm, *Hybrid Patched Conic Simulation of Lunar Return Trajectories*, TRW Systems, 05952-6187-R000, November 1967.
[36] F. R. Moulton, *Periodic Orbits*, Carnegie Institution of Washington, Washington, 1920.

6 Geocentric, Lunar, and Interplanetary Communications

Light is a phenomenon of vibrations like sound, but simply on a different scale of size, the vibrations being ever so much smaller and more rapid.

HUYGHENS [14]

6.1 COMMUNICATIONS ANALYSIS

Consider a space vehicle attempting to contact a specified Earth ground station. When will the ship be able to transmit to Earth? What are the earliest and latest times of transmission? For accurate tracking of interplanetary vehicles, when can a ground station obtain range, range-rate, and visual observations? These questions and others are analyzed in this chapter. More specifically, the analysis to be investigated is concerned with the geometry of the pertinent bodies involved in the communications or visual contact procedure. As may be obvious, the line of sight between ground station and vehicle must be clear of interfering barriers; for example, the station is positioned on one side of the Earth whereas the space vehicle is below the horizon, or the Moon is between the space vehicle and the station. Other geometrical constraints of this kind are numerous and are fundamental to lunar and interplanetary communications analysis. In this chapter the astrodynamical controlling equations of communications analysis are developed for several of the most important geometrical configurations. Whenever possible, the closed or semiclosed solution to the problem at hand is sought. References [1], [2], [3], [4], and [5] can be consulted for more specific details of this ever-growing problem. In passing, it is well to note that some geocentric communications problems, that is, geocentric rise-set and eclipsing, are not discussed here as these subjects have been treated in [1].

6.1.1 Basic Assumptions

Throughout this chapter a consistent effort is made to make the geometry as exact as possible. This includes the adoption of a realistic model for the shape of the Earth, the assumption that the positions or ephemerides of the Moon and planets are known as functions of time (Sections 1.3, 1.4, and 1.5), and the belief that position or state of the interplanetary space vehicle can be computed as accurately as is desired. These three items are the key to accurate communications analysis. The problem of multiple station communication is not discussed in this chapter; it is assumed that the analysis developed here can be repeated for any desired number of stations in the communications net.

Even though the geocentric model for the Earth, or any planet, is discussed in detail in [1], it is helpful to state the required geometrical relationships and the definition of symbols used in this chapter. Figure 6.1 illustrates the

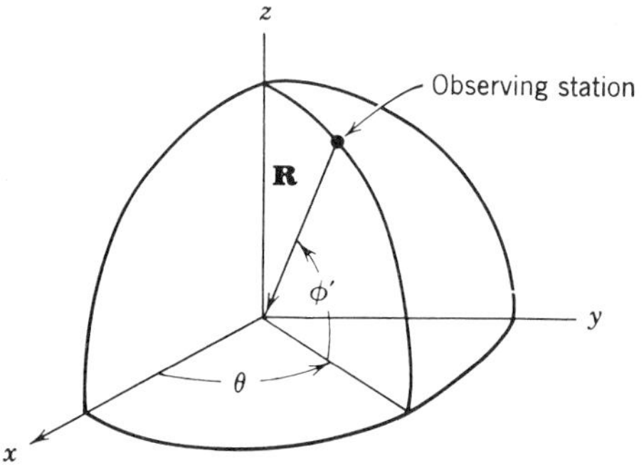

FIGURE 6.1 Ground station on a geometrically oblate planet.

Earth (or central planet) located in the right ascension-declination coordinate system ([1], Chapter 4). Consider a ground station with coordinates ϕ' and θ defined by $\phi' \equiv$ station geocentric latitude and $\theta \equiv$ station local sidereal time. The vector **R**, commonly called the station coordinate vector, can be shown to have [1], the three components

$$\mathbf{R} = \begin{bmatrix} X \\ Y \\ Z \end{bmatrix} = \begin{bmatrix} -G_1 \cos\phi \cos\theta \\ -G_1 \cos\phi \cos\theta \\ -G_2 \sin\phi \end{bmatrix}, \tag{6.1}$$

where

$$G_1 \equiv \frac{a_p}{[1 + (2f - f^2)\sin^2\phi]^{1/2}} + H,$$

$$G_2 \equiv \frac{(1 - f)^2 a_p}{[1 + (2f - f^2)\sin^2\phi]^{1/2}} + H,$$

with the geodetic latitude ϕ defined by

$$\phi = \tan^{-1}\left[\frac{\tan \phi'}{(1-f)^2}\right], \qquad -\frac{\pi}{2} \leq \phi \leq \frac{\pi}{2},$$

and $a_p \equiv$ equatorial radius of planet, $f \equiv$ flattening of adopted ellipsoid, and $H \equiv$ station elevation above and measured normal to the surface of the adopted ellipsoid.

The model for the location of a typical observing ground station defined by (6.1) is simple and provides sufficient accuracy for the communications problems to be discussed in later sections.

6.1.2 Light Time Correction

Electromagnetic radiation has a fixed speed in vacuum of approximately $c = 186{,}000$ mi/sec. It takes light about 8 minutes to reach the Earth from the Sun. If, for example, a vehicle in the vicinity of Mars and in opposition to the Earth attempts to contact an Earth ground station, a serious error in the communication window could occur without the inclusion of the light time correction. This compensation can be included quite easily once the uncorrected transmission times have been determined by means of the relationship

$$t_c = t + \frac{\rho}{c}, \tag{6.2}$$

where t_c is the corrected transmission time, t is the uncorrected time, and ρ is the slant range vector magnitude from the vehicle to the station.

6.2 RISE AND SET OF ONE SATELLITE WITH RESPECT TO ANOTHER

The desirability of having a simple analytic method for predicting when two or more satellites are mutually visible to each other is important in satellite communications systems and in space warfare. Mutually visible satellites are defined in this section as two satellites that can maintain direct line of sight between each other for a certain length of time. Actually, the analysis to be developed is primarily concerned with the rise and set time of a given satellite with respect to another, that is, the time of loss or gain of direct line of sight. For satellites of low eccentricity a single transcendental equation can be obtained which permits the relative viewing windows to be determined by a one-parameter iteration.

6.2.1 Development of Rise-Set Function

Consider the geometry defined in Figure 6.2. As illustrated, satellites 1 and 2 are in a near state of relative rise or set. Indeed, if the vector **S**, which emanates from the dynamical center of the Earth, had a magnitude equal to or less than the radius of the Earth and if it were perpendicular to **c**, the chord length vector between the satellites, it is evident that the satellites would not have direct line-of-sight communication. Owing to atmospheric

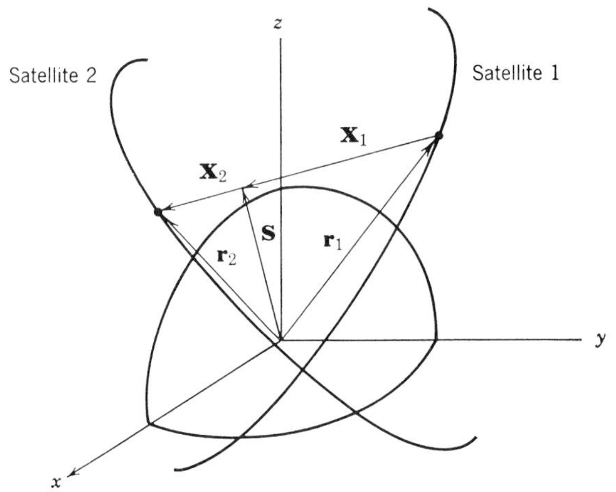

FIGURE 6.2 Relative rise-set geometry.

interference, however, a realistic analysis would let the magnitude of **S** be slightly larger than a_e, the radius of the Earth. Letting Δ be the thickness of the atmosphere or suitable bias factor, it follows that

$$S^2 = \mathbf{S} \cdot \mathbf{S} = (a_e + \Delta)^2. \tag{6.3}$$

Examination of Figure 6.2 allows the two fundamental vector closure equations to be written:

$$\mathbf{S} + \mathbf{X}_2 = \mathbf{r}_2, \tag{6.4}$$

$$\mathbf{S} - \mathbf{X}_1 = \mathbf{r}_1, \tag{6.5}$$

where $\mathbf{r}_i$, $i = 1, 2$, are the position vectors of the satellites and $\mathbf{X}_i$, $i = 1, 2$, are two unknown vectors. The magnitudes of $\mathbf{r}_2$ and $\mathbf{r}_1$ can be determined at once by dotting (6.4) and (6.5) on themselves:

$$r_2^2 = (\mathbf{S} + \mathbf{X}_2) \cdot (\mathbf{S} + \mathbf{X}_2) = S^2 + X_2^2 + 2\mathbf{S} \cdot \mathbf{X}_2 \tag{6.6}$$

$$r_1^2 = (\mathbf{S} - \mathbf{X}_1) \cdot (\mathbf{S} - \mathbf{X}_1) = S^2 + X_1^2 - 2\mathbf{S} \cdot \mathbf{X}_1. \tag{6.7}$$

At relative rise and set of satellite 1 with respect to satellite 2, $\mathbf{S} \cdot \mathbf{X}_1 = \mathbf{S} \cdot \mathbf{X}_2 = 0$ and (6.6) and (6.7) can be written as

$$X_2 = [r_2^2 - S^2]^{1/2}, \tag{6.8}$$

$$X_1 = [r_1^2 - S^2]^{1/2}. \tag{6.9}$$

Adding (6.8) and (6.9) yields

$$X_2 + X_1 = c = [r_2^2 - S^2]^{1/2} + [r_1^2 - S^2]^{1/2} \tag{6.10}$$

and, since c, the chord length, can be determined from

$$c = [(\mathbf{r}_2 - \mathbf{r}_1) \cdot (\mathbf{r}_2 - \mathbf{r}_1)]^{1/2} = [r_2^2 + r_1^2 - 2\mathbf{r}_1 \cdot \mathbf{r}_2]^{1/2}, \tag{6.11}$$

by substituting (6.11) into (6.10) and squaring twice, it is possible to obtain the relative rise-set function

$$R \equiv (\mathbf{r}_2 \cdot \mathbf{r}_1)^2 - r_2^2 r_1^2 + (r_2^2 + r_1^2)S^2 - 2S^2\mathbf{r}_2 \cdot \mathbf{r}_1, \tag{6.12}$$

where S is obtained from (6.3). The rise-set function defined by (6.12) can be used to predict explicitly whether or not satellites are visible to one another. The sign of R associated with visibility can be obtained by constructing a case in which direct line-of-sight visibility is an impossibility. By placing the satellites at opposition, that is, $\mathbf{r}_2 \cdot \mathbf{r}_1 = -r_2 r_1$, it is possible to reduce R to

$$R = (r_2 + r_1)^2 S^2 \geq 0. \tag{6.13}$$

Hence a negative value of R implies direct line-of-sight communication whereas in opposite fashion a positive value of R denotes nonvisibility.

6.2.2 Reduction of Rise-Set Function to a Two-Parameter Function

In terms of the orbital eccentricity e, semiparameter p, and true anomaly v, the equation of each orbit can be expressed by the relationship [2]

$$r_i = \frac{p_i}{1 + e_i \cos v_i}, \qquad i = 1, 2. \tag{6.14}$$

Hence, by substituting (6.14) into (6.12) and evaluating the dot products, R can be written as a function of v_1 and v_2 as

$$R(1 + e_1 \cos v_1)^2(1 + e_2 \cos v_2)^2 = R^*, \tag{6.15}$$

with R^* defined as

$$R^* \equiv [p_2^2(1 + e_1 \cos v_1)^2 + p_1^2(1 + e_2 \cos v_2)^2 - 2p_1 p_2(1 + e_1 \cos v_1) \times (1 + e_2 \cos v_2) \cos (v_2 - v_1)]S^2 - p_2^2 p_1^2 \sin^2 (v_2 - v_1). \tag{6.16}$$

Since R^* has the same mathematical properties as R, the use of R can be abandoned and the analysis continued utilizing R^* as the modified rise-set function. It should be noticed that (6.16) is only valid for two satellites in the same orbital plane. If desired, an equivalent relationship can be derived for satellites in different orbital planes by letting $\mathbf{r}_i = x_{\omega i}\mathbf{P} + y_{\omega i}\mathbf{Q}$, $i = 1, 2$ (See Equation 6.28). However, for purposes of illustrating the technique, the planar case is discussed.

6.2.3 Reduction of Rise-Set Function to a One-Parameter Function

Because communication satellite orbits, by their nature, will be low eccentricity orbits, a further reduction of R^* is possible. Consider the general expansion of v as a series [6] in the eccentricity e and mean anomaly M, namely,

$$v = M + 2e \sin M + \tfrac{5}{4}e^2 \sin 2M + \frac{e^3}{12}(13 \sin 3M - 3 \sin M) + \frac{e^4}{96}(103 \sin 4M - 44 \sin 2M) + \cdots. \tag{6.17}$$

By introducing the definition $M \equiv M_0 + n(t - t_0)$, where n is the mean motion, t is the universal time, and the subscript zero denotes an epoch reference time, (6.17) can be written as

$$v = M_0 + n\tau + 2e \sin M_0 \cos n\tau + 2e \cos M_0 \sin n\tau + \tfrac{5}{4}e^2 \times \sin 2M_0 \cos 2n\tau + \tfrac{5}{4}e^2 \cos 2M_0 \sin 2n\tau, \tag{6.18}$$

where $\tau \equiv t - t_0$, $n \equiv k\sqrt{\mu}\, a^{-3/2}$, $k \equiv$ planetary constant, $\mu \equiv$ sum of masses of planet and satellite, and $a \equiv$ semimajor axis of the satellite. By selecting the same epoch universal time for both satellites, it follows that

$$v_i = \bar{v}_i \equiv M_{0i} + n_i\tau + 2e_i \sin M_{0i} \cos n_i\tau + 2e \cos M_{0i} \sin n_i\tau + \cdots, \qquad i = 1, 2, \tag{6.19}$$

and (6.16) can now be written as

$$R^* = [p_2^2(1 + e_1 \cos \bar{v}_1) + p_1^2(1 + e_2 \cos \bar{v}_2)^2 - 2p_2p_1(1 + e_1 \cos \bar{v}_1)(1 + e_2 \cos \bar{v}_2) \cos(\bar{v}_2 - \bar{v}_1)](a_e + H)^2 - p_2^2p_1^2 \sin^2(\bar{v}_2 - \bar{v}_1), \tag{6.20}$$

so that $R^* = R^*(\tau)$. Letting $R^* = 0$, and solving the transcendental relationship defined by (6.20) by any standard iterative procedure for the zero of R^* yields τ, the total time interval before relative rise or set. Successive zeros of (6.20) determine visibility windows for the satellite system.

6.2.4 The High Eccentricity Case

If the relative rise and set points of satellites with high eccentricity, $e > 0.2$, are desired, the reduction of the analysis to a single transcendental equation is not practical. In this case, for specified times t, for each satellite, Kepler's equation would be solved and the satellite position vectors and magnitudes determined, that is, $\mathbf{r}_i$, $i = 1, 2$. It is convenient at this point to use the rise-set function R defined by (6.12) instead of R^*, (6.20), to determine if the satellites are mutually visible.

6.3 LUNAR ECLIPSE OF A SATELLITE OF THE EARTH

Certain geometric eclipse constraints on nuclear detection, communication, and abort orbits deal with the selection of orbits that are not eclipsed by either the Earth or the Moon for specified periods of time. Usually the effect of lunar eclipsing is neglected. However, certain astronomical situations can occur which cause a satellite to be eclipsed by the Moon for considerable periods of time. This eclipse interval can in turn cause severe damage to orbiting satellites. The eclipsing of satellites by the Earth can be conveniently handled by the analysis developed in ([1], Chapter 5). Lunar eclipses of satellites are discussed in this section. With this objective in mind, exact expressions for geometric eclipsing are developed in the form of a shadow function. This shadow function can be transformed further into a quartic equation in the cosine of the true anomaly of the satellite or conic section relative to the Earth. The analysis developed herein is due to Escobal and Robertson [2].

6.3.1 Development of Lunar Shadow Function

Examination of Figure 6.3 allows the following vector equation, referenced to the geocentric right ascension-declination coordinate system [1], to be written:

$$\mathbf{r} + \mathbf{d} + \mathbf{a}_{☾} = \mathbf{r}_{☾}, \tag{6.21}$$

where $\mathbf{r} \equiv$ radius vector from Earth to satellite, $\mathbf{d} \equiv$ vector from orbit to the lunar surface, $\mathbf{a}_{☾} \equiv$ vector from the lunar surface to the center of the Moon, and $\mathbf{r}_{☾} \equiv$ radius vector from Earth to Moon.

The geometry of eclipse, however, requires that two conditions be imposed on the vector closure defined by (6.21), namely,

$$\mathbf{R}_{☾} \cdot \mathbf{a}_{☾} = 0 \tag{6.22}$$

and

$$\mathbf{R}_{☾} \cdot \mathbf{d} = R_{☾} d. \tag{6.23}$$

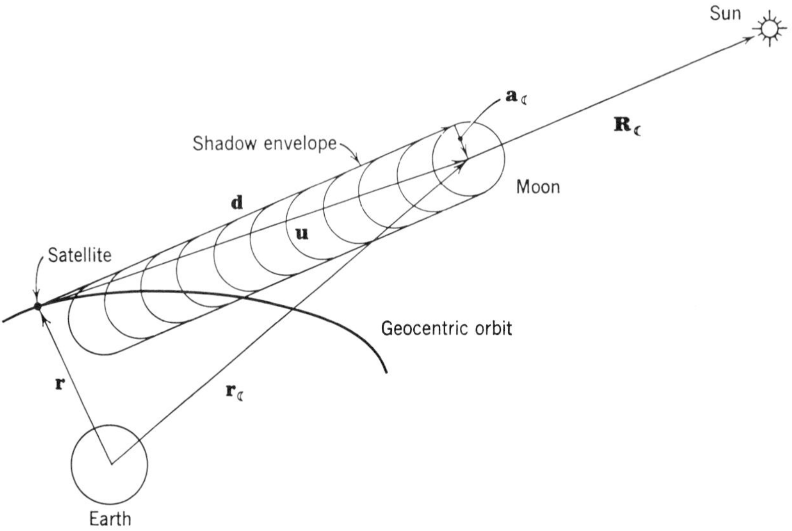

FIGURE 6.3 Orbit geometry.

Equation 6.22 requires that the twilight belt on the surface of the Moon be at right angles to $\mathbf{R}_{\leftmoon}$, the radius vector from the center of the Moon to the Sun. Furthermore, (6.23) requires that on entrance or exit from the cylinder generated by the Moon's shadow the **d** vector be parallel to the Sun's position vector. With these constraints and relationship 6.21, it follows from (6.22) that

$$\mathbf{R}_{\leftmoon} \cdot \mathbf{r}_{\leftmoon} - \mathbf{R}_{\leftmoon} \cdot \mathbf{r} = R_{\leftmoon} d. \tag{6.24}$$

For (6.24) to be of practical value, the unknown magnitude of **d** at the entrance and exit shadow points must be determined. From Figure 6.3 it follows that $\mathbf{u} \equiv \mathbf{r}_{\leftmoon} - \mathbf{r}$, or

$$u^2 = (\mathbf{r}_{\leftmoon} - \mathbf{r}) \cdot (\mathbf{r}_{\leftmoon} - \mathbf{r}),$$

but, since **d**, $\mathbf{a}_{\leftmoon}$, and **u** form a right triangle,

$$d^2 = (\mathbf{r}_{\leftmoon} - \mathbf{r}) \cdot (\mathbf{r}_{\leftmoon} - \mathbf{r}) - \mathbf{a}_{\leftmoon} \cdot \mathbf{a}_{\leftmoon}$$

or

$$d = [r_{\leftmoon}^2 + r^2 - 2\mathbf{r}_{\leftmoon} \cdot \mathbf{r} - \mathbf{a}_{\leftmoon}^2]^{1/2}. \tag{6.25}$$

Substituting (6.25) into (6.24) yields the lunar shadow function S,

$$S \equiv \mathbf{R}_{\leftmoon} \cdot \mathbf{r}_{\leftmoon} - \mathbf{R}_{\leftmoon} \cdot \mathbf{r} - R_{\leftmoon}[r_{\leftmoon}^2 + r^2 - a_{\leftmoon}^2 - 2\mathbf{r}_{\leftmoon} \cdot \mathbf{r}]^{1/2}, \tag{6.26}$$

where $S = 0$ implies either exit or entrance conditions.

6.3.2 Reduction of the Shadow Function to a Quartic Polynomial

The standard orientation vectors **P** and **Q**, where **P** is a unit vector from the dynamical center which points at perigee of the orbit and **Q** is advanced to **P** by a right angle in the plane and direction of motion, that is,

$$\begin{aligned} P_x &= \cos\omega\cos\Omega - \sin\omega\sin\Omega\cos i,\\ P_y &= \cos\omega\sin\Omega + \sin\omega\cos\Omega\cos i,\\ P_z &= \sin\omega\sin i,\\ Q_x &= -\sin\omega\cos\Omega - \cos\omega\sin\Omega\cos i,\\ Q_y &= -\sin\omega\sin\Omega + \cos\omega\cos\Omega\cos i,\\ Q_z &= \cos\omega\sin i, \end{aligned} \tag{6.27}$$

where $\omega \equiv$ the argument of perigee, $\Omega \equiv$ longitude of the ascending node, and $i \equiv$ orbit inclination, can be used to transform the shadow function as follows. Let the standard orbital mapping ([1], Chapter 3)

$$\mathbf{r} = x_\omega \mathbf{P} + y_\omega \mathbf{Q} \tag{6.28}$$

with x_ω, y_ω defined as functions of the radius vector magnitude r and true anomaly v through

$$x_\omega \equiv r\cos v, \qquad y_\omega \equiv r\sin v,$$

be introduced to simplify S. Hence, by direct substitution,

$$\begin{aligned} S \equiv{} & \mathbf{R}_☾ \cdot \mathbf{r}_☾ - \mathbf{R}_☾ \cdot \mathbf{P} r\cos v - \mathbf{R}_☾ \cdot \mathbf{Q} r \sin v \\ & - R_☾[r_☾^2 + r^2 - a_☾^2 - 2\mathbf{r}_☾ \cdot \mathbf{P} r\cos v - 2\mathbf{r}_☾ \cdot \mathbf{Q} r \sin v]^{1/2}. \end{aligned} \tag{6.29}$$

Furthermore, since the magnitude of the radius vector is given in terms of the orbital eccentricity e and semiparameter p by the well-known formula

$$r = \frac{p}{1 + e\cos v}, \tag{6.30}$$

it follows that

$$\begin{aligned} S \equiv{} & \mathbf{R}_☾ \cdot \mathbf{r}_☾ - \frac{\mathbf{R}_☾ \cdot \mathbf{P} p\cos v}{1 + e\cos v} - \frac{\mathbf{R}_☾ \cdot \mathbf{Q} p\sin v}{1 + e\cos v} \\ & - R_☾\left[r_☾^2 - a_☾^2 + \frac{p^2}{(1 + e\cos v)^2} - \frac{2\mathbf{r}_☾ \cdot \mathbf{P} p\cos v}{1 + e\cos v} - \frac{2\mathbf{r}_☾ \cdot \mathbf{Q} p\sin v}{1 + e\cos v}\right]^{1/2}. \end{aligned} \tag{6.31}$$

Functionally, by assuming that $\mathbf{R}_☾$ and $\mathbf{r}_☾$ are slowly varying functions of time which can be treated as constant vectors[1] and introducing the definitions

$$\beta_1 \equiv \mathbf{R}_☾ \cdot \mathbf{r}_☾, \qquad \beta_2 \equiv -\mathbf{R}_☾ \cdot \mathbf{P}p, \qquad \beta_3 \equiv -\mathbf{R}_☾ \cdot \mathbf{Q}p$$
$$\beta_4 \equiv (r_☾^2 - a_☾^2)R_☾^2, \qquad \beta_5 \equiv -2\mathbf{r}_☾ \cdot \mathbf{P}pR_☾^2, \qquad \beta_6 \equiv -2\mathbf{r}_☾ \cdot \mathbf{Q}pR_☾^2,$$
$$\beta_7 \equiv p^2R_☾^2,$$

the shadow function can be written as

$$S \equiv \beta_1 + \beta_2 \frac{\cos v}{1 + e\cos v} + \beta_3 \frac{\sin v}{1 + e\cos v}$$
$$- \left[\beta_4 + \beta_5 \frac{\cos v}{1 + e\cos v} + \beta_6 \frac{\sin v}{1 + e\cos v} + \frac{\beta_7}{(1 + e\cos v)^2}\right]^{1/2}. \tag{6.32}$$

Note that $S = S(v)$ and a double squaring of (6.32) yields the resolvent quartic of the lunar eclipse problem,

$$S^* \equiv a_0 \cos^4 v + a_1 \cos^3 v + a_2 \cos^2 v + a_3 \cos v + a_4, \tag{6.33}$$

where

$$a_0 \equiv \gamma_1^2 + \gamma_4^2, \qquad a_1 \equiv 2\gamma_1\gamma_2 + \gamma_5,$$
$$a_2 \equiv \gamma_2^2 + 2\gamma_1\gamma_3 - \gamma_4^2 + \gamma_6^2, \qquad a_3 \equiv 2\gamma_2\gamma_3 - \gamma_5, \qquad a_4 \equiv \gamma_3^3 - \gamma_6^2,$$

with the understanding that the a coefficients may be computed by means of the auxiliary variables

$$\gamma_1 \equiv (\beta_1^2 - \beta_4)e^2 + (2\beta_1\beta_2 - \beta_5)e + \beta_2^2 - \beta_3^2,$$
$$\gamma_2 \equiv 2(\beta_1^2 - \beta_4)e + 2\beta_1\beta_2 - \beta_5,$$
$$\gamma_3 \equiv \beta_1^2 + \beta_3^2 - \beta_4 - \beta_7,$$
$$\gamma_4 \equiv (2\beta_1\beta_3 - \beta_6)e + 2\beta_2\beta_3,$$
$$\gamma_5 \equiv 2(2\beta_1\beta_3 - \beta_6)[(2\beta_1\beta_3 - \beta_6)e + 2\beta_2\beta_3],$$
$$\gamma_6 \equiv 2\beta_1\beta_3 - \beta_6.$$

It should be noticed that $S^* = 0$, which can be solved in closed form ([1], Appendix 3), is the condition for orbital entrance or exit from the Moon's shadow. Furthermore, from the basic definition of S and (6.23), it follows that, for an orbit, S changes sign from negative to positive on entrance to the shadow.

This fact can best be established by examining a case in which satellite

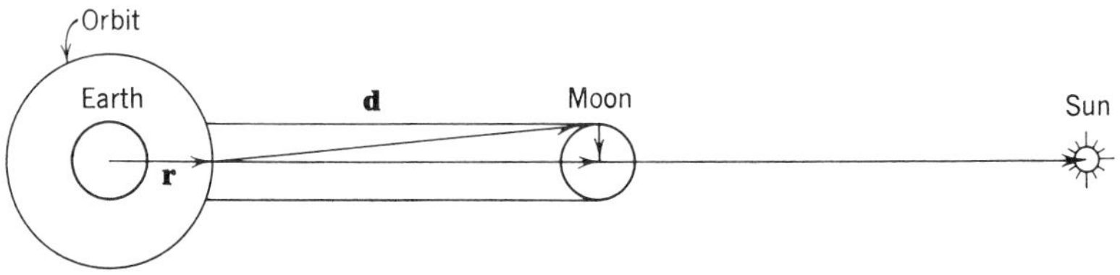

FIGURE 6.4 The satellite in a planar conjunction eclipse.

[1] See Section 6.3.4.

eclipsing is a certainty. Consider the case in which Earth, satellite, and Moon are in the same plane. As illustrated in Figure 6.4, let the satellite be at conjunction between the Earth and Moon and the Moon be at conjunction between the Earth and Sun. By construction, the satellite is eclipsed and it is possible to establish the sign of S in the following fashion. Since the shadow function is defined by (6.26), for the case of planar conjunction by evaluating the corresponding dot products, it is easy to show that

$$S = R_☾\{r_☾ - r - [(r_☾ - r)^2 - a_☾^2]^{1/2}\} > 0.$$

It therefore follows that S is positive if a satellite is eclipsed by the Moon and negative if no eclipse is evident. This statement could be established also by considering opposition class eclipses.

A final remark deals with the rejection of spurious roots associated with S^*. Since both opposition and conjunction eclipses can occur, the roots of S^* must be checked directly by substitution into S, (6.31), and only the roots

$$\cos v_i, \sin v_i = \pm[1 - \cos^2 v_i]^{1/2}, \qquad i = 1, 2, 3, 4, \tag{6.34}$$

which cause S to vanish accepted. Once these roots are known, the entrance and exit universal times can be determined more accurately, if desired, and the coefficients of the quartic updated. From this point on, the coefficients of the entrance quartic are different from the exit quartic; that is, two quartics must be carried forward in the analysis.

6.3.3 The Limiting Excursion of the Moon

The preceding analysis would not be performed if it could be proved a priori that a lunar eclipse of the geocentric satellite could not possibly occur. It is a simple exercise to show from Figure 6.5 that the limiting excursion of the Moon, the limiting angle a_l between the Moon's and the Sun's radius vectors, can be approximated by

$$a_l \simeq \tan^{-1}\left[\frac{(R_\odot - r_☾)a}{R_\odot r_☾}\right], \qquad 0 \leq a_l \leq \frac{\pi}{2}, \tag{6.35}$$

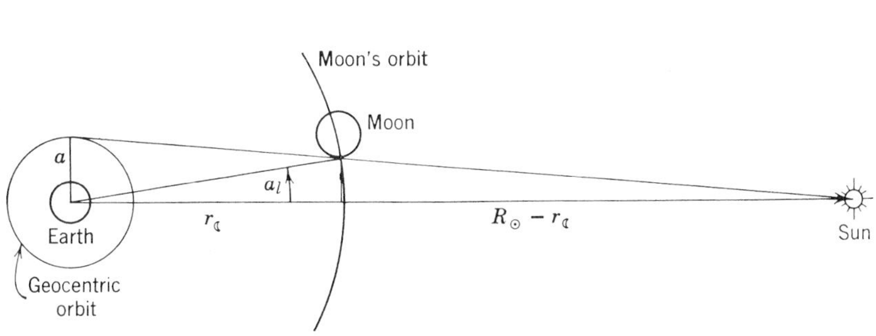

FIGURE 6.5 Maximum excursion of the Moon.

where a is the semimajor axis of the nearly circular geocentric orbit, $R_{\odot}$ is the distance between Earth and Sun, and $r_{☾}$ is the distance between Earth and Moon. Hence, assuming constant values for $R_{\odot}$, $r_{☾}$, and a rough circular orbit, the limiting a_l can be obtained. It is only when the Moon is in the vicinity of a_l that the previously developed quartic need be solved or the shadow function interrogated.

6.3.4 Time Variation of the Quartic Coefficients

In order to obtain a closed solution to the lunar eclipse problem it was assumed that over a length of time the lunar quartic coefficients β_i remain constant. The two main reasons for the time variation of these coefficients are due to the changes in $\mathbf{r}_{☾}$ and $\mathbf{R}_{☾}$, with the change in $\mathbf{r}_{☾}$ being most severe. However, if the quartic coefficients are evaluated at the time corresponding to a_l, the limiting excursion of the Moon, the approximation should be quite good. A second refinement may, however, be necessary and the quartic solved with updated coefficients.

The optimum strategy to compute lunar eclipses would probably utilize the shadow function S to provide a rough approximation to the point of actual eclipse via a stepping procedure. Once in the vicinity of an actual eclipse, the quartic can be solved in closed form as discussed in [1].

6.3.5 Umbra-Penumbra Effects

The preceding analysis has made the tacit assumption that the lunar shadow envelope illustrated in Figure 6.3 is a perfect cylinder. If umbra and penumbra effects [9] are of interest, however, it is fairly easy to modify the shadow function by means of the angles

$$\cos\delta_u = \frac{[R_{☾}^2 - (a_{\odot} - a_{☾})^2]^{1/2}}{R_{☾}}, \qquad \sin\delta_u = \frac{a_{\odot} - a_{☾}}{R_{☾}},$$

$$\cos\delta_p = \frac{[R_{☾}^2 - (a_{\odot} + a_{☾})^2]^{1/2}}{R_{☾}}, \qquad \sin\delta_p = \frac{a_{\odot} + a_{☾}}{R_{☾}}, \tag{6.36}$$

where $a_{\odot}$ is the radius of the Sun and the subscripts u and p denote umbra and penumbra, respectively. For the Earth-Moon system these angles are derived in ([1], Chapter 5). It therefore follows from the geometric considerations of Figure 6.6 that (6.22) and (6.24) must be modified. Equations 6.22 and 6.23, where the subscript i denotes umbra or penumbra conditions and the upper sign signifies penumbra entrance, now become

$$\mathbf{R}_{☾} \cdot (\mathbf{r}_{☾} - \mathbf{r} - \mathbf{d}) = \mp R_{☾} a_{☾} \sin\delta_i, \tag{6.37}$$

$$\mathbf{R}_{☾} \cdot \mathbf{r}_{☾} - \mathbf{R}_{☾} \cdot \mathbf{r} = R_{☾}[d\cos\delta_i \mp a_{☾} \sin\delta_i]. \tag{6.38}$$

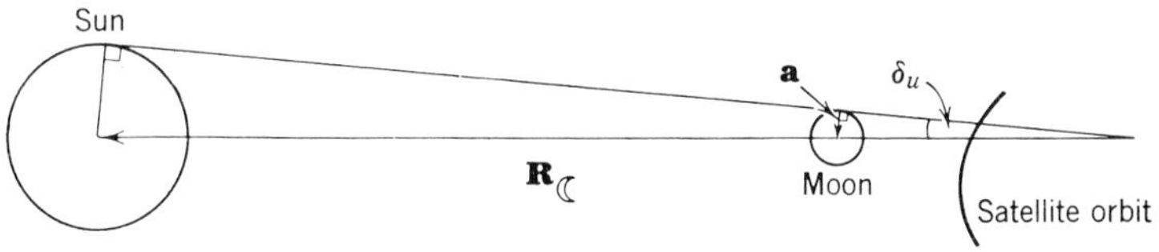

FIGURE 6.6 Geometry of an umbral eclipse.

The exact shadow function therefore should be modified to read

$$S \equiv \mathbf{R}_{\leftmoon} \cdot \mathbf{r}_{\leftmoon} - \mathbf{R}_{\leftmoon} \cdot \mathbf{r} - R_{\leftmoon}[(r_{\leftmoon}^2 + r^2 - a_{\leftmoon}^2 - 2\mathbf{r}_{\leftmoon} \cdot \mathbf{r})^{1/2} \cos \delta_i \mp a_{\leftmoon} \sin \delta_i]. \tag{6.39}$$

These umbra-penumbra effects can be absorbed into the β_j, $j = 1, 2, \ldots,$ 7, coefficients of (6.32) by letting

$$\begin{aligned}
\beta_1 &\equiv \mathbf{R}_{\leftmoon} \cdot \mathbf{r}_{\leftmoon} \pm R_{\leftmoon} a_{\leftmoon} \sin \delta_i, \\
\beta_4 &\equiv (r_{\leftmoon}^2 - a_{\leftmoon}^2) R_{\leftmoon}^2 \cos^2 \delta_i, \\
\beta_5 &\equiv -2\mathbf{r}_{\leftmoon} \cdot \mathbf{P} p R_{\leftmoon}^2 \cos^2 \delta_i, \\
\beta_6 &\equiv -2\mathbf{r}_{\leftmoon} \cdot \mathbf{Q} p R_{\leftmoon}^2 \cos^2 \delta_i, \\
\beta_7 &\equiv p^2 R_{\leftmoon}^2 \cos^2 \delta_i,
\end{aligned} \tag{6.40}$$

and retaining the previous values for β_2 and β_3.

A final remark on the umbra effects should be noted. It should be understood that a second test must be applied to the solutions of the umbra entrance-exit quartic in order to eliminate all spurious roots. Not only must the resulting trigonometric functions of true anomaly satisfy the shadow function, but also geometric considerations of total darkness require that the magnitude of $\mathbf{d}$ be no greater than $a_{\leftmoon}/\tan \delta_u$. It is, of course, assumed that, in every case where solutions of the corresponding entrance-exit quartic are to be checked for validity, $\mathbf{d} \cdot \mathbf{R}_{\leftmoon} \geq 0$. Otherwise no lunar eclipse could possibly occur.

6.3.6 Significance of Lunar Eclipses

An example of the numerical significance of lunar eclipses of geocentric satellites was obtained by examining the eclipse of the Sun which occurred over South America on November 12, 1966. A satellite in a nearly circular 60,000-nautical mile orbit with standard classical elements $e = 0.126$, $i = 36°.42$, $\Omega = 21°.15$, and $\omega = 114°.16$ was investigated [10]. The orientation of the orbital plane and the initial position of the spacecraft was chosen to produce an eclipse. Computations performed by means of the previous analysis yielded an umbral eclipse with duration 0.256 hr and a penumbral eclipse which lasted for 1.785 hr. Longer or shorter eclipses can, of course, occur as a function of the particular satellite elements.

6.4 RISE AND SET OF A SATELLITE OF THE MOON WITH RESPECT TO AN EARTH GROUND STATION

Future space missions, specifically the conquest of the Moon, as envisioned by the Apollo project, present special communications problems. This section investigates the optical or radar visibility of a satellite in an orbit about the Moon from a specified station on the Earth. The analysis is investigated from an astrodynamical point of view to yield a single transcendental equation in the eccentric anomaly of the lunicentered satellite which defines the viewing times.

This problem becomes rather complex and should be separated into two different but related conditions that ensure visibility of the lunar satellite. The first of these conditions, called here the critical viewing time condition, when the satellite is not obscured by the bulge of the Earth, is the problem attacked in this section. The second condition, the lunar occultation situation, when the satellite is hidden by the Moon, can be treated by a slight modification of eclipse theory as discussed in ([1], Chapter 5).[2] Both conditions must be satisfied in order to ensure a direct optical contact between the specified terrestrial station and the orbiting lunar satellite. The analysis developed in this section is due to Escobal [3].

6.4.1 Critical Viewing Time Condition

Consider the position of the Moon defined in the geocentric inertial coordinate system ([1], Chapter 4) by vector $\mathbf{r}_{☾}$, the position vector of the Moon's center with respect to the geocenter. The coordinates of $\mathbf{r}_{☾}$ can be obtained from Section 1.4 or 1.5. From the selenocenter the position vector of a lunar satellite, $\mathbf{r}_s$, can be introduced and the slant range vector $\boldsymbol{\rho}$ defined with the aid of Figure 6.7 as

$$\boldsymbol{\rho} = \mathbf{R} + \mathbf{r}_{☾} + \mathbf{r}_s. \tag{6.41}$$

Note that $\mathbf{R}$ is the station coordinate vector discussed in Section 6.1.1. Figure 6.8 affords further clarification of the vector construction.

[2] The occultation problem is also considered in Section 6.6.

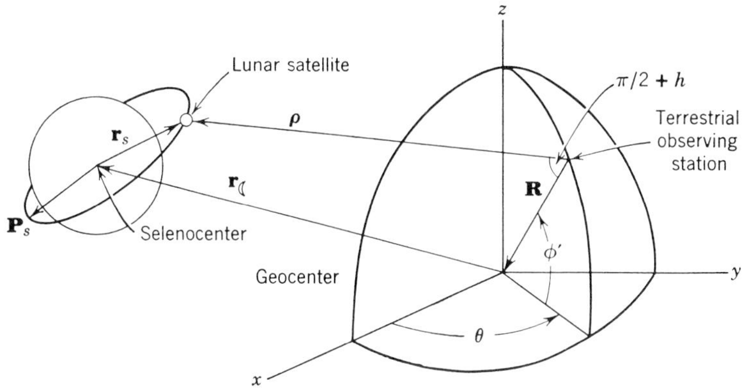

FIGURE 6.7 Orbit geometry.

From an observing station located at point A it is always possible to observe satellite S, if S is itself not occulted by the Moon. However, as the station moves into position A' because of the sidereal rotation of the Earth, a critical situation is reached. This critical situation of course is caused by the line-of-sight interference of the Earth, which has moved into a position between the lunar satellite and the observing station. For a spherical Earth the critical angle occurring at A' caused by line-or-sight interference is $\pi/2$, but for the sake of generality, and since the observing station may have defined observational constraints because of mountains, etc., let the critical angle be $\pi/2 + h$, where h is the minimum acceptable elevation angle of the

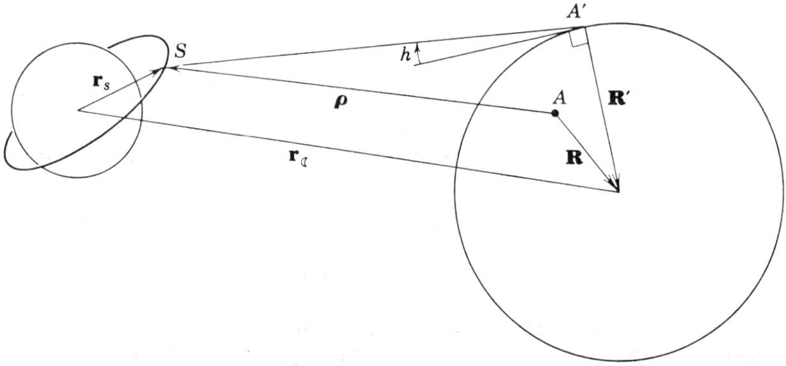

FIGURE 6.8 Constraining geometry.

observing station. For an oblate planet, h must be augmented by the angle ζ, where[3]

$$\zeta = \cos^{-1}\left[\frac{G_1 \cos^2 \phi + G_2 \sin^2 \phi}{R}\right], \qquad 0 \le \zeta \le \frac{\pi}{2}. \tag{6.42}$$

In light of the preceding discussion the critical situation occurs when

$$\boldsymbol{\rho} \cdot \mathbf{R} = \rho R \cos\left(\frac{\pi}{2} + h\right) = -\rho R \sin h. \tag{6.43}$$

Substituting for $\boldsymbol{\rho}$ from (6.41) results in the relation

$$(\mathbf{R} + \mathbf{r}_{☾} + \mathbf{r}_s) \cdot \mathbf{R} = -\rho R \sin h \tag{6.44}$$

or

$$(\mathbf{r}_{☾} + \mathbf{r}_s) \cdot \mathbf{R} = -R^2 - \rho R \sin h. \tag{6.45}$$

6.4.2 Reduction of Critical Viewing Condition to a One-Parameter Function

Let the two unit vectors $\mathbf{P}_s$, $\mathbf{Q}_s$ be introduced; $\mathbf{P}_s$ is a vector emanating from the selenocenter which points toward periapsis of the lunicentered orbit and $\mathbf{Q}_s$ is advanced to $\mathbf{P}_s$ by a right angle in the plane and direction of motion. Furthermore, if the lunicentered orbit is defined by the elements a_s, e_s, i_s, Ω_s, ω_s, and T_s, where a_s is the semimajor axis of the lunicentered vehicle, e_s is the eccentricity of the lunicentered vehicle, i_s is the orbital inclination of the lunicentered vehicle with respect to the Earth's equatorial plane, Ω_s is the longitude of the ascending node referred to the intersection of the planes of motion of the vehicle and the Earth's equator, ω_s is the argument of perilune referred to the intersection of the planes of motion of the vehicle and the Earth's equator, and T_s is the time of perifocal passage of the lunicentered vehicle, then $\mathbf{P}_s$ and $\mathbf{Q}_s$ are given by (6.27) as

$$\mathbf{P}_s = \mathbf{P}(\omega_s, \Omega_s, i_s), \qquad \mathbf{Q}_s = \mathbf{Q}(\omega_s, \Omega_s, i_s). \tag{6.46}$$

It should be noted that i, ω, Ω could be referred to the equatorial plane of the Moon and then rotated at the lunicenter to yield i_s, ω_s, and Ω_s, that is, the orientation angles of the lunicentered orbit referred to the fundamental plane of the Earth.

The $\mathbf{P}_s$ and $\mathbf{Q}_s$ unit vectors permit the following relation to be written [1]:

$$\mathbf{r}_s = x_{\omega s}\mathbf{P}_s + y_{\omega s}\mathbf{Q}_s, \tag{6.47}$$

[3] The angle ζ, as given by (6.42) and derived in [1], represents an actual upper bound which assumes that the space vehicle will always be visible. Indeed, for a given station, if $h > \zeta$ then h would be used in the computations. On the other hand, if no value of h is specified or $\zeta > h$, then ζ is the critical parameter and it would be used in place of h in the forthcoming analysis.

where

$$x_{\omega s} = a_s(\cos E - e_s), \qquad y_{\omega s} = a_s(1 - e_s^{\,2})^{1/2} \sin E$$

with $E =$ the eccentric anomaly of the lunicentered vehicle. Equation 6.47 is a mapping from two space to three space, and it expresses the rectangular coordinates of the lunicentered vehicle in the geocentric coordinate frame illustrated in Figure 6.7. By consequence of this mapping, (6.45) can be written as

$$\{\mathbf{r}_{\text{☾}} + a_s(\cos E - e_s)\mathbf{P}_s + a_s[(1 - e_s^{\,2})^{1/2} \sin E]\mathbf{Q}_s\} \cdot \mathbf{R} = -R^2 - \rho R \sin h. \tag{6.48}$$

Under the assumption that $\mathbf{r}_{\text{☾}}$ does not change by an appreciable amount in a period of about 12 hours, the approximate maximum upper time bound on the lunicentered satellite's time of observation, (6.48) reduces to a function of E and θ. The sidereal time θ enters through $\mathbf{R}$ by virtue of (6.1). The slight variation in $\mathbf{r}_{\text{☾}}$ is treated a little later in the analysis. To eliminate the time dependency of the $\mathbf{R}$ vector, consider the introduction of Kepler's equation,

$$t = \left[\frac{E - e_s \sin E}{n_s}\right] + T_s, \tag{6.49}$$

along with the exact relationship

$$\theta = \theta_0 + \dot{\theta}(t - t_0), \tag{6.50}$$

where t is the universal time, n_s is the mean motion $= k_{\text{☾}}(\mu)^{1/2} a_s^{-3/2}$, $k_{\text{☾}}$ is the gravitational constant of the Moon, μ is the sum of the masses of the lunicentered vehicle and Moon, θ_0 is the epoch local sidereal time, t_0 is the universal time corresponding to θ_0, and $\dot{\theta}$ is the sidereal rate of change, a constant. Evidently, utilizing (6.49) and (6.50), it is possible to define the local sidereal time θ in (6.1) as $\hat{\theta}$, where

$$\theta \equiv \hat{\theta} = \theta_0 + (T_s - t_0)\dot{\theta} + \frac{\dot{\theta}}{n_s}(E - e_s \sin E), \tag{6.51}$$

so that, as a function of E, (6.1) becomes

$$\hat{\mathbf{R}} = \begin{bmatrix} \hat{X} \\ \hat{Y} \\ \hat{Z} \end{bmatrix} = \begin{bmatrix} -G_1 \cos \phi \cos \hat{\theta} \\ -G_1 \cos \phi \sin \hat{\theta} \\ -G_2 \sin \phi \end{bmatrix}. \tag{6.52}$$

These are the components of vector $\hat{\mathbf{R}}$, which is equivalent to $\mathbf{R}$, but is a function of E instead of θ. The controlling equation of the critical viewing time regime, (6.48), therefore can be expanded as

$$F \equiv \{x_☾ + a_s(\cos E - e_s)P_{xs} + a_s[(1 - e^2)^{1/2} \sin E]Q_{xs}\} \cos \phi \cos \hat{\theta}$$

$$+ \{y_☾ + a_s(\cos E - e_s)P_{ys} + a_s[(1 - e^2)^{1/2} \sin E]Q_{ys}\} \cos \phi \sin \hat{\theta}$$

$$+ \{z_☾ + a_s(\cos E - e_s)P_{zs} + a_s[(1 - e^2)^{1/2} \sin E]Q_{zs}\} \frac{G_2}{G_1} \sin \phi$$

$$- \frac{1}{G_1} (\rho R \sin h + R^2). \tag{6.53}$$

The F function, (6.53), is the sought-for transcendental equation in the eccentric anomaly of the lunar satellite. Since geometrically an obtuse angle exists between $\boldsymbol{\rho}$ and $\mathbf{R}$ when the satellite is visible, it follows that $F < 0$ is associated with the visibility condition. Hence, as F varies from negative to positive, the satellite and Moon are setting with respect to the terrestrial station. A change of F from positive to negative, in opposite fashion, characterizes the satellite-Moon system as rising with respect to the Earth station. As the lunar satellite progresses through its period in eccentric anomaly, that is, 0 to 2π, the F function changes sign every few periods.

Equation 6.53 which defines F, is fully rigorous with the exception of the Keplerian approximation dominating the motion of the lunar satellite[4]. Keplerian motion for a lunar satellite is in effect a good approximation because the usual secular drift rates are not very prominent. Solution of (6.53) is accomplished by a direct search technique in which the condition $F = 0$ is the precise visibility condition. Hence a starting value of E is chosen and (6.49) is employed to yield the universal time. An interpolation into a suitable ephemeris yields $\mathbf{r}_☾$, and (6.50), which defines the local sidereal time, can be used to compute $\mathbf{R}$. Equation 6.41 immediately yields $\boldsymbol{\rho}$. A check is now made to see if $F = 0$; if it does, a critical condition has been found; if $F \neq 0$, increments are added to E and the procedure repeated. As soon as the critical eccentric anomalies have been determined, $\mathbf{r}_{☾i}$ for $i = 1, 2$ can be recalculated, and a more correct value can be used in (6.53). In passing, it is well to note that, when $h \neq 0$, (6.53) becomes even more functionally dependent on E since $\boldsymbol{\rho}$ must be obtained from $\mathbf{R}$, $\mathbf{r}_☾$, and $\mathbf{r}_s$; that is,

$$\rho = (R^2 + r_☾^2 + r_s^2 + 2\mathbf{R} \cdot \mathbf{r}_☾ + 2\mathbf{R} \cdot \mathbf{r}_s + 2\mathbf{r}_☾ \cdot \mathbf{r}_s)^{1/2}. \tag{6.54}$$

Once the rise eccentric anomaly (or the set eccentric anomaly) of the satellite-Moon system has been found, it is not necessary to continue the point-by-point critical viewing time regime check for the next x revolutions. In essence, since the period of the lunar satellite is known, a lower bound x

[4] If desired, the secular variation in the orbital elements can be included with relative ease.

can be found easily, where x is the number of revolutions from, for example, rise of the satellite-Moon system to the revolution before set of the satellite-Moon system. It is at this point that the checking procedure is reinitiated.

In closing, it should be evident that, if $\mathbf{r}_s$ is taken to be zero, that is, if the lunar satellite radius vector is neglected, (6.53) will yield the rise and set time ([1], Chapter 5) of the center of the Moon with respect to a terrestrial station.

6.5 RISE AND SET OF AN INTERPLANETARY SPACE VEHICLE

Tracking and communication with manned and unmanned interplanetary vehicles or any other interplanetary object require exact knowledge of available communication windows. The accessibility of such windows involves three respective motions, the diurnal rotation of the Earth about its axis, the Earth's motion about the Sun, and the space vehicle's motion along its heliocentric orbit. In addition, the physical features of the Earth ground station, such as meridian oblateness and elevation above the geoid, must be taken into account. To solve the astrodynamical problem of interplanetary rise and set times, that is, the determination of the times at which a space vehicle is visible from an observing Earth ground station, a rise-set function is developed. The rise-set function defines the range of times when the vehicle is visible or nonvisible. Furthermore, it answer's the question of space vehicle visibility by means of an examination of the sign of the function. The analysis in this section was performed by Escobal and Affatati [4].

6.5.1 Development of the Interplanetary Rise-Set Function

In the heliocentric system $(x_\epsilon, y_\epsilon, z_\epsilon)$ the geometry of Figure 6.9 allows the following vector equation to be written:

$$\mathbf{R}_\oplus - \mathbf{R}_\epsilon + \boldsymbol{\rho}_\epsilon - \mathbf{r}_\epsilon = \mathbf{0}, \tag{6.55}$$

where $\mathbf{R}_\oplus$ is the heliocentric position vector of the Earth, $\mathbf{R}_\epsilon$ is the geocentric position vector of the observing station, $\boldsymbol{\rho}_\epsilon$ is the topocentric slant range vector, and $\mathbf{r}_\epsilon$ is the heliocentric position vector of the space vehicle.

In the geocentric right ascension-declination coordinate system (x, y, z) the station coordinate vector $\mathbf{R}$ is expressed by (6.1). The vector $\mathbf{R}$, referenced to the ecliptic coordinate system, can be obtained by a rotation through the angle ϵ, the obliquity of the ecliptic, as follows:

$$\mathbf{R}_\epsilon = \begin{bmatrix} X_\epsilon \\ Y_\epsilon \\ Z_\epsilon \end{bmatrix} = \begin{bmatrix} X \\ Y \cos \epsilon + Z \sin \epsilon \\ -Y \sin \epsilon + Z \cos \epsilon \end{bmatrix}. \tag{6.56}$$

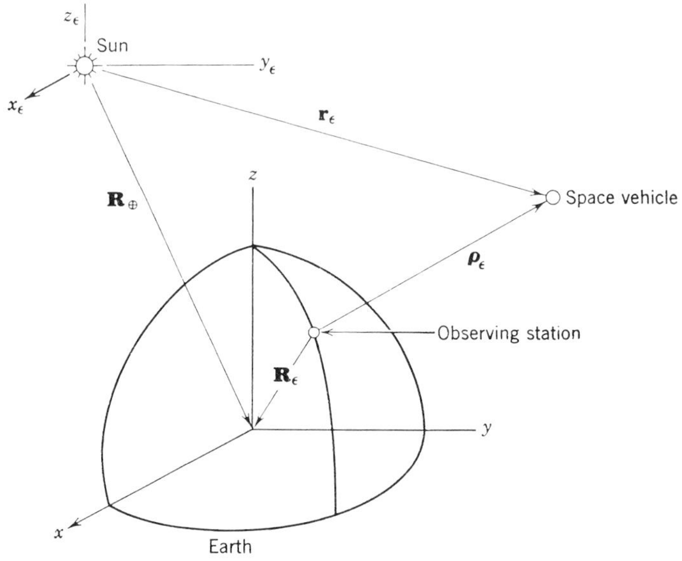

FIGURE 6.9 Heliocentric rise-set geometry.

Therefore, combining (6.1) and (6.56),

$$\mathbf{R}_\epsilon = \begin{bmatrix} -G_1 \cos\phi \cos\theta \\ -G_1 \cos\phi \cos\epsilon \sin\theta - G_2 \sin\phi \sin\epsilon \\ +G_1 \cos\phi \sin\epsilon \sin\theta - G_2 \sin\phi \cos\epsilon \end{bmatrix}. \tag{6.57}$$

More compactly, by introducing the definitions

$$S_0 \equiv -G_1 \cos\phi, \qquad S_1 \equiv -G_1 \cos\phi \cos\epsilon,$$
$$S_2 \equiv -G_2 \sin\phi \sin\epsilon, \qquad S_3 \equiv G_1 \cos\phi \sin\epsilon,$$
$$S_4 \equiv -G_2 \sin\phi \cos\epsilon, \tag{6.58}$$

it is possible to write (6.57) as

$$\mathbf{R}_\epsilon = \begin{bmatrix} S_0 \cos\theta \\ S_1 \sin\theta + S_2 \\ S_3 \sin\theta + S_4 \end{bmatrix}. \tag{6.59}$$

Under the assumption that $\epsilon \simeq 23°.45$ is nonvarying, the S_i coefficients can be evaluated from the station location data and treated as constants.

With the preceding understanding it is possible to return to (6.55) and obtain

$$\boldsymbol{\rho}_\epsilon = \mathbf{r}_\epsilon + \mathbf{R}_\epsilon - \mathbf{R}_\oplus, \tag{6.60}$$

$$\rho_\epsilon^2 = r_\epsilon^2 + R_\epsilon^2 + R_\oplus^2 + 2\mathbf{r}_\epsilon \cdot \mathbf{R}_\epsilon - 2\mathbf{r}_\epsilon \cdot \mathbf{R}_\oplus - 2\mathbf{R}_\epsilon \cdot \mathbf{R}_\oplus. \tag{6.61}$$

The vector relationship defined by (6.60) can now be used to obtain the rise-set function by imposing the condition

$$\mathbf{R} \cdot \boldsymbol{\rho} = R\rho \cos\left(\frac{\pi}{2} + h\right) = -R\rho \sin h \tag{6.62}$$

relative to the Earth-centered frame. This condition is geometrically evident from Figure 6.10. In (6.62) h is the minimum allowable elevation angle of

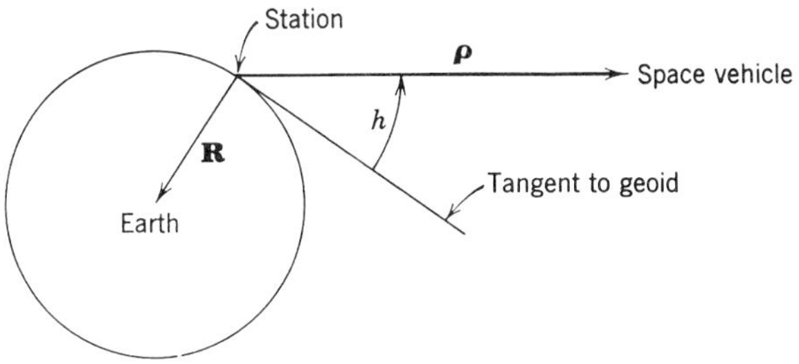

FIGURE 6.10 Constraining geometry.

the space vehicle with respect to the station. Actually, if h is the true minimum allowable elevation angle, to compensate for the geometric oblateness of the Earth this angle should be augmented by ζ where[5]

$$\zeta = \cos^{-1}\left(\frac{G_1 \cos^2 \phi + G_2 \sin^2 \phi}{R}\right), \qquad 0 \leq \zeta \leq \frac{\pi}{2}.$$

If it is noticed that scalar products are preserved under any transformation, (6.62) can be written in the equivalent fashion as

$$\mathbf{R}_\epsilon \cdot \boldsymbol{\rho}_\epsilon = -R\rho \sin h. \tag{6.63}$$

It is now possible to utilize (6.60) to yield the interplanetary rise set function F as

$$F \equiv \mathbf{R}_\epsilon \cdot (\mathbf{r}_\epsilon + \mathbf{R}_\epsilon - \mathbf{R}_\oplus) + R_\epsilon[r_\epsilon^2 + R_\epsilon^2 + R_\oplus^2 + 2\mathbf{r}_\epsilon \cdot \mathbf{R}_\epsilon - 2\mathbf{r}_\epsilon \cdot \mathbf{R}_\oplus - 2\mathbf{R}_\epsilon \cdot \mathbf{R}_\oplus]^{1/2} \sin h. \tag{6.64}$$

[5] See footnote 3.

The sign of F associated with visibility can be obtained by constructing a case in which the space vehicle is visible by definition. Consider the geometric construction illustrated in Figure 6.11. When the dot products in (6.64) are evaluated, it follows that

$$F = R_\epsilon(r_\epsilon + R_\epsilon - R_\oplus) + R_\epsilon[(r_\epsilon + R_\epsilon - R_\oplus)^2]^{1/2} \sin h$$

$$= R_\epsilon(r_\epsilon + R_\epsilon - R_\oplus)(1 + \sin h) < 0; \quad (6.65)$$

hence a negative rise-set function implies that the space vehicle is visible and a positive rise-set function denotes a condition of no visibility.

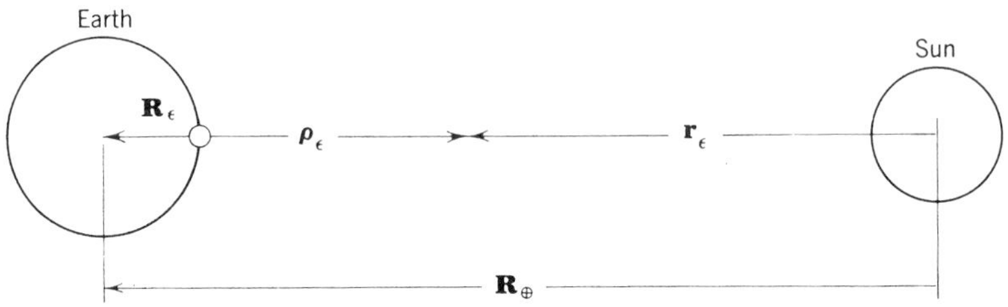

FIGURE 6.11 Space vehicle in conjunction.

6.5.2 Quartic Solution of the Rise-Set Function

Although the sign of F answers the question of visibility at any time, if the rise-set times are desired the F function must be solved for its critical zeros. This is most easily accomplished by treating F as a strict function of the eccentric anomaly E of the interplanetary vehicle. By considering a typical interplanetary transfer orbit, that is, a Hohmann transfer with a total eccentric anomaly change of 180° and an interplanetary trip time of approximately 200 days, it can be shown that the rate of change of E is a fraction of a degree per day. For any typical interplanetary flight the rise and set of the space vehicle occur within a degree of any epoch value of the eccentric anomaly. This indicates a high frequency variation of F with respect to E. Because of this high frequency oscillation in F an iteration technique to solve for the zeros of F would present a severe problem in the step size selection of E and would therefore require a lengthy iteration to determine the critical visibility points. To avoid these problems an approximate closed form solution to the F function is sought by the realistic assumption that the vehicle and Earth motions about the Sun are negligible for one diurnal revolution of the Earth.

Since the primary sinusoidal shape of F is due to the rotation of the Earth, that is, the orbital variations of the space vehicle cause a very slow variance in the amplitude of F, consider letting $\mathbf{r}_\epsilon$ and $\mathbf{R}_\oplus$ assume constant values at E_0, the eccentric anomaly corresponding to the local interrogation time. Functionally, since $F = F(\mathbf{R}_\epsilon, \mathbf{R}_\oplus, \mathbf{r}_\epsilon)$, the previously stated reasoning reduces the rise-set function to the tractable form

$$\bar{F} = F(\mathbf{R}_\epsilon, \mathbf{R}_\oplus^*, \mathbf{r}_\epsilon^*) = F(\sin\theta, \cos\theta),$$

where the superscript signifies that the vectors are held constant at their E_0 values. The explicit evaluation of F can be attained by use of (6.59) so that (6.64) becomes

$$F \approx s_1 \cos\theta + s_2 \sin\theta + s_3 + s_4[s_5 + s_1 \cos\theta + s_2 \sin\theta]^{1/2}, \qquad (6.66)$$

where

$$s_1 \equiv S_0(x_\epsilon - X_\oplus),$$

$$s_2 \equiv S_1(y_\epsilon - Y_\oplus) + S_3 z_\epsilon,$$

$$s_3 \equiv S_2(y_\epsilon - Y_\oplus) + S_4 z_\epsilon + R_\epsilon^2,$$

$$s_4 \equiv R_\epsilon \sin h,$$

$$s_5 \equiv r_\epsilon^2 + R_\oplus^2 + R_\epsilon^2 + 2S_4 z_\epsilon + 2S_2(y_\epsilon - Y_\oplus) - 2\mathbf{r}_\epsilon \cdot \mathbf{R}_\oplus.$$

It should be understood that the s coefficients vary very slowly with time in comparison to θ. Basically, if the s_i are assumed to be constant, this is equivalent to holding the vehicle and the center of the Earth fixed in inertial space. Setting (6.66) equal to zero and squaring twice yields

$$a_0 \cos^4\theta + a_1 \cos^3\theta + a_2 \cos^2\theta + a_3 \cos\theta + a_4 = 0 \qquad (6.67)$$

with the understanding that the transformation of algebraic forms is given via the auxiliary coefficients

$$\omega_0 \equiv s_1^2 - s_2^2,$$

$$\omega_1 \equiv 2s_1(s_3 - s_4^2),$$

$$\omega_2 \equiv s_2^2 + s_3^2 - s_5 s_4^2,$$

$$\omega_3 \equiv -2s_1 s_2,$$

$$\omega_4 \equiv -2s_2(s_3 - s_4^2),$$

and the direct coefficients

$$a_0 \equiv \omega_0^2 + \omega_3^2,$$

$$a_1 \equiv 2(\omega_0\omega_1 + \omega_3\omega_4),$$

$$a_2 \equiv \omega_1^2 - \omega_3^2 + \omega_4^2 + 2\omega_0\omega_2,$$

$$a_3 \equiv 2(\omega_1\omega_2 - \omega_3\omega_4),$$

$$a_4 \equiv \omega_2^2 - \omega_4^2.$$

Hence, given the interrogation eccentric anomaly E_0, the a_i can be directly computed and the quartic solved in closed form [1] for all real roots in the interval $-1 \leq \cos\theta_i \leq 1$. The rejection of spurious roots can be accomplished by substituting

$$\cos\theta_i,\ \sin\theta_i = \pm[1 - \cos\theta_i]^{1/2}, \qquad i = 1, 2, \ldots, r,$$

directly into (6.66) and eliminating all roots[6] such that $F \neq 0$. Once the unique $\cos\theta_i$, $\sin\theta_i$ have been obtained, the rise-set θ_i are known and the corresponding eccentric anomalies can be obtained as outlined in the next section.

6.5.3 Linearization of Kepler's Equation

Over a small range of the eccentric anomaly E about E_0 it is possible to expand Kepler's equation [1], that is,

$$M = E - e\sin E, \tag{6.68}$$

where $M \equiv n(t - T)$, n is the mean motion of the space vehicle, T is the time of perihelion passage, and e is the orbital eccentricity, by means of a Taylor expansion as follows:

$$M = M_0 + M_0'(E - E_0) + \frac{M_0''(E - E_0)^2}{2} + \cdots \tag{6.69}$$

with M_0' obtained from (6.68) as

$$M_0' = 1 - e\cos E_0. \tag{6.70}$$

By retaining two terms, the linearized version of Kepler's equation becomes

$$t = \alpha + \beta E, \tag{6.71}$$

where

$$\alpha \equiv t_0 + \frac{E_0}{n}(e\cos E_0 - 1), \qquad \beta \equiv \frac{1 - e\cos E_0}{n}.$$

Equation 6.71 is valid only over a very small variation of E, for example, about 3 degrees. For purposes of solution of the rise-set function, for local values of E corresponding to a rise or set this variation is more than sufficient. For extreme accuracy, quadratic terms could be carried if desired.

Proceeding formally, it follows that the variation of the sidereal time as a function of E can be obtained through the exact relation

$$\theta = \theta_g + \dot\theta(t - t_g) - \lambda_W, \tag{6.72}$$

[6] In a machine routine the possible combinations of "$\cos\theta_i$ and $\sin\theta_i$" can be substituted into F, and the so-called best zeros of F, that is, the zeros corresponding to the minimums of the absolute values of F, are chosen as the true roots of (6.67). This overcomes inherent numerical difficulties.

where θ_g is the epoch sidereal time at Greenwich, $\dot{\theta}$ is the constant sidereal rate of change, λ_W is the west longitude of the observing station, and t is the universal time. Hence, by use of (6.71),

$$\theta = \alpha^* + \beta^* E, \tag{6.73}$$

where

$$\alpha^* \equiv \theta_g - \lambda_W + \dot{\theta}(\alpha - t_g), \qquad \beta^* \equiv \dot{\theta}\beta.$$

The linear relation (6.73) can be used to obtain E_i from θ_i without the iterative solution of Kepler's equation. At this point, if a refinement of the rise-set times is desired, the E_i can be used to update $\mathbf{r}_\epsilon$, the t_i can be used to update $\mathbf{R}_\oplus$, and the a_j coefficients of (6.67) recomputed for each E_i. This operation results in two distinct quartics for the rise and set eccentric anomaly. It should be stated that in most cases the secondary quartics are not actually necessary to obtain suitable values of t_i. Furthermore, just before or after this refinement, depending on the desired accuracy, the slant range from Earth station to vehicle, see (6.61), should be computed and the light time correction (Section 6.1.2) obtained.

6.5.4 Numerical Results

Figure 6.12 depicts the variation of the F function around $E = 46°$ for a typical interplanetary flight from Earth to Jupiter as viewed from a tracking station at Maui, Hawaii. The pertinent orbital and ground station features that were used to obtain the F function variation are listed in Tables 6.1 and 6.2, respectively.

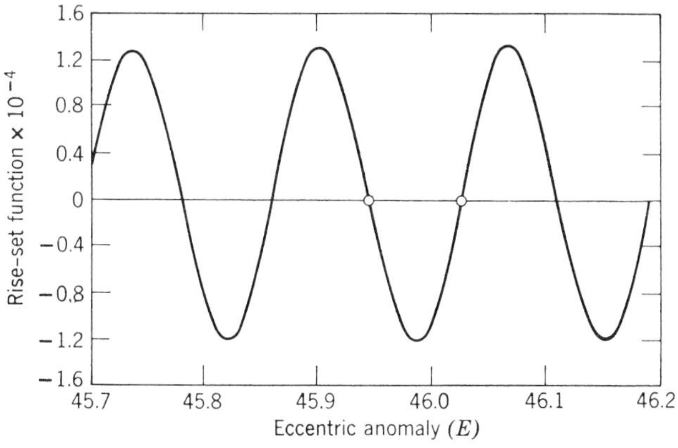

FIGURE 6.12 Rise-set function variance.

Table 6.1 Earth to Jupiter Flight

Characteristic	Numerical Value
Launch date	Julian date 2444188.0 or November 10, 1979
Trip time	542 Julian days
Central transfer angle	2.45834896 radians
Arrival date	Julian date 2444730.0
Semimajor axis	5.8440988 a.u.
Eccentricity	0.83063826
Inclination to ecliptic	0.03604020 radian
Longitude of the ascending node	0.82899855 radian
Argument of perihelion	6.23689151 radian
Time of perihelion passage	Julian date 2444186.040777

Figure 6.12 illustrates the actual rise-set function and the closed form analytic solution, that is, rise and set times of the quartic around $E_0 = 46°$. These solutions are indicated by heavy dots.

A comparison of the roots computed after solving and resolving the quartic with the actual rise-set times is listed in Table 6.3. Examination of

Table 6.2 Maui, Hawaii, Radar Site

East longitude	203°44′23″.40
Latitude	20°42′36″.00 N
Elevation	9997.440 ft
Elevation angle (approximate)	0.08726646 radian

the previous numerical results indicates that the quartic resolvent of the rise-set problem provides sufficient accuracy on the first iteration. With this understanding the analysis for the rise-set times can be accomplished in closed form.

Table 6.3 Rise Set-Times

		Closed Form Solution	
	Iterative Solution	First Solution	Second Solution
Rise time	45.992382	45.992383	45.992382
Set time	46.0724110	46.0724106	46.0724110

6.6 RISE AND SET OF A SATELLITE ABOUT A DISTANT PRIMARY

The problem of determining the rise-set times of a vehicle in orbit about a distant primary (the times at which a vehicle has direct line-of-sight contact with respect to a specified ground station) involves four distinct motions.

They are the rotation of the Earth about its axis, the orbital translation of the Earth, the orbital translation of the distinct primary, and the orbital movement of the satellite about the distant primary. Analytically it is possible to develop a single algebraic function whose zeros characterize either a rise or a set of the satellite with respect to a given Earth ground station. However, if this approach is taken, it is virtually impossible to obtain a closed solution of this type of astrodynamical problem. To obtain a closed solution to such a problem, the problem itself must be considered in two parts. A little thought will reveal that one part of the total problem involves the rise and set of the distant primary, for example, the planet Mars, with respect to a specified Earth ground station, and the other part involves the occulation of the satellite (the departure of the orbiting satellite behind the distant primary). In this section the rise and the set of the distant primary and the occultation of the satellite are discussed. It is shown that under suitable approximations both of these problems can be solved in closed form by the solution of quartic equations. Once each of these segments of the total problem is solved, it is demonstrated that, by overlaying the solutions, the total rise-set problem, that is, the interaction, can be treated by superposition. The rise-set analysis of the distant primary is treated first, the occultation analysis second, and finally the interaction is discussed. Rise-set analysis of this type was developed by Escobal, Stern, and Affatati [5].

6.6.1 Rise and Set Times of a Planet Relative to an Earth Ground Station

Section 6.5 treats the problem of the rise and set of an interplanetary body with respect to an Earth ground station. From the geometry displayed in Figure 6.13 a relationship can be developed to determine whether line-of-sight

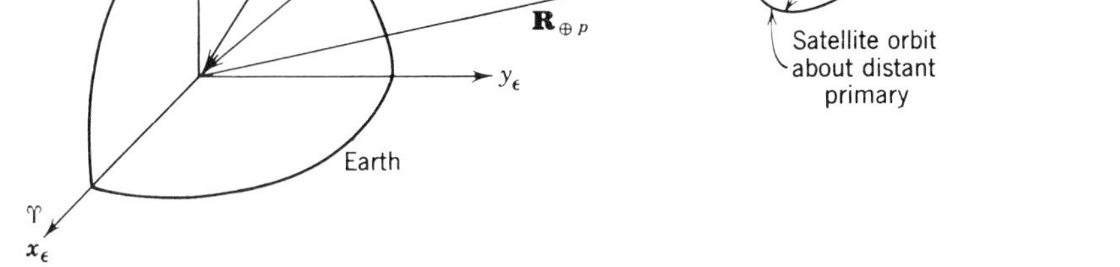

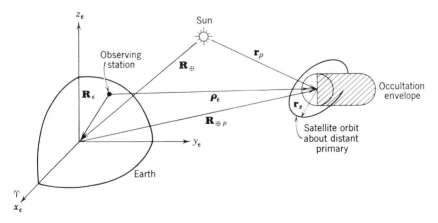

FIGURE 6.13 Geometry of station, Sun, and satellite about a distant primary.

visibility is possible. This relationship was developed in the preceding section by use of an elevation constraint which was described by the equation

$$\mathbf{R}_\epsilon \cdot \boldsymbol{\rho}_\epsilon = -R\rho \sin h, \tag{6.74}$$

where $\mathbf{R}_\epsilon$ is that station coordinate vector, $\boldsymbol{\rho}_\epsilon$ is the line-of-sight vector to the distant primary, and h is the minimum allowable elevation angle.[7] By employing the vector geometry displayed in Figure 6.13 the following vector equation, referenced to the ecliptic, becomes obvious:

$$\boldsymbol{\rho}_\epsilon = \mathbf{r}_p + \mathbf{R}_\epsilon - \mathbf{R}_\oplus, \tag{6.75}$$

where $\mathbf{r}_p$ is the heliocentric position vector of the distant primary (Section 1.3) and $\mathbf{R}_\oplus$ is the heliocentric position vector of the Earth. Substitution of (6.75) into (6.74) yields the interplanetary rise-set function F as

$$F \equiv \mathbf{R}_\epsilon \cdot (\mathbf{r}_p + \mathbf{R}_\epsilon - \mathbf{R}_\oplus) + R_\epsilon[r_p{}^2 + R_\epsilon{}^2 + R_\oplus{}^2 + 2\mathbf{r}_p \cdot \mathbf{R}_\epsilon$$
$$-2\mathbf{r}_p \cdot \mathbf{R}_\oplus - 2\mathbf{R}_\epsilon \cdot \mathbf{R}_\oplus]^{1/2} \sin h. \tag{6.76}$$

The function defined by (6.76) accounts for the Earth's motion about the Sun, the Earth's rotation about its axis, and the orbital motion of the distant primary. It was shown in Section 6.5 that a negative rise-set function implies that an interplanetary body is visible from the Earth ground station of interest, while a positive value of F denotes a condition of no visibility. It was also demonstrated that for suitable but realistic assumptions the rise-set times can be obtained in closed form. This is accomplished by assuming the Earth's position vector $\mathbf{R}_\oplus$ and the objective planet vector $\mathbf{r}_p$ to be invariant with respect to time during a diurnal rotation of the Earth about its axis.[8] With these assumptions a quartic polynomial in the cosine of the sidereal time θ can be obtained. This polynomial is given by

$$a_0 \cos^4 \theta + a_1 \cos^3 \theta + a_2 \cos^2 \theta + a_3 \cos \theta + a_4 = 0, \tag{6.77}$$

where the coefficients are defined by (6.67). Under the assumption that the a_j, $j = 0, 1, \ldots, 4$ are constants, (6.77) can be solved for its roots, namely the rise-set sidereal times. To find the respective planetary eccentric anomalies associated with the roots θ_k of (6.77), Kepler's equation [1] can be linearized, as in Section 6.5.3, and the following relation obtained:

$$E_k = \frac{\theta_k - \alpha^*}{\beta^*}, \tag{6.78}$$

[7] See footnote 3.

[8] In this section the position vector of the objective planet, r_p, is substituted for the position vector of the vehicle.

where

$$\alpha^* = \theta_g - \lambda_W + \dot{\theta}\left[t_0 - t_g + \frac{E_0}{n_p}(e_p \cos E_0 - 1)\right]$$

$$\beta^* = \dot{\theta}\left[\frac{1 - e_p \cos E_0}{n_p}\right],$$

with the understanding that θ_g is the epoch sidereal time at Greenwich, $\dot{\theta}$ is the constant sidereal rate of change, λ_W is the west longitude of the observing station, t is the universal time, n_p is the mean motion of the distant primary, e_p is the orbital eccentricity of the distant primary, E_0 is the epoch eccentric anomaly about which Kepler's equation is linearized, and t_0 is the time associated with E_0.

The linear relationship (6.78) can be used to obtain E_k from θ_k without the iterative solution of Kepler's equation. At this point, if a refinement of the rise-set times is desired, E_k can be used to update $\mathbf{r}_p$, the t_k associated with E_k can be used to update $\mathbf{R}_\oplus$, and the a_j coefficients of (6.77) can be recomputed for each E_k. This operation results in two distinct quartics for the rise and set eccentric anomalies. The numerical data obtained in Section 6.5.4 indicate that secondary quartics are not actually necessary to obtain sufficiently accurate values of E_k. By utilizing this method the rise-set function of Mars, as illustrated in Figure 6.14, was obtained for a ground

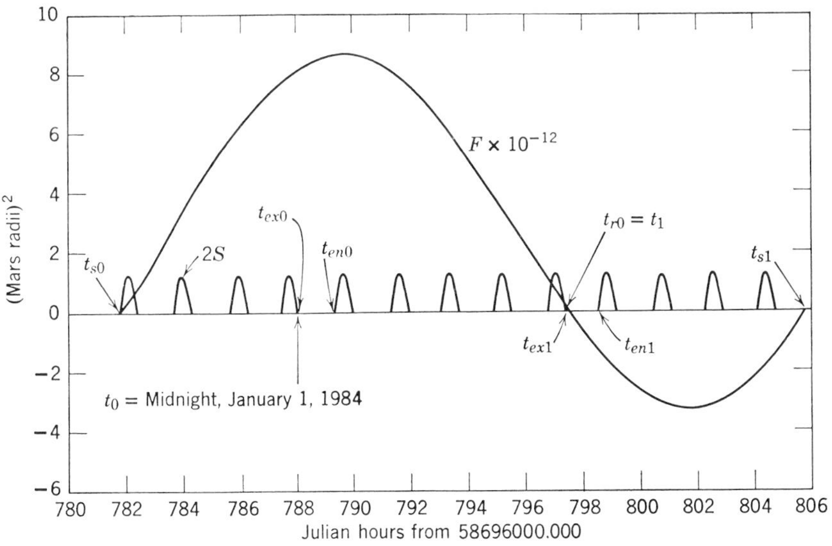

FIGURE 6.14 Visibility of Martian satellite.

station located at Maui, Hawaii. The oscillatory function F in Figure 6.14 shows the approximate rise and set times of the planet Mars for January 1, 1984. Other points of interest peculiar to Figure 6.14 will be discussed presently.

6.6.2 Occultation of Satellite by Distant Primary

Closely associated with interference of the line-of-sight visibility of the planetocentric satellite by rotation of the observing station on the Earth is the process of satellite occultation due to interference of the primary ([1], Chapter 5). As illustrated in Figure 6.13, this condition is caused by the rotation of the satellite behind the primary which controls the planetocentric orbital motion. Therefore, as the satellite enters the shaded domain, depicted in Figure 6.13, line-of-sight visibility is not possible. The process of entering the shaded occultation cylinder bears a one-to-one correspondence to the satellite eclipse geometry wherein the Sun is producing an actual physical shadow behind the planetary body. It should be evident from the geometry of Figure 6.13 that, if the Sun were shifted to the location of the observing ground station, the occultation problem would be transformed into a problem in the eclipse theory. Since the eclipse problem has been shown to be solvable in closed form [1] by means of the solution of a quartic equation in the true anomaly of the planetocentric satellite, it appears feasible to transform the occultation problem into a fictitious eclipse problem.

If it is understood that nearly all the contribution to $\boldsymbol{\rho}_\epsilon$ (6.75), comes from $\mathbf{r}_p$ and $\mathbf{R}_\oplus$, it is possible to determine numerically $\boldsymbol{\rho}_s$ at the time of planetary rise from Section 6.6.1 and assume that this vector is not varying with time for, say, one rise-set cycle of the distant primary[9]. With this understanding, since the planetocentric orbital elements i, Ω, and ω are assumed to be known, it is possible to calculate, from the standard expressions given by (6.27), the orientation vectors $\mathbf{P}$ and $\mathbf{Q}$, where $\mathbf{P}$ is a unit vector emanating from the dynamical center which points at periapsis of the planetocentric orbit and $\mathbf{Q}$ is advanced from $\mathbf{P}$ by a right angle in the plane and direction of motion.

Making the usual assumption that ϵ, the obliquity of the ecliptic, is constant, the $\mathbf{P}$ and $\mathbf{Q}$ vectors referred to the ecliptic [1] are given directly by

$$\begin{aligned} P_{x\epsilon} &= P_x, & Q_{x\epsilon} &= Q_x, \\ P_{y\epsilon} &= P_y \cos\epsilon + P_z \sin\epsilon, & Q_{y\epsilon} &= Q_y \cos\epsilon + Q_z \sin\epsilon, \\ P_{z\epsilon} &= -P_y \sin\epsilon + P_z \cos\epsilon, & Q_{z\epsilon} &= -Q_y \sin\epsilon + Q_z \cos\epsilon. \end{aligned} \tag{6.79}$$

[9] The subscript s is used to signify shadow. Actually the vector $\boldsymbol{\rho}_s$ is equivalent to $\boldsymbol{\rho}_\epsilon$ except that it does not include the effect of $\mathbf{R}_\epsilon$.

The artificial shadow (see Figure 6.13) that is utilized to calculate the occultation times of the satellite is assumed to be cylindrical with radius equal to the radius of the distant planet, and generated by a vector equal to the vector emanating from the topocentric observing station and directed toward the dynamical center of the distant planet. The auxiliary quantities ξ and β utilized to determine the occultation times are extracted from ([1], Chapter 5) as follows:

$$\beta = -\frac{(\mathbf{R}_{\oplus p} + \mathbf{R}_\epsilon)\cdot \mathbf{P}_\epsilon}{\rho_\epsilon} = -\frac{\boldsymbol{\rho}_\epsilon \cdot \mathbf{P}_\epsilon}{\rho_\epsilon}, \tag{6.80}$$

$$\xi = -\frac{(\mathbf{R}_{\oplus p} + \mathbf{R}_\epsilon)\cdot \mathbf{Q}_\epsilon}{\rho_\epsilon} = -\frac{\boldsymbol{\rho}_\epsilon \cdot \mathbf{Q}_\epsilon}{\rho_\epsilon}, \tag{6.81}$$

where

$$\rho_\epsilon^{\,2} = R_{\oplus p}^{\;2} + R_\epsilon^{\,2} + 2\mathbf{R}_\epsilon \cdot \mathbf{R}_{\oplus p}.$$

Even at the position of closest approach between Mars and Earth, the effect of the station vector $\mathbf{R}_\epsilon$ on the preceding calculations is extremely small because $R_\epsilon \ll \rho_\epsilon$. Consequently the generator of the artificial shadow, to a high degree of accuracy, can be taken as the $\mathbf{R}_{\oplus p}$ vector, that is, $\boldsymbol{\rho}_\epsilon \simeq \mathbf{R}_{\oplus p}$.

In strict analogy with eclipse theory ([1], Chapter 5) it is now possible to form the A_l coefficients in terms of the known planetocentric orbital parameters a, p, and e, where a is the semimajor axis, e is the orbital eccentricity, $p = a(1 - e^2)$, with a_p equal to the radius of the objective planet, as

$$A_0 = \left[\left(\frac{a_p}{p}\right)^4 e^4 - 2\left(\frac{a_p}{p}\right)^2(\xi^2 - \beta^2)e^2 + (\beta^2 + \xi^2)^2\right],$$

$$A_1 = \left[4\left(\frac{a_p}{p}\right)^4 e^3 - 4\left(\frac{a_p}{p}\right)^2(\xi^2 - \beta^2)e\right],$$

$$A_2 = \left[6\left(\frac{a_p}{p}\right)^4 e^2 - 2\left(\frac{a_p}{p}\right)^2(\xi^2 - \beta^2) - 2\left(\frac{a_p}{p}\right)^2(1 - \xi^2)e^2 + 2(\xi^2 - \beta^2)(1 - \xi^2) - 4\beta^2\xi^2\right],$$

$$A_3 = \left[4\left(\frac{a_p}{p}\right)^4 e - 4\left(\frac{a_p}{p}\right)^2(1 - \xi^2)e\right],$$

$$A_4 = \left[\left(\frac{a_p}{p}\right)^4 - 2\left(\frac{a_p}{p}\right)^2(1 - \xi^2) + (1 - \xi^2)^2\right].$$

It now follows from eclipse theory that the solution of

$$A_0 \cos^4 v + A_1 \cos^3 v + A_2 \cos^2 v + A_3 \cos v + A_4 = 0 \tag{6.82}$$

yields the occultation true anomalies, where only solutions that satisfy

$$\beta \cos v + \xi \sin v < 0 \tag{6.83}$$

and

$$S \equiv a_p^2(1 + e \cos v)^2 + p^2(\beta \cos v + \xi \sin v)^2 - p^2 = 0 \tag{6.84}$$

are of physical interest. It should also be noted from eclipse theory that, if a satellite is entering the occultation region, S must change sign from minus to plus. Exit from the occultation region, in opposite fashion, is characterized by S, changing sign from plus to minus.

The coefficients A_l of the quartic, (6.82), depend on ξ and β. These coefficients are in turn dependent on $\boldsymbol{\rho}_\epsilon$, which is a time-varying quantity. This complication can be overcome by noting that the variation of $\boldsymbol{\rho}_\epsilon$ is much less than the variation of $\mathbf{r}_s$. Consequently the quartic can be solved by first choosing $\boldsymbol{\rho}_\epsilon$ at the time of planetary rise and then updating $\boldsymbol{\rho}_\epsilon$ on the basis of the calculated occultation times. If extreme precision is desired, this iterative procedure can be repeated several times until the desired accuracy is obtained.

6.6.3 Interaction of the Effects of Planetary Rise and Set and Occultation

As has been noted previously, the problem of visibility of a planetocentric satellite depends on the combined effects of line-of-sight visibility as determined by the effect of Earth rotation on the observation station, and of occultation, caused by the distant primary. The prediction of each of these effects has been shown to be governed by a quartic polynomial equation. These two quartic polynomials can be coupled to establish a single equation for the determination of the times of visibility subject to the combined effect of the Earth rotation and occultation. Nevertheless the resulting equation is somewhat difficult to handle numerically. This fact suggests that perhaps the simplest approach to the combined problem is to solve each quartic polynomial individually, and then superimpose the two solutions, retaining only those times for which the satellite is neither occulted nor blocked from the line of sight because of the rotation of the observation station.

The primary objective of the visibility calculations is threefold: First, to determine if the satellite is visible at some specified interrogation time $t = t_0$; second, in the event of visibility, to determine the duration D of the visibility; and third, in the event of nonvisibility, to determine the delay time from $t = t_0$ until visibility is first attained. The logic associated with the calculations above is presented next. To begin the interaction logic, the occultation quartic can be solved in closed form to determine the local times of entrance t_{en0} and exit t_{ex0} of the artificial shadow function about the initial interrogation time $t = t_0$. Similarly, the line-of-sight quartic polynomial which accounts for the effect of the Earth rotation can be solved next to determine the local rise time t_{r0} and set times t_{s0} of the distant primary about the initial interrogation time $t = t_0$. Tests are next made to determine if the satellite is visible at $t = t_0$. If

$$t_0 < t_{s0} \quad \text{and} \quad t_0 < t_{en0}, \tag{6.85}$$

the satellite is visible. In this case the duration of the visibility, D_0, is given by

$$D_0 = \min\{t_{s0}, t_{en0}\} - t_0. \tag{6.86}$$

If the satellite is not visible at $t = t_0$, the calculations continue by computation of a new interrogation time,

$$t_1 = \max\{t_{ex0}, t_{r0}\}. \tag{6.87}$$

The visibility criterion at $t = t_1$ is similar to (6.85), that is,

$$t_1 < t_{s1}, \qquad t_1 < t_{en1},$$

where the subscript unity quantities refer to the local solutions of the two quartic polynomials about the interrogation time $t = t_1$. If the satellite is visible at $t = t_1$, the duration of visibility can be determined from (6.86) with the proper subscripts, namely,

$$D_1 = \min\{t_{s1}, t_{en1}\} - t_1. \tag{6.88}$$

In the event of nonvisibility at $t = t_1$ the process is repeated at $t = t_2$ and continued m times, if necessary, until visibility is attained. The expression for the nth interrogation time is given by

$$t_n = \max\{t_{ex(n-1)}, t_{r(n-1)}\}, \qquad n = 1, 2, 3, \ldots, m, \tag{6.89}$$

where the necessary condition for visibility at $t = t_n$ is

$$t_n < t_{s(n)}, \qquad t_n < t_{en(n)}. \tag{6.90}$$

The associated duration of visibility can now be taken as

$$D_n = \min\{t_{s(n)}, t_{e(n)}\} - t_n, \tag{6.91}$$

and the delay time before first visibility is simply given by

$$T_n = t_n - t_0. \tag{6.92}$$

6.6.4 Analysis of a Martian Satellite

As a specific example of the use of the previous results, an investigation was carried out to determine the visibility regions associated with a Martian satellite. The circular orbit of the Martian satellite was described by the orbital elements displayed in Table 6.4. The topocentric observation station

Table 6.4 Orbital Data

Semimajor axis, a	1.2000 Mars radii
Eccentricity, e	0.0000
Inclination, i	10.000°
Longitude of ascending node, Ω	200.00°

Time is measured from the point of intersection of the orbit with the Martian equator at midnight, January 1, 1984.

considered here is located at Maui, Hawaii; its data are collected in Table 6.5.

Figure 6.14 presents the results for both the line-of-sight visibility of Mars and the occultation of the Martian satellite. The oscillatory curve of F, scaled down by 10^{+12}, represents the visibility of the planet Mars with respect to the ground station. When $F < 0$, Mars is visible. Also displayed in Figure 6.14 are the regions of S, scaled up by a factor of 2, where the satellite is occulted. The satellite is occulted when (6.83) is satisfied and $S > 0$. The region of S for which both of these conditions hold is depicted in Figure 6.14.

Table 6.5 Maui, Hawaii, Radar Site

East longitude, λ_E	203°44′23″.40
Latitude, ϕ	20°42′36″.00 N
Elevation, H	9997.440 ft
Elevation angle (approximate), h	0.08726646 radian

Visibility calculations were carried out for an interrogation time $t_0 =$ midnight, January 1, 1984. The satellite is not visible at that date because $F > 0$. The calculations proceed with the determination of the subsequent interrogation time t_1, which is found to be $t_1 = t_0 + 9.50$ hr. The visibility criterion is satisfied at $t = t_1$; consequently the delay time to visibility is simply $T_1 = t_1 - t_0 = 9.50$ hr. Finally duration of visibility is next calculated as $D_1 = 1.151$ hr.

6.7 SUMMARY

Geocentric, lunar, and interplanetary communications analysis was discussed in this chapter from an astrodynamical point of view. The geocentric communications phase was de-emphasized to a certain degree because geocentric rise-set and eclipse studies have been discussed in detail in ([1], Chapter 5).

Relative rise-set times of two satellites about a central planet were discussed, and the analysis was developed in detail. It was shown how a rise-set function could be developed as a function of two parameters which would predict gain or loss of direct line of sight, as a result of central planet interference, for two planetocentric satellites. For low eccentricity satellites it was shown how the rise-set function could be reduced further to a one-parameter function whose zeros predict the critical viewing windows.

Eclipses of geocentric satellites by the shadow of the Moon were discussed next. An appropriate shadow function was developed whose sign indicated whether the Earth-centered trajectory was eclipsed or noneclipsed. By an appropriate algebraic transformation it was possible to develop a quartic equation, whose solution can be effected in closed form, to predict the actual shadow entrance and exit true anomalies. The shadow quartic was modified to permit the inclusion of umbra-penumbra effects.

Next the problem of the rise-set times of a satellite of the Moon relative to an Earth ground station was investigated. The rise-set function concept was utilized to yield a transcendental one-parameter function whose zeros yield the rise-set of the lunar satellite. It was indicated that, by shrinking the radius of the lunicentered orbit to zero, the rise and set times of the Moon could also be predicted.

The astrodynamical problem concerned with the visibility of a noneclipsed interplanetary vehicle from an Earth ground station was undertaken next. An interplanetary rise-set function was developed in order to provide a compact method of determining whether line-of-sight visibility is possible. The rise-set function is exact and includes the effects due to the geometric flattening of the Earth. It was shown that a negative value of the oscillatory rise-set function implies space vehicle visibility. The rise-set function accounts for the movement of the Earth about the Sun, the rotation of the Earth about its axis, and the motion of the space vehicle. It was shown that an accurate solution to the rise-set function is possible in closed form via suitable assumptions. The approximate solution reduces the problem of determining the rise-set times to the solution of a quartic equation. If further refinement is desired, the quartic can be solved again to provide high accuracy rise-set times. The quartic solution can also be used to provide an estimate of when the interrogation of the exact rise-set function should be made in the future. Owing to the construction of the rise-set function, it can be used to determine the relative visibility of any celestial object moving in a heliocentric orbit.

Finally, interplanetary communication analysis of a satellite about a distant planet in contact with an Earth ground station was investigated. It was shown that the determination of interplanetary line-of-sight analysis is a composite problem solvable by means of superposition. Two distinct problems must be analyzed in order to determine the rise and set of a satellite in orbit about an objective planet. These problems, specifically the rise-set of the planet with respect to an Earth ground station and the occultation of the satellite by the distant primary, can be solved in closed form under suitable approximations. For each of these problems a quartic equation can be derived which, on solution, yields the visibility window peculiar to each distinct problem. By superposition the composite visibility window can then be obtained. Some graphical and numerical results were displayed to test the theory.

EXERCISES

1. What are the three basic assumptions of communications analysis?
2. If Mars is in direct opposition to Earth, how long does it take light to travel from Mars to Earth? Use Table 1.1.
3. If the relative rise-set function R^* is never possessed of zeros, are the satellites visible or not visible?
4. Two geocentric satellites with orbital elements

$$a = 1.05 \text{ e.r.}, \qquad a = 1.2 \text{ e.r.},$$
$$e = 0.001, \qquad e = 0.02,$$
$$M_0 = 30.00°, \qquad M_0 = 120.00°,$$

   and reference epoch time t_0 = midnight, June 1, 1975, are to transmit a signal to each other for 30 minutes. Can they transmit? For what length of time is transmission possible or not possible? Assume the satellites are in the same plane and let $\Delta = 0.03$ e.r.
5. For geocentric satellites, what orientation of the satellite orbit yields the longest possible eclipse? Assume Keplerian motion in the analysis.
6. Devise a scheme which enables the longest possible lunar eclipse of a circular Earth satellite to be determined. Assume that the S function is known and the orbital elements of the satellite are fixed.
7. Establish that $S > 0$ for opposition class lunar eclipses.
8. Verify the validity of (6.35). Approximately how many degrees is a_l for an orbit with $a = 6$ e.r.?
9. What is the meaning and usefulness of the angle ξ, that is

$$\zeta = \cos^{-1} \frac{(G_1 \cos^2 \phi + G_2 \sin^2 \phi)}{R}, \qquad 0 \leq \zeta \leq \frac{\pi}{2}?$$

   Derive the formula.
10. Develop a quadratic approximation to Kepler's equation that could be substituted for (6.71). Why does the linearization procedure actually yield such accurate results?
11. In occultation and eclipse theory, why must

$$\beta \cos v + \xi \sin v < 0?$$

12. Prove that, if a satellite is entering the region of occultation, (6.84) must change sign from minus to plus.

13. Derive an algorithm that permits determination of whether or not the general quartic, that is,

$$a_0x^4 + a_1x^3 + a_2x^2 + a_3x + a_4 = 0,$$

has real or imaginary roots before solving the equation. (*Hint*: See [13].)

REFERENCES

[1] P. R. Escobal, *Methods of Orbit Determination*, John Wiley and Sons, New York, 1965.

[2] P. R. Escobal and R. A. Robertson, "Lunar Eclipse of a Satellite of the Earth," *Journal of Spacecraft and Rockets*, April 1967.

[3] P. R. Escobal, "Visibility of a Lunar Satellite from an Earth Ground Station," *AIAA Journal*, April 1965.

[4] P. R. Escobal and D. A. Affatati, "Rise and Set of An Interplanetary Space Vehicle," *Journal of the Astronautical Sciences*, July 1967.

[5] P. R. Escobal, G. S. Stern, and D. A. Affatati, "Rise and Set of a Satellite about a Distant Primary," *Journal of Spacecraft and Rockets*, November 1967.

[6] F. R. Moulton, *An Introduction to Celestial Mechanics*, Macmillan Company, New York, 1914.

[7] G. A. Crichton and H. L. Roth, *Relative Visibility between Orbiting Vehicles*, TRW Systems, 9892.3-176, November 1965.

[8] P. R. Escobal, "The Rise and Set of One Satellite with Respect to Another," TRW Systems, 3422.3-125, September 1965.

[9] T. R. Oppolzer, *Cannon of Eclipses*, Dover Publications, New York, 1962.

[10] R. R. Williams, *Lunar Eclipse Calculations for an Orbiting Spacecraft*, TRW Systems 05069-6021-R000, January 1967.

[11] *Explanatory Supplement to the Astronautical Ephemeris and the American Ephemeris and Nautical Almanac*, Her Majesty's Stationery Office, 1961.

[12] R. M. L. Baker, Jr., and M. W. Makemson, *An Introduction to Astrodynamics*, Academic Press, New York, 1960.

[13] W. S. Burnside and A. W. Panton, *The Theory of Equations*, Dover Publications, New York, 1960.

[14] I. B. Hart, *Makers of Science*, Oxford University Press, London, 1923.

7 Special Perturbations

With me everything turns into mathematics.

DESCARTES [10]

7.1 DEFINITION OF THE SPECIAL PERTURBATIONS PROCEDURE

7.1.1 Special, General, and Hybrid Perturbations

The term *special perturbations* is the name applied to a numerical procedure used in the determination of orbits. In one form or another this technique involves the numerical integration of the equations of motion of a body subject to complex perturbations. Special perturbation methods differ from *general perturbation* methods in that in the former techniques the accelerations of a vehicle or body are integrated numerically, whereas in the latter techniques series expansions of fundamental orbital elements are obtained and no numerical integration is required. A third technique presently in use is called the method of *hybrid perturbations*. In hybrid perturbations simultaneous use is made of special and general perturbations procedures in order to arrive at an optimal scheme for the computation of the future or past state of a celestial object. In this chapter several methods of special and hybrid computations are discussed: the method of Cowell, the method of Encke, and the hybrid patched conic technique.

7.1.2 Basic Procedures

The basic procedures of special perturbations are concerned with the generation of the next step or increment of the state variables of a given body under the influence of complex perturbations; that is,

$$\mathbf{q} = \mathbf{q}_0 + \int_{t=t_0}^{t=t_0+\Delta t} \left(\frac{d\mathbf{q}}{dt}\right) dt, \tag{7.1}$$

where $\mathbf{q}$ is state vector of the body under investigation and Δt is the step increment of the independent variable t. Usually the acceleration of the body under study is computed directly from the differential equations of motion, and a double integration by means of standard numerical integration formulas [16], for example, Gauss-Jackson, Runge-Kutta, Adams, is used to determine the position of the body. It is assumed that, once the procedure for taking a single step is known, the process can be repeated n times until the final desired state $\mathbf{q}_f$ at t_f has been reached.

The basic concept which characterizes special perturbation methods is a step-by-step procedure in which the next step always requires complete knowledge of the preceding step before the process can be continued. Inherent in this piecemeal stepping procedure is the inducement of errors due to the numerical nature of the process. Hence errors in roundoff and truncation result in an accumulated loss of significant figures. For this reason special perturbation procedures are not very applicable to long range prediction of the motion of a body. The interval of applicability of such procedures can be increased only at the expense of carrying larger and larger numbers of significant figures. In contrast, general perturbation procedures, owing to the closed form nature of the solutions, are not affected by such error buildup but are difficult to implement. Here, again, as in many physical problems, general and special perturbations have good and bad points which should be given careful consideration before the actual physical problem is attacked. In the case of special perturbations it can be stated generally that the main advantages of such techniques are simplicity of formulation, ability to include very complex perturbative influences, and compactness of storage location in modern electronic calculating machines.

As may be inferred from the preceding discussion, special perturbation techniques are most readily applicable to the inclusion of very complex and subtle perturbative influences. When, for example, it is desired to include or determine the effect of the planet Pluto on a space trajectory between the Earth and Mars, or to include the forces occasioned by solar radiation, it is mandatory to include all other realistic effects whose order of magnitude may be greater than those considered.

In this context the selection of coordinate systems in which the integration of the equations of motion is to be made is an extremely important point of consideration. Formulation of a meaningful integration program which is to be used for accurate space targeting and ephemeris generation usually requires the inclusion of the nutation and precession; that is, it must be determined whether the integration is to be handled in the mean of epoch, true of date, or mean of date coordinate system, etc. These coordinate systems are discussed in Chapter 8. Symbolically, using the analysis and algorithms to be developed in Chapter 8, it follows that, given initial conditions, for example, the position and velocity vectors $(\mathbf{r}, \dot{\mathbf{r}})$ of the body under study, it is possible to obtain the transformation

$$(\mathbf{r}, \dot{\mathbf{r}}) \rightarrow (\mathbf{r}, \dot{\mathbf{r}})_{\text{c.s.}} \tag{7.2}$$

which gives the position and velocity vectors referred to the appropriate coordinate system denoted by the subscript c.s. Usually, since the coordinates of perturbing planets[1] are required as inputs to special perturbation procedures, and since these coordinates are referred to standard coordinate systems and epochs, the initial conditions are transformed into such a system. Once this initial transformation has been made, the special perturbation procedures discussed here can be initiated.

7.1.3 The Perturbative Function

In special perturbation analysis it is usually convenient to refer the motion of the body under study to the body which induces the most predominant gravitational influence. With this understanding it is possible to adopt the relative form of the equations of motion[2] developed in ([1], Chapter 2), that is,

$$\frac{d^2\mathbf{r}}{dt^2} = -k^2(m_1 + m_2)\frac{\mathbf{r}}{r^3} + k^2\sum_{j=3}^{n} m_j\left(\frac{\mathbf{r}_{2j}}{r_{2j}^{\,3}} - \frac{\mathbf{r}_{1j}}{r_{1j}^{\,3}}\right) + \boldsymbol{\Sigma}_{r2} - \boldsymbol{\Sigma}_{r1}, \tag{7.3}$$

where the following associations are made.

- k The Gaussian or planetary constant. In this case a natural choice of k would be the k of the body to which motion is referred. Then it should be remembered that masses are measured as the ratio of each mass to the mass of the body to which motion is referred.
- m_j With k chosen as above, this is the mass ratio of any mass to the mass of the body to which motion is referred, that is, normalized masses denoted by $m_j \doteq m_j/m_1$.
- n Number of bodies in the system.
- 1 Refers to the central body (origin).
- 2 Refers to the body under consideration, for example, the space vehicle.
- j Refers to the existing jth body in the system not counting body 1 or 2.
- $\mathbf{r}_{2j}$ The distances between the jth body and the body under consideration.
- $\mathbf{r}_{1j}$ The distances between the jth body and the central body (origin).

[1] See [11], [12] for sources of planetary ephemeral data. Also see Sections 1.3, 1.4 and 1.5.

[2] At times it may be convenient to refer the body to the barycenter of all the bodies under consideration in a given system. This is not a conceptual difference, and the barycentric form of the equations of motion would then be used.

Notice that the directed distance of vector $\mathbf{r}_{2j}$ may be readily obtained if it is realized that

$$\mathbf{r}_{2j} = \mathbf{r}_j - \mathbf{r}_2 = \mathbf{r}_{1j} - \mathbf{r}_{12} = \mathbf{r}_{1j} - \mathbf{r}. \tag{7.4}$$

As discussed at length in ([1], Section 2.8), the quantity $\mathbf{\Sigma}_{r2} - \mathbf{\Sigma}_{r1}$ is the additional acceleration of other forces not of a gravitational nature. More explicitly,

$$\mathbf{\Sigma}_{\mathrm{r2}} - \mathbf{\Sigma}_{\mathrm{r1}} = \frac{\mathbf{T}}{m_v} + \frac{\mathbf{D}}{m_v} + \frac{\mathbf{L}}{m_v} + \frac{\mathbf{F}}{m_v}, \tag{7.5}$$

where, in the proper coordinate system and units, $\mathbf{T} \equiv$ thrust force, $\mathbf{D} \equiv$ drag force, $\mathbf{L} \equiv$ lift force, $\mathbf{F} \equiv$ additional forces such as solar radiation, and $m_v \equiv$ vehicle mass at time t.

Equation 7.3 can be written more compactly as

$$\frac{d^2\mathbf{r}}{dt^2} = -k^2(m_1 + m_2)\frac{\mathbf{r}}{r^3} + \mathbf{\Sigma}_{\mathrm{P}} + \mathbf{\Sigma}_{\mathrm{T}} + \mathbf{\Sigma}_{\mathrm{D}} + \mathbf{\Sigma}_{\mathrm{L}}, \tag{7.6}$$

where $\mathbf{\Sigma}_{\mathbf{P}}$ represents the planetary perturbative portion of (7.3) and the remaining $\mathbf{\Sigma}$'s represent the sums of all thrust, drag, and lift forces. The lead term of (7.6) represents the point mass attraction, that is, the two-body acceleration of the dominant central body. To include the aspherical effects of the central body, (7.6) can be written as

$$\frac{d^2\mathbf{r}}{dt^2} = \nabla\Phi + \mathbf{\Sigma}_{\mathrm{P}} + \mathbf{\Sigma}_{\mathrm{T}} + \mathbf{\Sigma}_{\mathrm{D}} + \mathbf{\Sigma}_{\mathrm{L}}, \tag{7.7}$$

where Φ is the aspherical potential discussed in [1]. For clarity a few terms of Φ are given by[3]

$$\Phi = \frac{k^2\mu}{r}\left[1 + \frac{J_2}{2r^3}(1 - 3\sin^2\delta) + \frac{J_3}{2r^3}(3 - 5\sin^2\delta)\sin\delta + \epsilon\right] \tag{7.8}$$

with $\mu \equiv$ mass of central body (in normalized masses) $= 1$, $k \equiv$ the Gaussian or planetary constant, $J_i =$ coefficient of the ith harmonic, and $\epsilon \equiv$ terms of higher order. The function Φ can be partitioned as follows:

$$\Phi = \frac{k^2\mu}{r} + R, \tag{7.9}$$

where R is by definition the gravitational perturbative function defined by virtue of (7.8) as

$$R \equiv \frac{k^2\mu}{r}\left[\frac{J_2}{2r^3}(1 - 3\sin^2\delta) + \frac{J_3}{2r^3}(3 - 5\sin^2\delta)\sin\delta + \epsilon\right]. \tag{7.10}$$

[3] The potential function including both zonal and tesseral harmonics may be used equally well. See Appendix 2.

Hence, inclusive of the gravitational anomalies of the central planet and planetary, thrust, drag and lift perturbations, (7.6) becomes

$$\frac{d^2\mathbf{r}}{dt^2} = -k^2\mu\frac{\mathbf{r}}{r^3} + \nabla R + \mathbf{\Sigma}_\mathrm{P} + \mathbf{\Sigma}_\mathrm{T} + \mathbf{\Sigma}_\mathrm{D} + \mathbf{\Sigma}_\mathrm{L}. \tag{7.11}$$

For convenience the following separation of accelerations can be made:

$$\frac{d^2\mathbf{r}}{dt^2} = \frac{d^2\mathbf{r}_c}{dt^2} + \frac{d^2\mathbf{r}_p}{dt^2}, \tag{7.12}$$

where

$$\frac{d^2\mathbf{r}_c}{dt^2} \equiv -k^2\mu\frac{\mathbf{r}}{r^3}$$

$$\frac{d^2\mathbf{r}_p}{dt^2} \equiv \nabla R + \mathbf{\Sigma}_\mathrm{P} + \mathbf{\Sigma}_\mathrm{T} + \mathbf{\Sigma}_\mathrm{D} + \mathbf{\Sigma}_\mathrm{L}$$

with the understanding that the subscript c denotes central accelerations and the subscript p denotes perturbative accelerations. At times the perturbative acceleration may be larger than the central acceleration, as when a thrust arc is being executed. This is not in any way a detraction since the previous associations are made merely for functional reasons. However, for most cases in planetocentric orbital analysis the central acceleration is the dominant term.

If the traditional astrodynamical time units discussed in ([1], Section 1) are used, it follows that (7.12) can be written as

$$\ddot{\mathbf{r}} = \ddot{\mathbf{r}}_c + \ddot{\mathbf{r}}_p, \tag{7.13}$$

where the dot notation denotes differentiation with respect to τ instead of t. Note that

$$\tau \equiv k(t - t_0), \tag{7.14}$$

where t = universal or more accurately ephemeris time and t_0 is the epoch time. In common usage $\ddot{\mathbf{r}}_p$ is referred to as the perturbative acceleration of a dynamical system.

7.2 THE METHOD OF COWELL

7.2.1 Applicability of Cowell's Method

The simplest of the special perturbation techniques is a method due to P. H. Cowell. This method, according to Baker [4], was used in the determination of the orbit of the eighth Satellite of Jupiter and was apparently developed after the more sophisticated technique of Encke. This approach was also used in predicting the return of Halley's comet from 1759 to 1910 [2].

One of the great advantages of Cowell's method is the rapidity with which the technique can be implemented on high speed computing machinery. Basically the method consists of a direct integration of the sum of the central and perturbative accelerations in a straightforward manner. Owing to this direct integration approach, which does of course minimize errors of implementation, numerical studies have shown that the Cowell method is about twelve times slower than other special perturbation methods. Furthermore, because of the rapid changes in the accelerations, the integration step size must usually be taken quite small. However, the method does not require as much core storage space in the computing machine as do other techniques.

7.2.2 Formulation of Cowell's Method

The formulation of Cowell's Method, as stated previously, is extremely simple. In brief, from (7.13), that is,

$$\ddot{\mathbf{r}} = \ddot{\mathbf{r}}_c + \ddot{\mathbf{r}}_p,$$

it follows that

$$\dot{\mathbf{r}}_{n+1} = \dot{\mathbf{r}}_n + \int_{\tau=\tau_n}^{\tau=\tau_n+\Delta\tau} \ddot{\mathbf{r}}\, d\tau, \qquad n = 1, 2, \ldots, q, \tag{7.15}$$

$$\mathbf{r}_{n+1} = \mathbf{r}_n + \int_{\tau=\tau_n}^{\tau=\tau_n+\Delta\tau} \dot{\mathbf{r}}\, d\tau, \qquad n = 1, 2, \ldots, q, \tag{7.16}$$

where $\mathbf{r}_n$ is the position vector from the dynamical center and $\dot{\mathbf{r}}_n$ is the velocity vector. Hence the accelerations $\ddot{\mathbf{r}}_c + \ddot{\mathbf{r}}_p$ are computed and, by the standard techniques of numerical analysis[4], integrated to yield the position and velocity vectors at the next time interval $\tau_n + \Delta\tau$.

7.3 THE METHOD OF ENCKE

7.3.1 The Fundamental Idea

A more refined special perturbations procedure bears the name of its discoverer, J. F. Encke. The technique devised by Encke has found favor in the computation of lunar trajectories. In this technique the only integration that is performed is the integration of the difference between the two-body or adopted reference orbit acceleration and the perturbative acceleration. The method of Encke, in essence, attempts to integrate smaller accelerations and thus retain more significant figures while increasing integration step size.

[4] A simple numerical integration scheme, Euler's method, is discussed in detail in [1, Section 2.8.3]; for other methods see [16].

The classical formulation of Encke's method has been recorded by Plummer [5], Brouwer and Clemence [2], Baker and Makemson [4], and many others. The formulation developed here, which may be new, is a modification of the classic procedure. It is due to Escobal [6].

7.3.2 Deviation from Two-Body Motion

In an Encke procedure a two-body reference orbit is usually adopted and deviations from Keplerian motion integrated to yield the correct orbital elements [9], [13]. Consider that a two-body orbit is defined at state $\mathbf{r}_n$, $\dot{\mathbf{r}}_n$, where $\mathbf{r}_n$ is the position vector from the dynamical center and $\dot{\mathbf{r}}_n$ is the velocity vector, as shown in Figure 7.1. From this point on, owing to existing

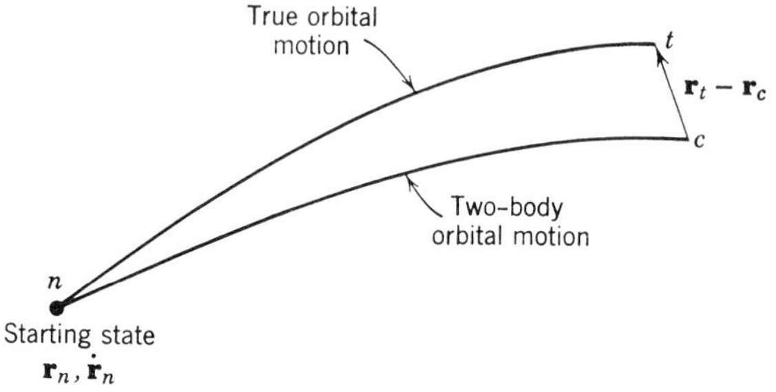

FIGURE 7.1 True and Keplerian orbits.

perturbations, the motion is from n to t. An approximation to this path is given by the Keplerian arc from n to c. Hence an error vector $\boldsymbol{\epsilon}$ defined as the difference between the true and Keplerian positions, that is,

$$\boldsymbol{\epsilon} = \mathbf{r}_t - \mathbf{r}_c, \tag{7.17}$$

can be formed. Evidently, by differentiating twice,

$$\ddot{\boldsymbol{\epsilon}} = \ddot{\mathbf{r}}_t - \ddot{\mathbf{r}}_c. \tag{7.18}$$

Now, if $\ddot{\mathbf{r}}_t$ is computed, that is,

$$\ddot{\mathbf{r}}_t = \ddot{\mathbf{r}}_c + \ddot{\mathbf{r}}_p,$$

where $\ddot{\mathbf{r}}_c$, the central or Keplerian acceleration, is obtained from

$$\ddot{\mathbf{r}}_c = -\frac{\mu \mathbf{r}_c}{r_c^3}, \tag{7.19}$$

the deviation acceleration becomes numerically known. It is therefore possible to write

$$\dot{\mathbf{r}}_{n+1} = (\dot{\mathbf{r}}_c)_n + \int_{\tau=\tau_n}^{\tau=\tau_n+\Delta\tau} \ddot{\boldsymbol{\epsilon}}\, d\tau,$$

$$\mathbf{r}_{n+1} = (\mathbf{r}_c)_n + \int_{\tau=\tau_n}^{\tau=\tau_n+\Delta\tau} \dot{\boldsymbol{\epsilon}}\, d\tau, \tag{7.20}$$

which, after integration of the error acceleration vector $\ddot{\boldsymbol{\epsilon}}$, yield the true positions and velocities of an orbit at modified time

$$\tau = k(t - t_0),$$

where $k \equiv$ Gaussian or planetary constant, $t \equiv$ universal or ephemeris time of desired position and velocity vectors, and $t_0 \equiv$ epoch time. This procedure obviously allows larger integration step sizes and preserves more accuracy than the previously described Cowell technique. However, the method is more difficult to implement.

The procedure described has a given disadvantage. The deviation vector $\boldsymbol{\epsilon}$ is not bounded and will continue to grow. It could happen that $\boldsymbol{\epsilon}$ might become so large that the advantages of this technique would be overcome. The usual attack is to rectify the two-body orbit; that is, at the point where $\boldsymbol{\epsilon}$ becomes too large, the process is stopped and a new epoch is adopted from which a new osculating orbit is computed. This process is commonly known as *rectification*. Consider, however, the use of a non two-body reference orbit which results in a *hybrid formulation*, that is, in a mixture of general and special perturbational techniques.

7.3.3 Effects Due to Oblateness

For planetocentric orbits, if a secularly perturbed reference orbit is adopted $\boldsymbol{\epsilon}$ will be bounded by definition. In a conservative system the secular effects due to the oblateness of a central body manifest themselves by:

$$\Omega = \Omega_n + \dot{\Omega}_n \tau, \qquad \omega = \omega_n + \dot{\omega}_n \tau, \qquad n = \bar{n}, \tag{7.21}$$

where the subscript n signifies the revolution number, $\Omega =$ longitude of the ascending node, $\omega =$ argument of perifocus, and $n =$ the mean motion[5]. For a first-order theory[6] the secular rate of change of the elements can be obtained from

$$\dot{\Omega} = -\frac{3}{2}\frac{J_2}{p^2}\frac{\bar{n}}{k}\cos i,$$

$$\dot{\omega} = +\frac{3}{2}\frac{J_2}{p^2}\left(2 - \tfrac{5}{2}\sin^2 i\right)\frac{\bar{n}}{k},$$

$$\bar{n} = n\left[1 + \tfrac{3}{2}J_2\,\frac{[1 - e^2]^{1/2}}{p^2}\left(1 - \tfrac{3}{2}\sin^2 i\right)\right], \tag{7.22}$$

[5] Obviously, if in this formulation $\dot{\Omega}_n = \dot{\omega}_n = 0$ and n is used in place of $\bar{n}$, the regular Encke technique with a Keplerian reference orbit results, that is, $J_2 = 0$.

[6] See ([1], Chapter 10) for secular expressions of the second order.

where $n = k[\mu]^{1/2}a^{-3/2}$, μ = sum of masses of celestial object and central mass, a = semimajor axis of orbit, e = orbital eccentricity, p = semi-parameter of orbit, i = orbital inclination to the equator, and J_2 = second harmonic coefficient.

A standard relation for the orbital position vector ([1], Section 3.4) is

$$\mathbf{r} = x_\omega \mathbf{P} + y_\omega \mathbf{Q}, \tag{7.23}$$

where, in terms of the Keplerian eccentric anomaly E,

$$x_\omega = a(\cos E - e), \qquad y_\omega = a[1 - e^2]^{1/2} \sin E; \tag{7.24}$$

the unit vectors **P**, **Q** will be defined presently.

In a conservative field, a, e, and i are true constants from a secular point of view and the perturbed eccentric anomaly is defined through Kepler's equation by

$$\bar{n}(t - T) = \bar{E} - e \sin \bar{E}. \tag{7.25}$$

Since, T, the time of perifocal passage, is known, the parameters x_ω and y_ω can be determined at once through (7.24) with $\bar{E}$ replacing E. In (7.23) the unit vector **P** emanating from the origin which points at perifocus is defined as

$$\begin{aligned} P_x &= \cos \omega \cos \Omega - \sin \omega \sin \Omega \cos i, \\ P_y &= \cos \omega \sin \Omega + \sin \omega \cos \Omega \cos i, \\ P_z &= \sin \omega \sin i, \end{aligned} \tag{7.26}$$

and, similarly, **Q** advanced to **P** by a right angle in the plane and direction of motion is given in terms of Ω, ω, and i as

$$\begin{aligned} Q_x &= -\sin \omega \cos \Omega - \cos \omega \sin \Omega \cos i, \\ Q_y &= -\sin \omega \sin \Omega + \cos \omega \cos \Omega \cos i, \\ Q_z &= \cos \omega \sin i. \end{aligned} \tag{7.27}$$

The first derivatives of **P** and **Q** are given by virtue of (7.21), (7.26), and (7.27) as

$$\begin{aligned} \dot{P}_x &= -P_y\dot{\Omega}_n + Q_x\dot{\omega}_n, & \dot{Q}_x &= -Q_y\dot{\Omega}_n - P_x\dot{\omega}_n, \\ \dot{P}_y &= P_x\dot{\Omega}_n + Q_y\dot{\omega}_n, & \dot{Q}_y &= Q_x\dot{\Omega}_n - P_y\dot{\omega}_n, \\ \dot{P}_z &= Q_z\dot{\omega}_n, & \dot{Q}_z &= -P_z\dot{\omega}_n, \end{aligned} \tag{7.28}$$

while the second derivatives are obtained from $\dot{\mathbf{P}}$ and $\dot{\mathbf{Q}}$ as

$$\begin{aligned} \ddot{P}_x &= -\dot{P}_y\dot{\Omega}_n + \dot{Q}_x\dot{\omega}_n, & \ddot{Q}_x &= -\dot{Q}_y\dot{\Omega}_n - \dot{P}_x\dot{\omega}_n, \\ \ddot{P}_y &= \dot{P}_x\dot{\Omega}_n + \dot{Q}_y\dot{\omega}_n, & \ddot{Q}_y &= \dot{Q}_x\dot{\Omega}_n - \dot{P}_y\dot{\omega}_n, \\ \ddot{P}_z &= \dot{Q}_z\dot{\omega}_n, & \ddot{Q}_z &= -\dot{P}_z\dot{\omega}_n. \end{aligned} \tag{7.29}$$

Collecting terms, the second derivatives of the unit vectors become

$$\ddot{P}_x = -(\dot{\Omega}_n^2 + \dot{\omega}_n^2)P_x - 2\dot{\Omega}_n\dot{\omega}_n Q_y,$$
$$\ddot{P}_y = -(\dot{\Omega}_n^2 + \dot{\omega}_n^2)P_y + 2\dot{\Omega}_n\dot{\omega}_n Q_x,$$
$$\ddot{P}_z = -\dot{\omega}_n^2 P_z, \tag{7.30}$$

$$\ddot{Q}_x = -(\dot{\Omega}_n^2 + \dot{\omega}_n^2)Q_x + 2\dot{\Omega}_n\dot{\omega}_n P_y,$$
$$\ddot{Q}_y = -(\dot{\Omega}_n^2 + \dot{\omega}_n^2)Q_y - 2\dot{\Omega}_n\dot{\omega}_n P_x,$$
$$\ddot{Q}_z = -\dot{\omega}_n^2 Q_z. \tag{7.31}$$

The secularly perturbed reference acceleration vector $\ddot{\mathbf{r}}_s$ can now be written in terms of $\mathbf{P}$ and $\mathbf{Q}$ in the form

$$\ddot{\mathbf{r}}_s = \frac{d^2}{d\tau^2}(x_\omega \mathbf{P} + y_\omega \mathbf{Q}) \tag{7.32}$$

and, by double differentiation of the right side of (7.32), it follows that

$$\ddot{\mathbf{r}}_s = x_\omega \ddot{\mathbf{P}} + 2\dot{x}_\omega \dot{\mathbf{P}} + \ddot{x}_\omega \mathbf{P} + y_\omega \ddot{\mathbf{Q}} + 2\dot{y}_\omega \dot{\mathbf{Q}} + \ddot{y}_\omega \mathbf{Q}. \tag{7.33}$$

The orbit plane accelerations can be eliminated by virtue of Newton's second law as

$$\ddot{\mathbf{r}}_\omega = -\frac{\mu \mathbf{r}_\omega}{r^3}. \tag{7.34}$$

and the velocities likewise eliminated by the standard expressions extracted from ([1], Chapter 3), that is

$$\dot{x}_\omega = -\frac{\bar{n}[\mu a]^{1/2}}{n r_s}\sin \bar{E}, \qquad \dot{y}_\omega = \frac{\bar{n}[\mu p]^{1/2}}{n r_s}\cos \bar{E}, \tag{7.35}$$

where $r_s = a(1 - e\cos \bar{E})$.

7.3.4 The Problem of Small Differences

The previous section has shown how it is possible to generate a reference orbit that contains the secular perturbations due to an aspherical potential Φ defined by (7.8). It is therefore possible to express the error acceleration more exactly than by (7.18) as $\ddot{\boldsymbol{\epsilon}} = \ddot{\mathbf{r}}_t - \ddot{\mathbf{r}}_s$ or $\ddot{\boldsymbol{\epsilon}} = \nabla\Phi - \ddot{\mathbf{r}}_s$, that is,

$$\ddot{\boldsymbol{\epsilon}} = -\frac{\mu \mathbf{r}}{r^3} - \ddot{\mathbf{r}}_s + \nabla R, \tag{7.36}$$

where $R \equiv$ gravitational perturbative function, (7.10), which is equal to $\Phi - V$, with V defined as the two-body potential.

Now, since $\ddot{\mathbf{r}}_s$ can be regrouped as

$$\ddot{\mathbf{r}}_s = \ddot{x}_\omega \mathbf{P} + \ddot{y}_\omega \mathbf{Q} + 2\dot{x}_\omega \dot{\mathbf{P}} + 2\dot{y}_\omega \dot{\mathbf{Q}} + x_\omega \ddot{\mathbf{P}} + y_\omega \ddot{\mathbf{Q}}$$

or by (7.34) as

$$\ddot{\mathbf{r}}_s = -\frac{\mu}{r^3}(x_\omega \mathbf{P} + y_\omega \mathbf{Q}) + 2\dot{x}_\omega \dot{\mathbf{P}} + 2\dot{y}_\omega \dot{\mathbf{Q}} + x_\omega \ddot{\mathbf{P}} + y_\omega \ddot{\mathbf{Q}},$$

it follows that

$$\ddot{\mathbf{r}}_s = -\frac{\mu \mathbf{r}_s}{r_s^{\,3}} + 2\dot{x}_\omega \dot{\mathbf{P}} + 2\dot{y}_\omega \dot{\mathbf{Q}} + x_\omega \ddot{\mathbf{P}} + y_\omega \ddot{\mathbf{Q}}. \tag{7.37}$$

This acceleration can be written compactly as

$$\ddot{\mathbf{r}}_s \equiv -\frac{\mu \mathbf{r}_s}{r_s^{\,3}} + \mathbf{s}, \tag{7.38}$$

where the additive secular acceleration is given by

$$\mathbf{s} \equiv 2\dot{x}_\omega \dot{\mathbf{P}} + 2\dot{y}_\omega \dot{\mathbf{Q}} + x_\omega \ddot{\mathbf{P}} + y_\omega \ddot{\mathbf{Q}}.$$

The deviation vector $\ddot{\boldsymbol{\epsilon}}$, (7.36), now becomes

$$\ddot{\boldsymbol{\epsilon}} = -\mu \frac{\mathbf{r}}{r^3} + \mu \frac{\mathbf{r}_s}{r_s^{\,3}} - \mathbf{s} + \nabla R. \tag{7.39}$$

The dominant terms of (7.39) are the first two leading terms on the right side. It follows, however, that, if (7.36) or (7.39) is used to determine $\ddot{\boldsymbol{\epsilon}}$ numerically, a great number of significant figures have to be carried in the computations. The reason is that the first two terms are the difference of nearly equal quantities. In order to avoid this problem it is possible to introduce

$$\frac{r^2}{r_s^{\,2}} = 1 + 2q, \tag{7.40}$$

where q is a small quantity that makes this equation hold true, that is,

$$q \equiv \frac{r^2 - r_s^{\,2}}{2r_s^{\,2}}. \tag{7.41}$$

However, q can be expanded as

$$q = \frac{r_s^{-2}(x^2 + y^2 + z^2 - x_s^{\,2} - y_s^{\,2} - z_s^{\,2})}{2}$$

or equivalently as

$$q = r_s^{-2}(xx_s - x_s^{\,2} + yy_s - y_s^{\,2} + zz_s - z_s^{\,2} - xx_s$$
$$+ \tfrac{1}{2}x^2 - yy_s + \tfrac{1}{2}y^2 - zz_s + \tfrac{1}{2}z^2 + \tfrac{1}{2}x_s^{\,2} + \tfrac{1}{2}y_s^{\,2} + \tfrac{1}{2}z_s^{\,2}$$

so that

$$q = r_s^{-2}[x_s\epsilon_x + y_s\epsilon_y + z_s\epsilon_z + \tfrac{1}{2}(\epsilon_x^{\,2} + \epsilon_y^{\,2} + \epsilon_z^{\,2})]. \tag{7.42}$$

Finally in terms of $\tilde{\boldsymbol{\epsilon}}$ defined by

$$\tilde{\boldsymbol{\epsilon}} \equiv \mathbf{r}_s + \tfrac{1}{2}\boldsymbol{\epsilon} \tag{7.43}$$

it is possible to compute q from

$$q = r_s^{-2}\boldsymbol{\epsilon} \cdot \tilde{\boldsymbol{\epsilon}}. \tag{7.44}$$

Furthermore (7.40) can be used to obtain

$$\frac{r_s^3}{r^3} = (1 + 2q)^{-3/2} \equiv 1 - fq, \tag{7.45}$$

where from a direct expansion of $(1 + 2q)^{-3/2}$ the following well-known formula [4] is obtained:

$$f \equiv 3\left[1 - \tfrac{5}{2}q + \frac{5 \cdot 7}{2 \cdot 3}q^2 - \frac{5 \cdot 7 \cdot 9}{2 \cdot 3 \cdot 4}q^3 + \cdots\right]. \tag{7.46}$$

The basic devices are now available for the transformation of (7.39) in order to avoid the difference calculation of two nearly equal numbers. Hence, substituting for r^3 in (7.39) and using relationship (7.45) yields

$$\ddot{\boldsymbol{\epsilon}} = -\frac{\mu \mathbf{r}}{r_s^3}(1 - fq) + \frac{\mu \mathbf{r}_s}{r_s^3} + \nabla R - \mathbf{s} \tag{7.47}$$

or

$$\ddot{\boldsymbol{\epsilon}} = \frac{\mu}{r_s^3}(fq\mathbf{r} - \boldsymbol{\epsilon}) + \nabla R - \mathbf{s}, \tag{7.48}$$

As may be evident, the only complication encountered over and above the standard Encke procedure is the inclusion of the factor $\mathbf{s}$ defined by (7.38).

7.3.5 Inclusion of Drag

The analysis here was restricted to the inclusion of gravitational effects occasioned by an oblate force field. Because this restriction results in the absence of drag and other forces, the reference orbit always stays close to the true orbit. The inclusion of drag effects can also be handled if, for example, the mean rates, denoted by a caret, $\widehat{da}/d\tau$, $\widehat{de}/d\tau$, $\widehat{di}/d\tau$ are known. These rates can be determined by the procedure of Sterne [8], [1]. Since the only secular effects of drag are that it affects a, e, and i, it follows that

$$a = a_0 + \frac{\widehat{da}}{d\tau}\tau, \qquad e = e_0 + \frac{\widehat{de}}{d\tau}\tau, \qquad i = i_0 + \frac{\widehat{di}}{d\tau}\tau; \tag{7.49}$$

the effects of oblateness are represented by (7.21). Equations 7.49 must therefore be carried in the analysis and in the differentiation of the secular position vector. The analysis is lengthy but straightforward.

7.3.6 Algorithm for Encke's Method with Shifting Reference Orbit

The modified Encke computational procedure for elliptic orbits with moderate eccentricity follows. Given the true position and velocity vectors $\mathbf{r}_n$, $\dot{\mathbf{r}}_n$ at time τ_n, determine at time $\tau_{n+1} = \tau_n + \Delta\tau$, where $\Delta\tau$ is the integration step size, the new state vectors $\mathbf{r}_{n+1}$, $\dot{\mathbf{r}}_{n+1}$ by computing[7]

$$\bar{n} = n\left[1 + \tfrac{3}{2}J_2 \frac{[1 - e^2]^{\frac{1}{2}}}{p^2}(1 - \tfrac{3}{2}\sin^2 i)\right],$$

$$\Omega = \Omega_n + \dot{\Omega}_n\tau_{n+1},$$

$$\omega = \omega_n + \dot{\omega}_n\tau_{n+1}, \tag{7.50}$$

$$\begin{aligned}
P_x &= \cos\omega\cos\Omega - \sin\omega\sin\Omega\cos i,\\
P_y &= \cos\omega\sin\Omega + \sin\omega\cos\Omega\cos i,\\
P_z &= \sin\omega\sin i,\\
Q_x &= -\sin\omega\cos\Omega - \cos\omega\sin\Omega\cos i,\\
Q_y &= -\sin\omega\sin\Omega + \cos\omega\cos\Omega\cos i,\\
Q_z &= \cos\omega\sin i;
\end{aligned}\tag{7.51}$$

$$\begin{aligned}
\dot{P}_x &= -P_y\dot{\Omega}_n + Q_x\dot{\omega}_n,\\
\dot{P}_y &= P_x\dot{\Omega}_n + Q_y\dot{\omega}_n,\\
\dot{P}_z &= Q_z\dot{\omega}_n,\\
\dot{Q}_x &= -Q_y\dot{\Omega}_n - P_x\dot{\omega}_n,\\
\dot{Q}_y &= Q_x\dot{\Omega}_n - P_y\dot{\omega}_n,\\
\dot{Q}_z &= -P_z\dot{\omega}_n;
\end{aligned}\tag{7.52}$$

$$\begin{aligned}
S &= -(\dot{\Omega}_n^2 + \dot{\omega}_n^2),\\
P &= 2\dot{\Omega}_n\dot{\omega}_n,\\
Q &= -\dot{\omega}_n^2;
\end{aligned}\tag{7.53}$$

$$\begin{aligned}
\ddot{P}_x &= SP_x - PQ_y,\\
\ddot{P}_y &= SP_y + PQ_x,\\
\ddot{P}_z &= QP_z,\\
\ddot{Q}_x &= SQ_x + PP_y,\\
\ddot{Q}_y &= SQ_y - PP_x,\\
\ddot{Q}_z &= QQ_z.
\end{aligned}\tag{7.54}$$

[7] It should be noted that $\tau_{n+1} \equiv k(t_{n+1} - t_n)$.

Solve Kepler's equation[8] for $\bar{E}$,

$$\bar{n}(t_{n+1} - T) = \bar{E} - e \sin \bar{E},$$

and continue calculating with

$$x_\omega = a(\cos \bar{E} - e),$$
$$y_\omega = a[1 - e^2]^{1/2} \sin \bar{E},$$
$$r_s = [x_\omega^2 + y_\omega^2]^{1/2}; \tag{7.55}$$

$$\dot{x}_\omega = -\frac{\bar{n}[\mu a]^{1/2}}{nr_s} \sin \bar{E},$$

$$\dot{y}_\omega = \frac{\bar{n}[\mu p]^{1/2}}{nr_s} \cos \bar{E}; \tag{7.56}$$

$$(\mathbf{r}_s)_{n+1} = x_\omega \mathbf{P} + y_\omega \mathbf{Q},$$
$$(\dot{\mathbf{r}}_s)_{n+1} = \dot{x}_\omega \mathbf{P} + \dot{y}_\omega \mathbf{Q},$$
$$(\mathbf{s})_{n+1} = 2\dot{x}_\omega \dot{\mathbf{P}} + 2\dot{y}_\omega \dot{\mathbf{Q}} + x_\omega \ddot{\mathbf{P}} + y_\omega \ddot{\mathbf{Q}}. \tag{7.57}$$

Obtain the deviation vector $\boldsymbol{\epsilon}$ from

$$\boldsymbol{\epsilon} = \mathbf{r}_n - (\mathbf{r}_s)_n, \tag{7.58}$$

$$\tilde{\boldsymbol{\epsilon}} = (\mathbf{r}_s)_n + \tfrac{1}{2}\boldsymbol{\epsilon},$$

$$q = \frac{\boldsymbol{\epsilon} \cdot \tilde{\boldsymbol{\epsilon}}}{(r_s^2)_n}; \tag{7.59}$$

$$f = 3\left(1 - \tfrac{5}{2}q + \frac{5 \cdot 7}{2 \cdot 3} q^2 - \frac{5 \cdot 7 \cdot 9}{2 \cdot 3 \cdot 4} q^3 + \cdots\right), \qquad q^n < \delta. \tag{7.60}$$

Determine the gradient of the perturbative function R evaluated with $\mathbf{r}_n$, $\dot{\mathbf{r}}_n$ and obtain the deviation acceleration from[9]

$$\ddot{\boldsymbol{\epsilon}} = \frac{\mu}{(r_s^3)_n}(fq\mathbf{r}_n - \boldsymbol{\epsilon}) + \nabla R - \mathbf{s}_n.$$

Finally integrate to yield

$$\dot{\mathbf{r}}_{n+1} = (\dot{\mathbf{r}}_s)_n + \int_{\tau_n}^{\tau_n+\Delta\tau} \ddot{\boldsymbol{\epsilon}}\, d\tau, \tag{7.61}$$

$$\mathbf{r}_{n+1} = (\mathbf{r}_s)_n + \int_{\tau_n}^{\tau_n+\Delta\tau} \dot{\boldsymbol{\epsilon}}\, d\tau. \tag{7.62}$$

[8] Much more rapid results can be obtained by taking E as the independent variable and thus avoiding the iterative solution of Kepler's equation.

[9] If $s = 0$, the standard Encke technique with a Keplerian orbit as the reference orbit results.

7.3.7 Remarks on Zero and Low Eccentricity Orbits

In some instances it may be convenient to have an Encke formulation which is valid for zero or low eccentricity orbits. It should be evident from the discussion in [1, Chapter 3] that the element ω becomes badly defined for eccentricities approaching zero. In this light a new formulation for the representation of the reference position and velocity vectors must be adopted. Hence, in the algorithm of the preceding section, (7.57) must be replaced with an equivalent set that permits the position and velocity vectors to be obtained free of indeterminacies. The collection of formulas to be stated permits a completely free variance of the orbital eccentricity for elliptic motion. An analogous set of equations can be developed for hyperbolic motion if desired. The algorithm permits an indeterminacy-free representation of the reference orbit position and velocity vectors but does not include a shifting reference orbit. If desired, the shifting reference orbit, including secular perturbations, can be derived as in the previous section. For compactness the following chain calculation is extracted from ([1], Appendix 1).

The computation commences by performing the common calculations,

$$r_n^2 = \mathbf{r}_n \cdot \mathbf{r}_n,$$

$$D_n = \frac{\mathbf{r}_n \cdot \dot{\mathbf{r}}_n}{\sqrt{\mu}},$$

$$\frac{1}{a} = \frac{2}{r_n} - \frac{\dot{\mathbf{r}}_n \cdot \dot{\mathbf{r}}_n}{\mu}. \tag{7.63}$$

For an elliptic or circular orbit it is possible to determine

$$C_e = 1 - \frac{r_n}{a},$$

$$S_e = \frac{D_n}{\sqrt{a}}, \tag{7.64}$$

and obtain the difference in mean motions as

$$M_{n+1} - M_n = \frac{\tau_{n+1}\sqrt{\mu}}{a^{3/2}}. \tag{7.65}$$

Kepler's modified equation can now be solved iteratively in terms of $g \equiv (E_{n+1} - E_n)/2$ to yield the corresponding difference in eccentric anomalies, that is,

$$g_{i+1} = g_i - \frac{g_i + S_e \sin^2 g_i - C_e \sin g_i \cos g_i - \frac{1}{2}(M_{n+1} - M_n)}{1 + 2S_e \sin g_i \cos g_i - C_e(1 - 2\sin^2 g_i)}, \tag{7.66}$$

for $i = 1, 2, \ldots, \nu$, until $|g_{i+1} - g_i| < \delta$, where δ is a tolerance. Finally, obtain the Keplerian position and velocity vectors at state $n + 1$ from

$$
\begin{aligned}
E_{n+1} - E_n &= 2g, \\
C &= a[1 - \cos(E_{n+1} - E_n)], \\
S &= \sqrt{a}\sin(E_{n+1} - E_n), \\
f &= 1 - \frac{C}{r_n}, \\
g &= \frac{1}{\sqrt{\mu}}(r_n S + D_n C), \\
r_{n+1} &= r_n + \left(1 - \frac{r_n}{a}\right)C + D_n S, \\
\dot{f} &= -\frac{\sqrt{\mu}}{r_{n+1}r_n}S, \\
\dot{g} &= 1 - \frac{C}{r_{n+1}}, \\
(\mathbf{r}_c)_{n+1} &= f\mathbf{r}_n + g\dot{\mathbf{r}}_n, \\
(\dot{\mathbf{r}}_c)_{n+1} &= \dot{f}\mathbf{r}_n + \dot{g}\dot{\mathbf{r}}_n.
\end{aligned}
\tag{7.67}
$$

The standard Encke equations, (7.59) through (7.62), where the s is replaced by subscript c, can now be utilized to obtain the true position and velocity vectors at $\tau = \tau_n + \Delta\tau$.

7.4 THE HYBRID PATCHED CONIC

7.4.1 Preliminary Remarks

Chapter 5 presented the pertinent equations of the patched conic technique. The method of patched conics has found extensive use in industry because of its simplicity and ease of implementation. Unfortunately, one of the disadvantages of this method has been the sizable errors resulting from the neglect of perturbative effects. The intent of this section is to introduce a technique which can be used to remove the inherent disadvantages of using the patched conic method while maintaining its simplicity and speed advantages. The technique, called the hybrid patched conic, is due to Escobal [17], and its extension to Stern [18]. In order to be definite the analysis in this section is aimed at the lunar trajectory problem.

7.4.2 Equations of Motion

As derived in ([1], Chapter 2), the relative equations of motion[10] of an n-body system can be written as

$$\ddot{\mathbf{r}} = -\mu\frac{\mathbf{r}}{r^3} + \sum_{j=3}^{n} m_j\left(\frac{\mathbf{r}_{2j}}{r_{2j}^{\ 3}} - \frac{\mathbf{r}_{1j}}{r_{1j}^{\ 3}}\right), \tag{7.68}$$

where the dots denote differentiation with respect to $\tau = k(t - t_0)$ and the origin is assumed to be fixed at the dynamical center of the Earth. More compactly,

$$\ddot{\mathbf{r}} = -\mu\frac{\mathbf{r}}{r^3} + \ddot{\mathbf{r}}_p, \tag{7.69}$$

where the first term on the right is the standard two-body acceleration and $\ddot{\mathbf{r}}_p$, the perturbative acceleration due to the remaining $n - 2$ bodies, is defined by the summation in (7.68).

Following the technique utilized in the method of Encke, Section 7.3, the position vector $\mathbf{r}$ and velocity vector $\dot{\mathbf{r}}$ at any future or past time can be written in terms of the deviation vector $\boldsymbol{\epsilon} = \mathbf{r} - \mathbf{r}_c$, as

$$\dot{\mathbf{r}} = \dot{f}\mathbf{r}_0 + \dot{g}\dot{\mathbf{r}}_0 + \int \ddot{\boldsymbol{\epsilon}}\, d\tau,$$

$$\mathbf{r} = f\mathbf{r}_0 + g\dot{\mathbf{r}}_0 + \iint \ddot{\boldsymbol{\epsilon}}\, d\tau\, d\tau, \tag{7.70}$$

where the first two terms of these two equations represent the Keplerian (or patched conic) motion about the central body and for elliptic motion are defined by ([1], Chapter 3)

$$f \equiv 1 - \frac{a}{r_0}[1 - \cos(E - E_0)],$$

$$g \equiv \tau - \frac{a^{3/2}}{\sqrt{\mu}}[E - E_0 - \sin(E - E_0)],$$

$$\dot{f} \equiv -\frac{\sqrt{\mu a}}{r r_0}\sin(E - E_0),$$

$$\dot{g} \equiv 1 - \frac{a}{r}[1 - \cos(E - E_0)].$$

The basic idea of the hybrid patched conic technique is rooted in the evaluation of the deviation acceleration $\ddot{\boldsymbol{\epsilon}}$. Since the integrands of (7.70) can be

[10] See also Section 7.1.3.

evaluated only if a functional form of $\ddot{\boldsymbol{\epsilon}}$ is assumed, the problem at hand is to find such a form. Studies of lunar trajectories have shown that, for escape trajectories toward the Moon, the deviation acceleration can always be represented by the expression[11]

$$\ddot{\boldsymbol{\epsilon}} = \mathbf{A} + \frac{\mathbf{B}}{(\tilde{\tau} - \tau)} + \frac{\mathbf{C}}{(\tilde{\tau} - \tau)^2} + \cdots, \tag{7.71}$$

where **A**, **B**, **C**, etc., are vectorial constants and $\tilde{\tau}$ is a scalar constant. If for the present it is assumed that the fundamental constants of (7.71) are known, it follows by direct integration that

$$\dot{\boldsymbol{\epsilon}} \equiv \int_0^{\tau} \ddot{\boldsymbol{\epsilon}}\, dt = \mathbf{A}\tau + \mathbf{B} \log\left(\frac{\tilde{\tau}}{\tilde{\tau} - \tau}\right) + \frac{\mathbf{C}\tau}{\tau(\tilde{\tau} - \tau)} + \cdots, \tag{7.72}$$

$$\boldsymbol{\epsilon} \equiv \int_0^{\tau} \dot{\boldsymbol{\epsilon}}\, dt = \frac{\mathbf{A}}{2}\tau^2 + \left[\mathbf{B} \log \tilde{\tau} + \mathbf{B} - \frac{\mathbf{C}}{\tilde{\tau}}\right]\tau$$

$$+ \mathbf{C} \log\left(\frac{\tilde{\tau}}{\tilde{\tau} - \tau}\right) + \mathbf{B}\tilde{\tau} \log\left(\frac{\tilde{\tau} - \tau}{\tilde{\tau}}\right) - \mathbf{B}\tau \log(\tilde{\tau} - \tau) + \cdots. \tag{7.73}$$

Equations 7.70 can therefore be written in the explicit analytic form

$$\dot{\mathbf{r}} = \dot{f}\mathbf{r}_0 + \dot{g}\dot{\mathbf{r}}_0 + \dot{\boldsymbol{\epsilon}}, \tag{7.74}$$

$$\mathbf{r} = f\mathbf{r}_0 + g\dot{\mathbf{r}}_0 + \boldsymbol{\epsilon}, \tag{7.75}$$

with the understanding that $\dot{\boldsymbol{\epsilon}}$ and $\boldsymbol{\epsilon}$ are defined by (7.72) and (7.73).

7.4.3 Evaluation of the Fundamental Constants

To accomplish the evaluation of the fundamental constants consider the following scheme. An ordinary patched conic trajectory, to accomplish a desired mission, is computed by means of any of the standard techniques. As illustrated in Figure 7.2, $\ddot{\boldsymbol{\epsilon}}$ now can be evaluated on the Keplerian orbit j times at corresponding values of the eccentric anomaly E, starting with E_0 and ending with E_m. Note that E_m is the value of the eccentric anomaly on the patched conic trajectory at MSI puncture. Since at any E_k the corresponding time variable is determined by

$$\tau_k = \frac{1}{n}\left[E_k - E_0 + 2S_e \sin^2\left(\frac{E_k - E_0}{2}\right) - C_e \sin(E_k - E_0)\right], \tag{7.76}$$

it follows that for specified values of E it will be possible to obtain the set

[11] For lunar return trajectories, the fit $\ddot{\boldsymbol{\epsilon}} = \mathbf{A} + \mathbf{B}\tau + \mathbf{C}\tau^2 + \cdots$ has been found to yield excellent results.

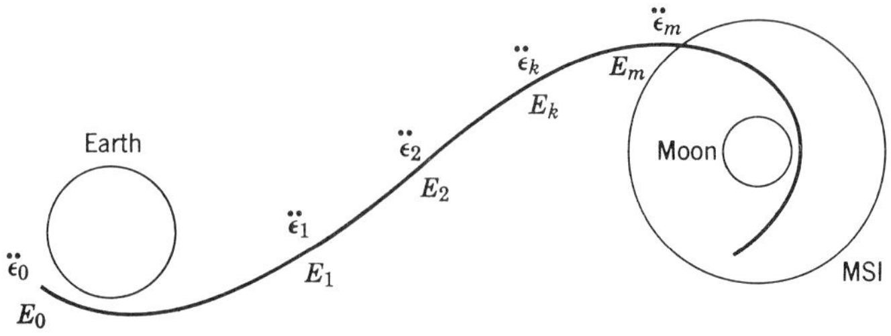

FIGURE 7.2 Evaluation of the deviation acceleration on the patched conic trajectory.

$(\ddot{\boldsymbol{\epsilon}}, \tau)$. Let it be assumed that the set $(\ddot{\boldsymbol{\epsilon}}, \tau)$ is obtained n times so that (7.71) can be written as

$$\mathbf{A} + \left[\frac{1}{\tilde{\tau} - \tau_j}\right]\mathbf{B} + \left[\frac{1}{(\tilde{\tau} - \tau_j)^2}\right]\mathbf{C} + \cdots = \ddot{\boldsymbol{\epsilon}}_j. \tag{7.77}$$

Equation 7.77 represents a linear system of equations which on solution yields the sought-for fundamental constants **A**, **B**, **C**, etc. In practice the constant $\tilde{\tau}$ should be taken at a convenient epoch, for example, in lunar trajectories a suitable value for $\tilde{\tau}$ is the predicted patched conic time of pericythion passage.

7.4.4 The First- and Second-Order Corrections

As mentioned previously, the hybrid patched conic technique assumes that the perturbative accelerations $\ddot{\mathbf{r}}_p$ can be evaluated on the adopted reference orbit, that is, on the patched conic trajectory. Mathematically this corresponds to performing a Taylor expansion about the reference orbit such that

$$\ddot{\boldsymbol{\epsilon}} = \ddot{\boldsymbol{\epsilon}}\Big|_{\boldsymbol{\epsilon}=0} + \left[\frac{\partial \ddot{\boldsymbol{\epsilon}}}{\partial \boldsymbol{\epsilon}}\right]_{\boldsymbol{\epsilon}=0} \boldsymbol{\epsilon} + \cdots, \tag{7.78}$$

where the expansion is taken about the point $\boldsymbol{\epsilon} = \mathbf{0}$, that is, on the reference orbit. The matrix denoted by $[\partial\ddot{\boldsymbol{\epsilon}}/\partial\boldsymbol{\epsilon}]$ can be written out more formally as

$$\left[\frac{\partial \ddot{\boldsymbol{\epsilon}}}{\partial \boldsymbol{\epsilon}}\right] \equiv \begin{bmatrix} \dfrac{\partial \ddot{\epsilon}_x}{\partial \epsilon_x} & \dfrac{\partial \ddot{\epsilon}_x}{\partial \epsilon_y} & \dfrac{\partial \ddot{\epsilon}_x}{\partial \epsilon_z} \\[2ex] \dfrac{\partial \ddot{\epsilon}_y}{\partial \epsilon_x} & \dfrac{\partial \ddot{\epsilon}_y}{\partial \epsilon_y} & \dfrac{\partial \ddot{\epsilon}_y}{\partial \epsilon_z} \\[2ex] \dfrac{\partial \ddot{\epsilon}_z}{\partial \epsilon_x} & \dfrac{\partial \ddot{\epsilon}_z}{\partial \epsilon_y} & \dfrac{\partial \ddot{\epsilon}_z}{\partial \epsilon_z} \end{bmatrix}. \tag{7.79}$$

As it has been used in practice, the method of hybrid patched conics makes two basic corrections. To commence the correction process the terms in series (7.78) except $\ddot{\boldsymbol{\epsilon}}\,|_{\boldsymbol{\epsilon}=0}$ are disregarded and the formulas of Section 7.4.2 are utilized to obtain the so-called first-order correction. Once this correction has been obtained, since $\boldsymbol{\epsilon}$ at the points for which (7.77) are solved to yield **A**, **B**, **C**, etc., is also known, it is possible to return to (7.78) and include the effects of the second term. Numerical results have shown that it suffices to carry only two terms in the $\ddot{\boldsymbol{\epsilon}}$ expansion (see Section 7.4.6).

For the lunar case the expression for the perturbative acceleration, defined by the summation in (7.68), becomes

$$\ddot{\mathbf{r}}_p = m_3\left(\frac{\mathbf{r}_{23}}{r_{23}^{\;3}} - \frac{\mathbf{r}_{13}}{r_{13}^{\;3}}\right) = \mu_☾\left(\frac{\mathbf{r}_{v☾}}{r_{v☾}^{\;3}} - \frac{\mathbf{r}_{\oplus☾}}{r_{\oplus☾}^{\;3}}\right); \tag{7.80}$$

the geometry is illustrated in Figure 7.3. To avoid subscripts let **R** represent

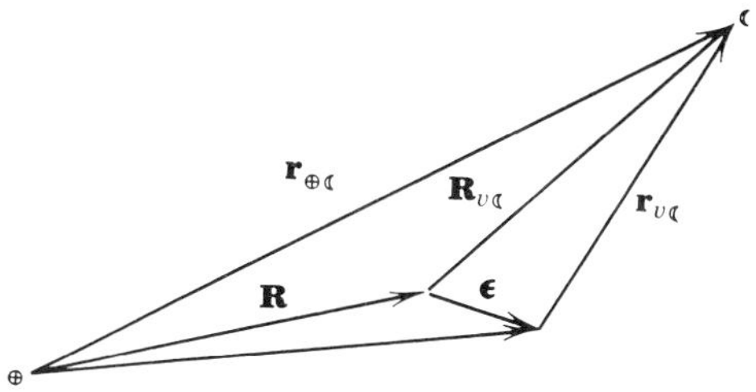

FIGURE 7.3 Three-body geometry.

the Keplerian position vector of the space vehicle and $\boldsymbol{\epsilon}$, as introduced previously, be the deviation $\mathbf{r} - \mathbf{R}$. With this understanding $\ddot{\mathbf{r}}_p$ on the Keplerian orbit can be written as

$$[\ddot{\mathbf{r}}_p]_k = \mu_☾\left(\frac{\mathbf{R}_{v☾}}{R_{v☾}^{\;3}} - \frac{\mathbf{r}_{\oplus☾}}{r_{\oplus☾}^{\;3}}\right) = -\mu_☾\left(\frac{\mathbf{R}_{☾v}}{R_{☾v}^{\;3}} + \frac{\mathbf{r}_{\oplus☾}}{r_{\oplus☾}^{\;3}}\right). \tag{7.81}$$

The total acceleration can therefore be written as

$$\ddot{\mathbf{r}} = -\mu_\oplus\frac{\mathbf{R}}{R^3} - \mu_☾\left(\frac{\mathbf{R}_{☾v}}{R_{☾v}^{\;3}} + \frac{\mathbf{r}_{\oplus☾}}{r_{\oplus☾}^{\;3}}\right), \tag{7.82}$$

where $\mu_\oplus$ and $\mu_☾$ are the masses of the Earth and the Moon, respectively.

By formal differentiation it follows that the elements of matrix (7.79) can be obtained as

$$\frac{\partial \ddot{\epsilon}_x}{\partial \epsilon_x} = -\frac{\mu_{\oplus}}{R^3} + 3\mu_{\oplus}\frac{X^2}{R^5} - \frac{\mu_{\leftmoon}}{R_{\leftmoon v}{}^3} + 3\mu_{\leftmoon}\frac{X_{\leftmoon v}^2}{R_{\leftmoon v}{}^5},$$

$$\frac{\partial \ddot{\epsilon}_x}{\partial \epsilon_y} = +3\mu_{\oplus}\frac{XY}{R^5} + 3\mu_{\leftmoon}\frac{X_{\leftmoon v}Y_{\leftmoon v}}{R_{\leftmoon v}{}^3},$$

$$\frac{\partial \ddot{\epsilon}_x}{\partial \epsilon_z} = +3\mu_{\oplus}\frac{XZ}{R^5} + 3\mu_{\leftmoon}\frac{X_{\leftmoon v}Z_{\leftmoon v}}{R_{\leftmoon v}{}^5},$$

$$\frac{\partial \ddot{\epsilon}_y}{\partial \epsilon_x} = +3\mu_{\oplus}\frac{XY}{R^5} + 3\mu_{\leftmoon}\frac{X_{\leftmoon v}Y_{\leftmoon v}}{R_{\leftmoon v}{}^5},$$

$$\frac{\partial \ddot{\epsilon}_y}{\partial \epsilon_y} = -\frac{\mu_{\oplus}}{R^3} + 3\mu_{\oplus}\frac{Y^2}{R^5} - \frac{\mu_{\leftmoon}}{R_{\leftmoon v}{}^3} + 3\mu_{\leftmoon}\frac{Y_{\leftmoon v}{}^2}{R_{\leftmoon v}{}^5},$$

$$\frac{\partial \ddot{\epsilon}_y}{\partial \epsilon_z} = +3\mu_{\oplus}\frac{YZ}{R^5} + 3\mu_{\leftmoon}\frac{Y_{\leftmoon v}Z_{\leftmoon v}}{R_{\leftmoon v}{}^5},$$

$$\frac{\partial \ddot{\epsilon}_z}{\partial \epsilon_x} = +3\mu_{\oplus}\frac{XZ}{R^5} + 3\mu_{\leftmoon}\frac{X_{\leftmoon v}Z_{\leftmoon v}}{R_{\leftmoon v}{}^5},$$

$$\frac{\partial \ddot{\epsilon}_z}{\partial \epsilon_y} = +3\mu_{\oplus}\frac{YZ}{R^5} + 3\mu_{\leftmoon}\frac{Y_{\leftmoon v}Z_{\leftmoon v}}{R_{\leftmoon v}{}^5},$$

$$\frac{\partial \ddot{\epsilon}_z}{\partial \epsilon_z} = -\frac{\mu_{\oplus}}{R^3} + 3\mu_{\oplus}\frac{Z^2}{R^5} - \frac{\mu_{\leftmoon}}{R_{\leftmoon v}{}^3} + 3\mu_{\leftmoon}\frac{Z_{\leftmoon v}{}^2}{R_{\leftmoon v}{}^5}.$$

7.4.5 Additional Comments on the Hybrid Patched Conic Technique

The hybrid technique, as outlined in preceding sections, has been particularly aimed at the lunar trajectory problem. As may be evident from the correction equations, the improvement provided by the technique results in times of puncture at the MSI slightly different from those provided by the patched conic equations. This problem can be most easily circumvented by letting the radius of the MSI change slightly as the first- and second-order corrections are made. Indeed, depending on when the points for the determination of fundamental constants are taken, the hybrid technique is valid for an extended distance into the MSI. Once inside the MSI, depending on the accuracy desired, either the previous technique can be applied to a Moon-centered orbit or a simple Keplerian orbit can be initiated at the final values of the corrected orbital position and velocity vectors. Since the hybrid technique works for an extended distance into the MSI, the latter approach usually suffices.

7.4.6 Numerical Results

Numerical studies of trajectories departing from parking orbits near the Earth and going to the vicinity of the lunar MSI have indicated that the hybrid patched conic technique has several advantages. First and foremost is the ease of implementation of the method. As may be evident from the equations developed in preceding sections, once the ordinary patched conic

Table 7.1 Trajectory Position Discrepancy

Time from Launch (hours)	Position Discrepancy between Integrated and Patched Conic Trajectory $\lvert\epsilon\rvert$ (percent)	Position Discrepancy between Integrated and Hybrid Patched Conic Trajectory $\lvert\epsilon\rvert$ (percent)
0	0%	0%
20.0000	0.025%	0.004%
40.0000	0.118%	0.006%
60.0000	0.329%	0.011%
80.0000	0.790%	0.016%
93.4677	1.470%	0.028%

trajectory is known, only a minimum amount of computation is required to yield an improved orbital state. Studies have indicated that only a doubling of computing time over and above the ordinary patched conic technique is incurred. Secondly, a marked increase in accuracy can be obtained by utilizing the proposed technique. A measure of the accuracy of the method is given in Tables 7.1 and 7.2. In particular, Table 7.1 compares the position error between a numerically integrated trajectory (Cowell's method) and a patched conic trajectory. It also shows the percent errors between the integrated and hybrid patched conic trajectories. Examination of Table 7.2,

Table 7.2 Trajectory Velocity Discrepancy

Time from Launch (hours)	Velocity Discrepancy between Integrated and Patched Conic Trajectory $\lvert\dot{\epsilon}\rvert$ (percent)	Velocity Discrepancy between Integrated and Hybrid Patched Conic Trajectory $\lvert\dot{\epsilon}\rvert$ (percent)
0	0%	0%
20.0000	0.110%	0.008%
40.0000	0.621%	0.025%
60.0000	2.21%	0.044%
80.0000	7.80%	0.170%
93.4677	23.50%	0.215%

which shows the respective errors for velocity components, clearly demonstrates the marked improvement of the method over the ordinary patched conic method. The interested reader can find more detailed numerical investigations in [18].

7.5 SUMMARY

In this chapter the definition of special perturbations was stated. It was seen that special perturbations is a procedure depending on numerical integration of the accelerations acting on a vehicle subjected to forces of other bodies and additional forces, such as thrust, drag, lift.

The total acceleration acting on a body was shown to be separable into two parts, the central or two-body acceleration and the so-called perturbative acceleration. The perturbative acceleration contains all other accelerations to be considered in the solution of a physical problem with the exception of the two-body acceleration.

In the method of Cowell the direct sum of the central and perturbative accelerations is integrated. The implementation of this technique is simple, but integration step size must be restricted.

The method of Encke integrates the acceleration difference between the acceleration on an adopted reference orbit and the total acceleration acting on the space vehicle or body. It was shown that it appeared feasible to adopt more sophisticated reference orbits than the usual Keplerian orbit. To this end a secularly perturbed orbit was adopted as the reference orbit. This device permitted greater integration step size. Even the regular Encke method permits greater step size than Cowell's method and therefore possesses greater computing speed, but is more complex to implement.

A new method for the approximate computation of lunar trajectories was introduced. The method, called the hybrid patched conic technique, relies on the use of an ordinary patched conic trajectory as a reference orbit. From this reference orbit it is possible to obtain sufficient information to permit analytic formulas to be obtained which can predict the error between patched conic and integrated trajectories with high accuracy. The method has the advantage of extreme rapidity and ease of implementation. This technique requires further investigation. It was shown numerically that good results are obtained when the method is used to compute Apollo-type trajectories.

EXERCISES

1. Why are special perturbation techniques restricted to short range prediction of the motion of a planet or space vehicle? Can anything be done to increase the range of applicability?

2. What is a hybrid perturbation technique? Give a specific example of such a technique.
3. Before commencing a special perturbation scheme, why must the initial conditions be referred to an appropriate coordinate system, that is, why will not any coordinate system suffice?
4. What is the main advantage of using a modified Encke procedure utilizing a secularly perturbed orbit?
5. Derive the time rates of change of **P** and **Q**, that is, (7.26), if ω, Ω, and i are varying with time.
6. Develop an algorithm for Encke's method if both secular drag and gravitational perturbations are to be included.
7. In the method of Encke, that is, the variation of the method that uses a secularly perturbed reference orbit, verify that

$$\frac{d\bar{E}}{dE} = \frac{\bar{n}r}{n\bar{r}}.$$

8. What is orbital rectification? When should it be performed in the Encke technique?
9. Would you expect the hybrid patched conic technique to work equally well for return trajectories from the lunar vicinity? Why?
10. Do you believe that adding more terms to the series defined by (7.71) will result in substantially improved orbits for the method of hybrid patched conics?

REFERENCES

[1] P. R. Escobal, *Methods of Orbit Determination*, John Wiley and Sons, New York, 1965.

[2] D. Brouwer and G. M. Clemence, *Methods of Celestial Mechanics*, Academic Press, New York and London, 1961.

[3] F. R. Moulton, *An Introduction to Celestial Mechanics*, The Macmillan Company, New York, 1914.

[4] R. M. L. Baker, Jr., and M. W. Makemson, *An Introduction to Astrodynamics* Academic Press, New York, 1960.

[5] H. C. Plummer, *An Introductory Treatise on Dynamical Astronomy*, Dover Publications, New York, 1960.

[6] P. R. Escobal, *Non Two-Body Reference Orbit for Encke's Method, AIAA Journal*, June 1966.

[7] W. T. Kyner and M. M. Bennett, JAS Specialist Meeting, Denver, July 1966.

[8] T. E. Sterne, *An Introduction to Celestial Mechanics*, Interscience, New York, 1960.

[9] J. E. Ball, M. L. Birkholz, and P. R. Escobal, *An Encke Special Perturbations Program for Geocentric, Earth-Satellite and Lunar/Cislunar Trajectories*, Lockheed California Co., LAC/421571, February 1964.

[10] J. R. Newman, *The World of Mathematics*, Volume 4, Simon & Schuster, 1956.

[11] *Explanatory Supplement to the Astronomical Ephemeris and the American Ephemeris and Nautical Almanac*, Her Majesty's Stationery Office, London, 1961.

[12] *Connaissance des Temps*, Le Bureau des Longitudes, Paris, France, 1965.

[13] F. Shaffer, R. K. Squires, and H. Wolf, *Interplanetary Encke Method Program Manual*, X-640-63-71, Goddard Space Flight Center, Greenbelt, Md., 1963.

[14] R. H. Battin, *Astronautical Guidance*, McGraw-Hill Book Company, New York, 1964.

[15] S. Pines and H. Wolf, *Generalized Perturbation Methods in Trajectory Analysis*, First International Symposium on Analytical Astrodynamics, UCLA, Los Angeles, June 27–29, 1961.

[16] F. B. Hildebrand, *Introduction to Numerical Analysis*, McGraw-Hill Book Company, New York, 1956.

[17] P. R. Escobal, *Inclusion of Planetary Perturbations into Patched Conic Programs*, TRW Systems, 3431.2-19, March 1967.

[18] P. R. Escobal and G. S. Stern, *The Hybrid Patched Conic Technique*, TRW Independent Research Report, 05952-6170-R000, July 1967.

8 Precise Lunar Coordinate Transformations

No illustration of astronomical precession can be devised more perfect than that presented by the properly balanced top . . . .

J. C. MAXWELL [35]

8.1 DETERMINATION OF AUXILIARY VARIABLES REQUIRED FOR LUNAR TRANSFORMATIONS

Lunar coordinate transformations are of fundamental importance in precise trajectory generation. The subject of this chapter augments and extends the special perturbation techniques discussed in Chapter 7. Specifically, this chapter is aimed at the precise location of landing areas relative to the lunar surface. Important effects such as precession and nutation are introduced and discussed from the point of view of determining the proper coordinate systems for the integration of the equations of motion relative to real world reference frames. The importance of these transformations stems from the fact that the neglect of such effects as precession and nutation can cause position and velocity errors to be incurred which are equally important as are other perturbative sources, for example, distant planets.

The transformations presented in this chapter are precise extensions of the algorithms stated in [1]. Most of the analysis was performed by Roth [13], [33], [34] and is patterned after the development of Gabbard [5], [9], [14].

8.1.1 Definition and Causes of Precession and Nutation

A top is spinning. As the top, illustrated in Figure 8.1, rotates about its axis of symmetry OS, it has a tendency to wobble and generate a cone with vertex angle ϕ from the vertical OV. It would appear at first that the weight of the top acting through the center of gravity should cause the half cone

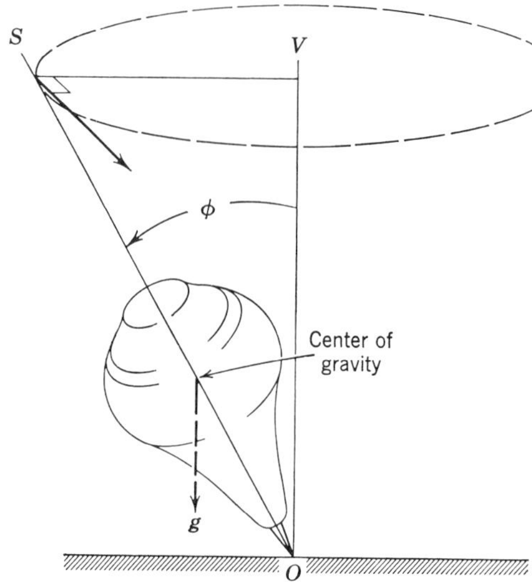

FIGURE 8.1 Spinning top.

angle ϕ to increase as a result of the moment created about the ground contact point O. Observations show, however, that the divergence of ϕ does not occur until the rotational rate of the top is very slow. The top continues to wobble about the vertical OV for a considerable length of time, or until rotation practically ceases. In fact, instead of falling to the ground, the axis of symmetry OS is instantaneously moving at right angles to the plane defined by OS and OV. The motion of axis OS is called precession. This dynamical interpretation of precession was first given by Newton. Let the Earth be substituted for the top. Of course, the correspondence of the top and the Earth is used as a superficial example, with different principles involved, in order to expedite the explanation of precession.

If the equatorial bulge of the Earth is thought of as an extra ring of matter surrounding a perfectly spherical Earth, it follows that the attraction of the Sun and Moon gives rise to a couple that has a tendency to turn the ring into the plane of the ecliptic[1] because both the perturbing bodies are in or near the ecliptic. In the analogy the ecliptic would be substituted for the ground and the perpendicular to the ecliptic would correspond to the vertical. In analogy to the top, if the Earth were not spinning, the couple caused by the Sun and Moon would turn the Earth over until equator and ecliptic were coincident.

[1] The plane traced on the celestial sphere by the apparent annual motion of the Sun around the Earth.

The spinning Earth, however, behaves like a gyrostat and its axis moves at right angles to the plane of the perturbing couple. Therefore, as the top, the axis of the Earth moves about the normal to the ecliptic and generates a cone.

The period of generation of this cone is approximately 26,000 years and corresponds to the equinox, or intersection of equator and ecliptic planes, movement during this cycle. As can be seen from Figure 8.2, this corresponds to the equinox, ♈, moving backward along the ecliptic.

The vernal equinox is also called the first point of Aries or the sign of the ram's horns, ♈. A relatively short while ago, the equinox did point

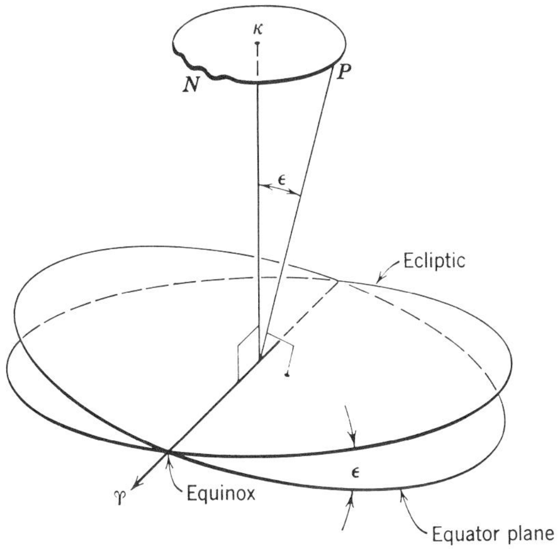

FIGURE 8.2 Regression of the equinox.

toward this particular point in the heavens. Today the precessional movement has drifted ♈ until it now points at the constellation of Pisces, the fish, denoted by ♓. The north pole of the celestial sphere describes a circle with center κ of radius ϵ called the obliquity of the ecliptic. The obliquity of the ecliptic is approximately equal to $23^{\circ}.5$.

At present the pole is near the star α Ursae Minoris, the pole or north star. In prehistoric times, about 3000 B.C., the polestar would have been α Draconis, whereas in A.D. 13000 Vega will be situated where the pole of the Earth should be pointing, barring unforeseen circumstances.

The sum of the major oscillatory precessional effects caused by the Moon and the minor oscillatory effects due to the Sun is called *nutation.* The orbit of the Moon, inclined to the ecliptic by about 5°, is constantly undergoing change in orientation. Actually, the nodes of the lunar orbit move around the ecliptic in 18.6 years. Hence, as a function of lunar position, a deviation from the ecliptic will occur. Over a period of time these deviations will cancel; however, at any specified time a periodic contribution will be available.

To reiterate, in classical astronomical theory the average secular motion of the equinox, which is due to the combined motion of the ecliptic and celestial equator, is termed *precession.* On the other hand, the periodic fluctuations are called nutation. The periodic fluctuations have a tendency to superimpose a sinuous curve on the circle generated by the movement of the Earth's pole in Figure 8.2, as illustrated by point N. The part of the total precession, which is due to the motion of the celestial equator, is commonly called *lunisolar precession.*

The remaining bodies of the solar system, owing to analogous physical principles, also cause precession to occur. This is called *planetary precession.* The effect of planetary precession is to perturb the ecliptic, thereby producing an additional component in the motion of the equinox. The combined effects of lunisolar and planetary precession, or the *general precession*, are treated in this chapter. The physical description of precession and nutation is modeled after the discussion in [12].

In analytical form, expressions for the backward regression of the vernal equinox can be justified by considering an expansion of the form

$$a_1 t + a_2 \sin 2L, \tag{8.1}$$

where a_1, a_2 are constants determined from dynamical theory and L is the longitude of the Sun. This relationship expresses the fact that during one orbital period the magnitude of the radius vector from the Earth to the Sun is not constant and thus affects the equatorial bulge of the Earth in different ways; hence the periodic trigonometric term. The first term of (8.1) denotes the steady uniform backward movement of the vernal equinox. This term is referred to as the secular term, in which time t appears explicitly, in opposition to the remaining periodic terms.

The Moon, owing to the ellipticity of its orbit, also affects the equatorial bulge of the Earth, and thus the approximate perturbing effects can be written in analogous fashion as

$$a_3 t + a_4 \sin 2☾, \tag{8.2}$$

where a_3, a_4 are constants determined from dynamical theory and ☾ is the longitude of the Moon. The assumption is made that the orbit of the Moon lies in the ecliptic plane. Since the orbit of the Moon is inclined to the ecliptic by about 5°, this effect must also be accounted for in the motion of the vernal equinox.

To compensate for the out of ecliptic situation or position of the Moon's orbital plane with respect to the ecliptic plane, consider the expression

$$a_5 \sin \Omega + a_6 \sin 2\Omega, \tag{8.3}$$

where a_5, a_6 are constants obtained from dynamical theory and Ω is the longitude of the ascending node of the lunar orbit. From observations it is possible to verify that the Moon's orbit plane is not fixed with respect to the ecliptic plane. In fact, the perpendicular to the Moon's orbit plane moves about the perpendicular to the ecliptic plane in a small circle with a period of 18.6 years. Expression 8.3 expresses the corresponding variance of Ω.

It follows by addition and combination of the constants in (8.1), (8.2), and (8.3) that the motion of the equinox on the ecliptic can be approximately expressed as

$$at + b \sin \Omega + c \sin 2\Omega + d \sin 2L + e \sin 2☾, \tag{8.4}$$

where a, b, c, d, and e are constants. In classical astronomy, at is the luni-solar precession in longitude, and the trigonometric terms are the nutation in longitude.

As previously noted, the variance of the perpendiculars to the Moon's orbit and ecliptic planes also cause a variance in ϵ, the obliquity of the ecliptic, to occur in a periodic fashion. These effects can be similarly represented by an expansion of the form

$$f \cos \Omega + g \cos 2\Omega + h \cos 2L + i \cos 2☾, \tag{8.5}$$

where f, g, h, and i are constants derivable from dynamical theory. These strictly periodic terms are called the nutation in obliquity, denoted by $\delta\epsilon$.

The expressions for precession and nutation can be developed with high precision by carrying a large number of terms. Perhaps the most complete version of these expansions is given in [3] and is employed in this chapter.

Also, it is well to mention that the dynamical analysis employed here is carried out in the standard notation, and nondimensional units of classical astrodynamical analysis, in which the unit of distance is the Earth's radius (e.r.), the unit of velocity is the velocity of circular satellite at the unit distance, and the unit of time is the classical modified time variable [1] defined by

$$\tau \equiv k(t - t_0), \tag{8.6}$$

where $k \equiv$ the planetary constant, $t \equiv$ universal or more precisely ephemeris time, and $t_0 \equiv$ epoch universal or ephemeris time.

It should be noted that k is, by definition and not by physical approximation, taken to be equal to 0.07436574 $(\text{e.r.})^{3/2}/\text{min}$. For a further discussion of this system of units, (see [1], Section 1).

8.1.2 Mean Angles Necessary for Selenographic Transformations

Selenographic transformations are fundamentally based on rotations which utilize small deviations from a set of mean angles. Through observations these mean angles can be obtained as functions of time that are representable as polynomials. In this section the mean angles are defined and the polynomial expansions stated. Expressions for the time rate of change of these angles, that is, the mean longitudes and the mean obliquity, are also obtained.

These mean longitudes and the mean obliquity are defined as follows.

$\Gamma \equiv$ mean longitude of the Sun's perigee measured in the ecliptic plane from the mean equinox of date.

$\Gamma' \equiv$ mean longitude of the Moon's perigee measured in the ecliptic plane from the mean equinox of date to the mean ascending node of the lunar orbit and then along the orbit.

$L \equiv$ mean longitude of the Sun measured in the ecliptic plane from the mean equinox of date.

$\Omega \equiv$ longitude of the mean ascending node of the lunar orbit measured in the ecliptic plane from the mean equinox of date.

☾ $\equiv$ geocentric mean longitude of the Moon measured in the ecliptic plane from the mean equinox of date to the mean ascending node of the lunar orbit, and then along the orbit.

$\epsilon \equiv$ angle between the mean celestial equator and the ecliptic (mean obliquity).

The mean longitude of the Sun's perigee, Γ, and the mean obliquity, ϵ, are slowly varying functions of time as compared to the other four mean longitudes. The nutations and librations are small periodic variations about the mean position. Some auxiliary angles that are frequently used are $g' \equiv$ ☾ $- \Gamma'$, $g \equiv L - \Gamma$, $\omega \equiv \Gamma - \Omega$, and $\omega' \equiv \Gamma' - \Omega$; they are defined more accurately in Section 8.1.5.

The expressions for the mean longitudes adopted in this chapter are due to Woolard [16]. They are the expressions utilized in the *Explanatory Supplement* [6]. The expression for the mean obliquity is due to Newcomb [2]. It is also adopted in [3] and [6]. These expressions and alternative expressions appear in *Connaissance des Temps* [3].

The previously defined parameters are obtained as functions of the independent variable

$$T_u = \frac{\text{J.D.} - (\text{J.D.})_{\text{Jan }0.5,\ 1900}}{36525.0}, \tag{8.7}$$

where J.D. is the Julian date at instant [1] with the epoch date taken as

$$(\text{J.D.})_{\text{Jan }0.5,\ 1900} = 2415020.0. \tag{8.8}$$

The variable T_u is the time in Julian centuries of 36,525.0 ephemeris days[2] from the epoch date. The derivative $\dot{T}_u$ of T_u, with respect to the modified time variable τ, is the constant value

$$\dot{T}_u \equiv \frac{dT_u}{d\tau} = (36{,}525.0 \times 24 \times 60 \times 0.074\,36574)^{-1}$$
$$\simeq 0.255\,666824 \times 10^{-6}. \tag{8.9}$$

The mean angles and the mean obliquity are then given, as in [1, Appendix 1], by the expressions

$$\Gamma = 281^\circ.220\,8333 + 1^\circ.719\,1750T_u + 0^\circ.452\,7778 \times 10^{-3}T_u^2 + 0^\circ.333\,3333 \times 10^{-5}T_u^3,$$
$$\Gamma' = 334^\circ.329\,5556 + 4069^\circ.034\,0333T_u - 0^\circ.103\,2500 \times 10^{-1}T_u^2 - 0^\circ.125\,0000 \times 10^{-4}T_u^3,$$
$$L = 279^\circ.696\,6778 + 36{,}000^\circ.768\,9250T_u + 0^\circ.302\,5000 \times 10^{-3}T_u^2,$$
$$\Omega = 259^\circ.183\,2750 - 1{,}934^\circ.142\,0083T_u + 0^\circ.207\,7778 \times 10^{-2}T_u^2 + 0^\circ.222\,2222 \times 10^{-5}T_u^3,$$
$$☾ = 270^\circ.434\,1639 + 481{,}267^\circ.883\,1417T_u - 0^\circ.113\,3333 \times 10^{-2}T_u^2 + 0^\circ.188\,8889 \times 10^{-5}T_u^3,$$
$$\epsilon = 23^\circ.452\,2944 - 0^\circ.130\,1250 \times 10^{-1}T_u - 0^\circ.163\,8889 \times 10^{-5}T_u^2 + 0^\circ.502\,7778 \times 10^{-6}T_u^3. \tag{8.10}$$

The time derivatives of the mean longitudes and the mean obliquity are readily obtained by differentiation as follows.

$$\dot{\Gamma} \equiv \frac{d\Gamma}{d\tau} = \frac{d\Gamma}{dT_u}\dot{T}_u = [1^\circ.719\,1750 + 0^\circ.905\,5556 \times 10^{-3}T_u + 1^\circ.000\,0000 \times 10^{-5}T_u^2]\frac{\pi}{180}\dot{T}_u,$$

$$\dot{\Gamma}' \equiv \frac{d\Gamma'}{d\tau} = \frac{d\Gamma'}{dT_u}\dot{T}_u = [4069^\circ.034\,0333 - 0^\circ.206\,5000 \times 10^{-1}T_u - 0^\circ.375\,0000 \times 10^{-4}T_u^2]\frac{\pi}{180}\dot{T}_u,$$

$$\dot{L} \equiv \frac{dL}{d\tau} = \frac{dL}{dT_u}\dot{T}_u = [36{,}000^\circ.768\,9250 + 0^\circ.605\,0000 \times 10^{-3}T_u]\frac{\pi}{180}\dot{T}_u,$$

$$\dot{\Omega} \equiv \frac{d\Omega}{d\tau} = \frac{d\Omega}{dT_u}\dot{T}_u = [-1{,}934^\circ.142\,0083 + 0^\circ.415\,5556 \times 10^{-2}T_u + 0^\circ.666\,6667 \times 10^{-5}T_u^2]\frac{\pi}{180}\dot{T}_u,$$

[2] In these expressions either universal or ephemeris time can be used.

$$\dot{☾} \equiv \frac{d☾}{d\tau} = \frac{d☾}{dT_u}\dot{T}_u = [481{,}267^\circ.883\,1417 - 0^\circ.226\,6667 \times 10^{-2}T_u$$

$$+ 0^\circ.566\,6667 \times 10^{-5}T_u^{\,2}]\frac{\pi}{180}\dot{T}_u,$$

$$\dot{\epsilon} \equiv \frac{d\epsilon}{d\tau} = \frac{d\epsilon}{dT_u}\dot{T}_u = [-0^\circ.130\,1250 \times 10^{-1} - 0^\circ.327\,7778 \times 10^{-5}T_u$$

$$+ 1^\circ.508\,3333 \times 10^{-6}T_u^{\,2}]\frac{\pi}{180}\dot{T}_u, \tag{8.11}$$

where $\dot{T}_u$ is defined by (8.9).

These values of Ω, ☾, ϵ, $\dot{\Omega}$, $\dot{☾}$, and $\dot{\epsilon}$ are used directly in the selenographic transformation discussed in Section 8.2. The values of all the mean angles, as well as the five corresponding derivatives, are also required in the determination of the expressions for the nutations and lunar librations.

8.1.3 Computation of the Nutations in Obliquity and Longitude

The nutation equations presented in this section compute the nutations in obliquity and longitude from the previously obtained values of the mean longitudes and T_u. The nutations are expressed as sines and cosines of various linear combinations of five mean angles.

Woolard's expressions [3], [16] for the nutation expansions contain a total of 109 sine and cosine terms. Since digital computers are quite slow computing trigonometric functions, the retention of the full expansions would appear to produce an excessively long computational time for this phase of the transformation. On the other hand, truncation of the series, as done in [10], entails an undesirable loss of accuracy. Both speed and accuracy can be attained by retaining all Woolard's terms while reducing the computing time by replacing all the trigonometric terms in Woolard's expansion by equivalent sums and products of the sines and cosines of the five fundamental mean longitudes.

The basic purpose of such a reorganization of the equations is to reduce the number of computations of trigonometric functions at the expense of increasing the number of additions and multiplications, thereby achieving a significant reduction in overall computation time for the nutation terms.

The expressions given by Woolard [16] for the long period and short period terms of the nutation in obliquity and longitude may be written in the form

$$d\epsilon = \sum_{i=1}^{8} A_i X_i + (1 - \lambda) \sum_{i=9}^{24} A_i X_i,$$

$$\Delta\epsilon = \sum_{i=1}^{7} B_i Y_i + (1 - \lambda) \sum_{i=8}^{16} B_i Y_i,$$

$$d\psi = \sum_{i=1}^{12} C_i Z_i + (1 - \lambda) \sum_{i=13}^{46} C_i Z_i,$$

$$\Delta\psi = \sum_{i=1}^{9} D_i W_i + (1 - \lambda) \sum_{i=10}^{23} D_i W_i, \tag{8.12}$$

where $d\epsilon$ = short period terms of the nutation in obliquity, $\Delta\epsilon$ = long period terms of the nutation in obliquity, $d\psi$ = short period terms of the nutation in longitude, and $\Delta\psi$ = long period terms of the nutation in longitude.

According to [3], [6] the short period terms of the nutation have periods of less than 40 days while the long period terms have periods of greater than 90 days. The total nutations in obliquity and longitude are given, respectively, by adding the short and long period effects as follows

$$\delta\epsilon = d\epsilon + \Delta\epsilon, \qquad \delta\psi = d\psi + \Delta\psi. \tag{8.13}$$

The X_i, Y_i, Z_i, and W_i terms in (8.12) correspond to sine and cosine functions. These functions and their corresponding coefficients are given in the following development. The expressions in (8.12) include all terms for which the coefficients have a magnitude of $0''.0002$ or more

The parameter λ in (8.12) has the value of either 0 or 1. When $\lambda = 0$, all the terms in (8.12) are retained. If $\lambda = 1$, then the series is truncated such that the only terms retained are of major order [10]. This technique facilitates checking against independent computations based on other expressions and provides a measure of the effects of the truncation employed in [10]. In passing, it is well to note that the unexpanded version of the following equation as displayed in Exercise 4 can be found in [3].

The sine and cosine functions are computed as strict multiplications, additions, and subtractions as follows (8.14):

$$X_1 = \cos 2☾ = 1 - 2\sin^2 ☾,$$

$$\sin 2☾ = 2 \sin ☾ \cos ☾,$$

$$X_2 = \cos (2☾ - \Omega) = X_1 \cos \Omega + \sin 2☾ \sin \Omega,$$

$$\sin (2☾ - \Omega) = \sin 2☾ \cos \Omega - X_1 \sin \Omega,$$

$$\cos g' = \cos ☾ \cos \Gamma' + \sin ☾ \sin \Gamma',$$

$$
\begin{aligned}
\sin g' &= \sin \leftmoon \cos \Gamma' - \cos \leftmoon \sin \Gamma', \\
X_3 &= \cos (2\leftmoon + g') = X_1 \cos g' - \sin 2\leftmoon \sin g', \\
\sin (2\leftmoon + g') &= \sin 2\leftmoon \cos g' + X_1 \sin g', \\
X_4 &= \cos (\leftmoon + \Gamma') = \cos \leftmoon \cos \Gamma' - \sin \leftmoon \sin \Gamma', \\
X_5 &= \cos (g' + \Omega) = \cos g' \cos \Omega - \sin g' \sin \Omega, \\
X_6 &= \cos (g' - \Omega) = \cos g' \cos \Omega + \sin g' \sin \Omega, \\
\sin (g' - \Omega) &= \sin g' \cos \Omega - \cos g' \sin \Omega, \\
X_7 &= \cos [2\leftmoon + (g' - \Omega)] \\
&= X_1 X_6 - \sin 2\leftmoon \sin (g' - \Omega), \\
\cos 4\leftmoon &= 1 - 2 \sin^2 2\leftmoon, \\
\sin 4\leftmoon &= 2X_1 \sin 2\leftmoon, \\
\cos 2L &= 1 - 2 \sin^2 L, \\
\sin 2L &= 2 \sin L \cos L, \\
X_9 &= \cos (4\leftmoon - 2L) = \cos 4\leftmoon \cos 2L + \sin 4\leftmoon \sin 2L, \\
\sin (4\leftmoon - 2L) &= \sin 4\leftmoon \cos 2L - \cos 4\leftmoon \sin 2L, \\
X_8 &= \cos [(4\leftmoon - 2L) - g'] \\
&= X_9 \cos g' + \sin (4\leftmoon - 2L) \sin g', \\
\sin [(4\leftmoon - 2L) - g'] &= \sin (4\leftmoon - 2L) \cos g' - X_9 \sin g', \\
X_{10} &= \cos [(2\leftmoon + g') + g'] \\
&= X_3 \cos g' - \sin (2\leftmoon + g') \sin g', \\
X_{11} &= \cos (2L + g') = \cos 2L \cos g' - \sin 2L \sin g', \\
\sin (2L + g') &= \sin 2L \cos g' + \cos 2L \sin g', \\
\cos \omega' &= \cos \Gamma' \cos \Omega + \sin \Gamma' \sin \Omega, \\
\sin \omega' &= \sin \Gamma' \cos \Omega - \cos \Gamma' \sin \Omega, \\
X_{12} &= \cos (\leftmoon + \omega') = \cos \leftmoon \cos \omega' - \sin \leftmoon \sin \omega', \\
\cos (2L - \leftmoon) &= \cos 2L \cos \leftmoon + \sin 2L \sin \leftmoon, \\
\sin (2L - \leftmoon) &= \sin 2L \cos \leftmoon - \cos 2L \sin \leftmoon, \\
X_{13} &= \cos [(2L - \leftmoon) - \omega'] = \cos (2L - \leftmoon) \cos \omega' \\
&\qquad + \sin (2L - \leftmoon) \sin \omega', \\
\sin [(2L - \leftmoon) - \omega'] &= \sin (2L - \leftmoon) \cos \omega' - \cos (2L - \leftmoon) \sin \omega', \\
\cos 2\Omega &= 1 - 2 \sin^2 \Omega, \\
\sin 2\Omega &= 2 \sin \Omega \cos \Omega, \\
X_{14} &= \cos (2L - \leftmoon - \omega' - 2\Omega) \\
&= X_{13} \cos 2\Omega + \sin (2L - \leftmoon - \omega') \sin 2\Omega,
\end{aligned}
$$

$$
\begin{aligned}
X_{15} &= \cos [(4☾ - 2L - g') - \Omega] \\
&= X_8 \cos \Omega + \sin (4☾ - 2L - g') \sin \Omega, \\
\cos g &= \cos L \cos \Gamma + \sin L \sin \Gamma, \\
\sin g &= \sin L \cos \Gamma - \cos L \sin \Gamma, \\
X_{16} &= \cos (2☾ - g) = X_1 \cos g + \sin 2☾ \sin g, \\
X_{17} &= \cos [(4☾ - 2L) + g'] \\
&= X_9 \cos g' - \sin (4☾ - 2L) \sin g', \\
X_{18} &= \cos [(2☾ - \Omega) - 2L] \\
&= X_2 \cos 2L + \sin (2☾ - \Omega) \sin 2L, \\
\sin [(2☾ - \Omega) - 2L] &= \sin (2☾ - \Omega) \cos 2L - X_2 \sin 2L, \\
X_{19} &= \cos (2☾ - 2L + \Omega) \\
&= \cos [(2☾ - \Omega - 2L) + 2\Omega] \\
&= X_{18} \cos 2\Omega - \sin (2☾ - \Omega - 2L) \sin 2\Omega, \\
X_{20} &= \cos [(4☾ - 2L) - \Omega] \\
&= X_9 \cos \Omega + \sin (4☾ - 2L) \sin \Omega, \\
X_{21} &= \cos (2☾ + g) = X_1 \cos g - \sin 2☾ \sin g, \\
X_{22} &= \cos [2L + (g' - \Omega)] \\
&= X_6 \cos 2L - \sin (g' - \Omega) \sin 2L, \\
X_{23} &= \cos (2☾ - \Omega + 2g') \\
&= \cos [(2☾ + g') + (g' - \Omega)] \\
&= X_3 X_6 - \sin (2☾ + g') \sin (g' - \Omega), \\
X_{24} &= \cos (2L + 2g') = \cos [(2L + g') + g'] \\
&= X_{11} \cos g' - \sin (2L + g') \sin g', \\
Y_1 &= \cos \Omega, \\
Y_2 &= \cos 2L, \\
Y_3 &= \cos 2\Omega, \\
Y_4 &= \cos (2L + g) = \cos 2L \cos g - \sin 2L \sin g, \\
\sin (2L + g) &= \sin 2L \cos g + \cos 2L \sin g, \\
Y_5 &= \cos (L + \Gamma) = \cos L \cos \Gamma - \sin L \sin \Gamma, \\
Y_6 &= \cos (2L - \Omega) = \cos 2L \cos \Omega + \sin 2L \sin \Omega, \\
\sin (2L - \Omega) &= \sin 2L \cos \Omega - \cos 2L \sin \Omega, \\
Y_7 &= \cos (\Gamma' + \omega') = \cos \Gamma' \cos \omega' - \sin \Gamma' \sin \omega', \\
\sin (\omega' + \Gamma') &= \sin \omega' \cos \Gamma' + \cos \omega' \sin \Gamma', \\
Y_8 &= \cos (\Omega + g) = \cos \Omega \cos g - \sin \Omega \sin g,
\end{aligned}
$$

$$
\begin{aligned}
Y_9 &= \cos(2L + 2g) = \cos[(2L + g) + g] \\
&= Y_4 \cos g - \sin(2L + g)\sin g, \\
Y_{10} &= \cos(\Omega - g) = \cos\Omega\cos g + \sin\Omega\sin g, \\
\cos 2\Gamma' &= 1 - 2\sin^2\Gamma', \\
\sin 2\Gamma' &= 2\sin\Gamma'\cos\Gamma', \\
Y_{11} &= \cos(2\Gamma' - 2L + \Omega) \\
&= Y_6\cos 2\Gamma' + \sin(2L - \Omega)\sin 2\Gamma', \\
\cos 2\Gamma &= 1 - 2\sin^2\Gamma, \\
\sin 2\Gamma &= 2\sin\Gamma\cos\Gamma, \\
Y_{13} &= \cos(2\Gamma - \Omega) = \cos 2\Gamma\cos\Omega + \sin 2\Gamma\sin\Omega, \\
\sin(2\Gamma - \Omega) &= \sin 2\Gamma\cos\Omega - \cos 2\Gamma\sin\Omega, \\
Y_{12} &= \cos(2\Gamma - \Omega + g) = \cos[(2\Gamma - \Omega) + g], \\
&= Y_{13}\cos g - \sin(2\Gamma - \Omega)\sin g, \\
Y_{14} &= \cos 2\Gamma', \\
Y_{15} &= \cos(2L - \omega' - \Gamma') = \cos[2L - (\omega' + \Gamma')] \\
&= Y_7\cos 2L + \sin 2L\sin(\omega' + \Gamma'), \\
Y_{16} &= \cos[(2L - \Omega) + g] \\
&= Y_6\cos g - \sin(2L - \Omega)\sin g, \\
Z_1 &= \sin 2☾, \\
Z_2 &= \sin g', \\
Z_3 &= \sin(2☾ - \Omega), \\
Z_4 &= \sin(2☾ + g'), \\
Z_5 &= \sin[(2L - ☾) - \Gamma'] \\
&= \sin(2L - ☾)\cos\Gamma' - \cos(2L - ☾)\sin\Gamma', \\
Z_6 &= \sin(☾ + \Gamma') = \sin☾\cos\Gamma' + \cos☾\sin\Gamma', \\
\cos(☾ + \Gamma') &= \cos☾\cos\Gamma' - \sin☾\sin\Gamma', \\
Z_7 &= \sin(2☾ - 2L) = \sin 2☾\cos 2L - X_1\sin 2L, \\
Z_8 &= \sin(\Omega + g') = \sin\Omega\cos g' + \cos\Omega\sin g', \\
Z_9 &= -\sin(g' - \Omega), \\
Z_{10} &= \sin(4☾ - 2L - g'), \\
Z_{11} &= \sin[(2☾ - \Omega) + g'] = Z_3\cos g' + X_2\sin g', \\
Z_{12} &= \sin 2g' = 2\sin g'\cos g', \\
Z_{13} &= \sin(4☾ - 2L), \\
Z_{14} &= \sin(2L + g'),
\end{aligned}
$$

$$
\begin{aligned}
Z_{15} &= \sin(2☾ + 2g') = \sin[(2☾ + g') + g'] \\
&= Z_4 \cos g' + X_3 \sin g', \\
Z_{16} &= \sin(2☾ - 2\Omega) = \sin[(2☾ - \Omega) - \Omega] \\
&= Z_3 \cos\Omega - X_2 \sin\Omega, \\
Z_{17} &= \sin(☾ + \omega') = \sin ☾ \cos\omega' + \cos ☾ \sin\omega', \\
Z_{18} &= \sin(2L - ☾ - \omega'), \\
Z_{19} &= \sin(☾ - 2L + \Omega + \Gamma') \\
&= \sin[(☾ + \Gamma') - (2L - \Omega)] \\
&= Z_6 Y_6 - \cos(☾ + \Gamma') \sin(2L - \Omega), \\
Z_{20} &= \sin(4☾ - 2L - \Omega - g') \\
&= \sin[(4☾ - 2L - g') - \Omega] \\
&= \sin(4☾ - 2L - g') \cos\Omega - X_8 \sin\Omega, \\
Z_{21} &= \sin(2☾ + g) = Z_1 \cos g + X_1 \sin g, \\
Z_{22} &= \sin(2L + g - ☾ - \Gamma') \\
&= \sin[(2L + g) - (☾ + \Gamma')] \\
&= \sin(2L + g) \cos(☾ + \Gamma') - Y_4 Z_6, \\
Z_{23} &= \sin(2L + 2g') = \sin[(2L + g') + g'] \\
&= \sin(2L + g') \cos g' + X_{11} \sin g', \\
Z_{24} &= \sin(2☾ - 2L + g') = \sin[(2☾ + g') - 2L] \\
&= \sin(2☾ + g') \cos 2L - X_3 \sin 2L, \\
Z_{25} &= \sin(2☾ - g) = Z_1 \cos g - X_1 \sin g, \\
Z_{26} &= \sin(4☾ - 2L + g') = \sin[(4☾ - 2L) + g'] \\
&= Z_{13} \cos g' + X_9 \sin g', \\
Z_{27} &= \sin(2☾ - 2L + \Omega) = \sin[2☾ - (2L - \Omega)] \\
&= Y_6 \sin 2☾ - X_1 \sin(2L - \Omega), \\
Z_{28} &= \sin(2L + g' - \Omega) = \sin[(2L + g') - \Omega] \\
&= \sin(2L + g') \cos\Omega - X_{11} \sin\Omega, \\
Z_{29} &= \sin(2L - 2☾ + \Omega) = \sin[2L - (2☾ - \Omega)] \\
&= X_2 \sin 2L - \sin(2☾ - \Omega) \cos 2L, \\
Z_{30} &= \sin(4☾ - 2L - \Omega) = \sin[(4☾ - 2L) - \Omega] \\
&= \sin(4☾ - 2L) \cos\Omega - X_9 \sin\Omega, \\
Z_{31} &= \sin(g' - g) = \sin g' \cos g - \cos g' \sin g, \\
\cos(☾ + \omega') &= \cos ☾ \cos\omega' - \sin ☾ \sin\omega',
\end{aligned}
$$

$$
\begin{aligned}
Z_{32} &= \sin(\Omega - \text{☾} - \omega') = \sin[\Omega - (\text{☾} + \omega')] \\
&= \sin\Omega\cos(\text{☾} + \omega') - Z_{17}\cos\Omega, \\
Z_{33} &= \sin(2\text{☾} - \Omega + 2g') \\
&= \sin[(2\text{☾} + g') + (g' - \Omega)] \\
&= X_6\sin(2\text{☾} + g') + X_3\sin(g' - \Omega), \\
Z_{34} &= \sin(2L + g - 2\text{☾}) = \sin[(2L + g) - 2\text{☾}] \\
&= X_1\sin(2L + g) - Y_4\sin 2\text{☾}, \\
Z_{35} &= \sin(\text{☾} - L) = \sin\text{☾}\cos L - \cos\text{☾}\sin L, \\
Z_{36} &= \sin(2\text{☾} + g' - 2\Omega) = \sin[(2\text{☾} + g') - 2\Omega] \\
&= \sin(2\text{☾} + g')\cos 2\Omega - X_3\sin 2\Omega, \\
Z_{37} &= \sin(g + g') = \sin g\cos g' + \cos g\sin g', \\
Z_{38} &= \sin(2\text{☾} + g' - g) = \sin[(2\text{☾} + g') - g] \\
&= \sin(2\text{☾} + g')\cos g - X_3\sin g, \\
Z_{39} &= \sin(\Omega + 2g') = \sin[(\Omega + g') + g'] \\
&= Z_8\cos g' + X_5\sin g', \\
Z_{40} &= \sin(2\text{☾} + g' + g) = \sin[(2\text{☾} + g') + g] \\
&= \sin(2\text{☾} + g')\cos g + X_3\sin g, \\
Z_{41} &= \sin(2g' - \Omega) = \sin[g' + (g' - \Omega)] \\
&= X_6\sin g' - Z_9\cos g', \\
Z_{42} &= \sin(2L - \Omega - g') = \sin[(2L - \Omega) - g'] \\
&= \sin(2L - \Omega)\cos g' - Y_6\sin g', \\
Z_{43} &= \sin(4\text{☾} - 2L - g' - g) \\
&= \sin[(4\text{☾} - 2L - g') - g] \\
&= \sin(4\text{☾} - 2L - g')\cos g - X_8\sin g, \\
Z_{44} &= \sin(4\text{☾} - 2L - g) = \sin[(4\text{☾} - 2L) - g] \\
&= \sin(4\text{☾} - 2L)\cos g - X_9\sin g, \\
Z_{45} &= \sin(2\Omega + g') = \sin 2\Omega\cos g' + \cos 2\Omega\sin g', \\
Z_{46} &= \sin(2\text{☾} + 3g') = \sin[(2\text{☾} + 2g') + g'] \\
&= Z_{15}\cos g' + X_{10}\sin g', \\
W_1 &= \sin\Omega, \\
W_2 &= \sin 2L, \\
W_3 &= \sin 2\Omega,
\end{aligned}
$$

$$
\begin{aligned}
W_4 &= \sin g, \\
W_5 &= \sin (2L + g), \\
W_6 &= \sin (L + \Gamma) = \sin L \cos \Gamma + \cos L \sin \Gamma, \\
W_7 &= \sin (2L - \Omega), \\
W_8 &= \sin (\omega' + \Gamma'), \\
W_9 &= \sin (2L - 2\Gamma') \\
&= \sin 2L \cos 2\Gamma' - \cos 2L \sin 2\Gamma', \\
W_{10} &= \sin (2L - 2\Omega) \\
&= \sin 2L \cos 2\Omega - \cos 2L \sin 2\Omega, \\
W_{11} &= \sin 2g = 2 \sin g \cos g, \\
W_{12} &= \sin (2L + 2g) = \sin [(2L + g) + g] \\
&= \sin (2L + g) \cos g + Y_4 \sin g, \\
W_{13} &= \sin (g + \Omega) = \sin g \cos \Omega + \sin \Omega \cos g, \\
W_{14} &= \sin 2\omega' = 2 \sin \omega' \cos \omega', \\
W_{15} &= \sin (\Omega - g) = \sin \Omega \cos g - \cos \Omega \sin g, \\
W_{16} &= \sin (2\Gamma' - 2L + \Omega) \\
&= Y_6 \sin 2\Gamma' - \cos 2\Gamma' \sin (2L - \Omega), \\
W_{17} &= \sin (L + \Gamma - \Omega) = \sin [(L + \Gamma) - \Omega] \\
&= W_6 \cos \Omega - Y_5 \sin \Omega, \\
\cos (2L + \Omega) &= \cos 2L \cos \Omega - \sin 2L \sin \Omega, \\
\sin (2L + \Omega) &= \sin 2L \cos \Omega + \cos 2L \sin \Omega, \\
W_{18} &= \sin (2L + \Omega - 2\Gamma') \\
&= \sin (2L + \Omega) \cos 2\Gamma' - \cos (2L + \Omega) \sin 2\Gamma', \\
W_{19} &= \sin (2\Gamma - \Omega), \\
W_{20} &= \sin (2L - \Omega + g) \\
&= \sin (2L - \Omega) \cos g + Y_6 \sin g, \\
W_{21} &= \sin 2\Gamma', \\
W_{22} &= \sin (L - \Gamma') = \sin L \cos \Gamma' - \cos L \sin \Gamma', \\
W_{23} &= \sin (\Gamma - \Gamma') = \sin \Gamma \cos \Gamma' - \cos \Gamma \sin \Gamma'.
\end{aligned}
\tag{8.14}
$$

The corresponding coefficients of the X_i, Y_i, Z_i, and W_i are displayed according to [16] in Table 8.1. They can be found in [3] also.

Table 8.1 Nutation Coefficients

$A_1 = 0''.0884 - 0''.00005T_u$	$A_9 = 0''.0014$	$A_{17} = A_{16}$
$A_2 = 0''.0183$	$A_{10} = 0''.0011$	$A_{18} = A_{16}$
$A_3 = 0''.0113 - 0''.00001T_u$	$A_{11} = -0''.0011$	$A_{19} = A_{16}$
$A_4 = -0''.0050$	$A_{12} = -0''.0010$	$A_{20} = A_{16}$
$A_5 = -0''.0031$	$A_{13} = 0''.0007$	$A_{21} = -A_{16}$
$A_6 = 0''.0030$	$A_{14} = -0''.0007$	$A_{22} = -A_{16}$
$A_7 = 0''.0023$	$A_{15} = 0''.0005$	$A_{23} = 0''.0002$
$A_8 = 0''.0022$	$A_{16} = 0''.0003$	$A_{24} = -A_{23}$
$B_1 = 9''.2100 + 0''.00091T_u$	$B_9 = 0''.0007$	
$B_2 = 0''.5522 - 0''.00029T_u$	$B_{10} = 0''.0005$	
$B_3 = -0''.0904 + 0''.00004T_u$	$B_{11} = 0''.0003$	
$B_4 = 0''.0216 - 0''.00006T_u$	$B_{12} = B_{11}$	
$B_5 = -0''.0093 + 0''.00003T_u$	$B_{13} = 0''.0002$	
$B_6 = -0''.0066$	$B_{14} = B_{13}$	
$B_7 = -0''.0024$	$B_{15} = -B_{14}$	
$B_8 = 0''.0008$	$B_{16} = -B_{14}$	
$C_1 = -0''.2037 - 0''.00002T_u$	$C_{24} = C_{23}$	
$C_2 = 0''.0675 + 0''.00001T_u$	$C_{25} = -C_{23}$	
$C_3 = -0''.0342 - 0''.00004T_u$	$C_{26} = -C_{23}$	
$C_4 = -0''.0261$	$C_{27} = -C_{23}$	
$C_5 = -0''.0149$	$C_{28} = 0''.0005$	
$C_6 = 0''.0114$	$C_{29} = -C_{28}$	
$C_7 = 0''.0060$	$C_{30} = -C_{28}$	
$C_8 = 0''.0058$	$C_{31} = 0''.0004$	
$C_9 = -0''.0057$	$C_{32} = C_{31}$	
$C_{10} = -0''.0052$	$C_{33} = -C_{31}$	
$C_{11} = -0''.0044$	$C_{34} = -C_{31}$	
$C_{12} = 0''.0028$	$C_{35} = -C_{31}$	
$C_{13} = -0''.0032$	$C_{36} = 0''.0003$	
$C_{14} = 0''.0026$	$C_{37} = -C_{36}$	
$C_{15} = -C_{14}$	$C_{38} = -C_{36}$	
$C_{16} = 0''.0025$	$C_{39} = 0''.0002$	
$C_{17} = 0''.0019$	$C_{40} = C_{39}$	
$C_{18} = -0''.0013$	$C_{41} = C_{39}$	
$C_{19} = 0''.0014$	$C_{42} = -C_{39}$	
$C_{20} = -0''.0009$	$C_{43} = -C_{39}$	
$C_{21} = 0''.0007$	$C_{44} = -C_{39}$	
$C_{22} = -C_{21}$	$C_{45} = -C_{39}$	
$C_{23} = 0''.0006$	$C_{46} = -C_{39}$	
$D_1 = -17''.2327 - 0''.01737T_u$	$D_{13} = -0''.0015$	
$D_2 = -1''.2729 - 0''.00013T_u$	$D_{14} = -0''.0010$	
$D_3 = 0''.2088 + 0''.00002T_u$	$D_{15} = D_{14}$	
$D_4 = 0''.1261 - 0''.00031T_u$	$D_{16} = -0''.0005$	
$D_5 = -0''.0497 + 0''.00012T_u$	$D_{17} = D_{16}$	
$D_6 = 0''.0214 - 0''.00005T_u$	$D_{18} = 0''.0004$	
$D_7 = 0''.0124 + 0''.00001T_u$	$D_{19} = -D_{18}$	
$D_8 = 0''.0045$	$D_{20} = 0''.0003$	
$D_9 = D_8$	$D_{21} = -D_{20}$	
$D_{10} = -0''.0021$	$D_{22} = -D_{20}$	
$D_{11} = 0''.0016 - 0''.00001T_u$	$D_{23} = -0''.0002$	
$D_{12} = -0''.0015 + 0''.00001T_u$		

8.1.4 Time Derivatives of the Nutation in Obliquity and Longitude

As previously noted, the twelve inputs required to compute the nutation expressions are the five mean longitudes Γ, Γ', L, Ω, and $☾$, their first time derivatives $\dot{\Gamma}$, $\dot{\Gamma}'$, $\dot{L}$, $\dot{\Omega}$, $\dot{☾}$, the time in Julian centuries, T_u, and the selection parameter λ. All of these variables are listed and discussed in Section 8.1.2. The immediate concern of this section is with the determination of the values of the intermediate variables

$$d\dot{\epsilon} \equiv \frac{d}{d\tau}(d\epsilon),$$

$$\Delta\dot{\epsilon} \equiv \frac{d}{d\tau}(\Delta\epsilon),$$

$$d\dot{\psi} \equiv \frac{d}{d\tau}(d\psi),$$

$$\Delta\dot{\psi} \equiv \frac{d}{d\tau}(\Delta\psi). \tag{8.15}$$

In essence, the formulas of this section define the values of the variables

$$\delta\dot{\epsilon} = d\dot{\epsilon} + \Delta\dot{\epsilon}, \qquad \delta\dot{\psi} = d\dot{\psi} + \Delta\dot{\psi}, \tag{8.16}$$

where

$$\delta\dot{\epsilon} \equiv \text{the time derivative of the nutation in obliquity (total)} = \frac{d}{d\tau}(\delta\epsilon),$$

$$\delta\dot{\psi} \equiv \text{the time derivative of the nutation in longitude (total)} = \frac{d}{d\tau}(\delta\psi).$$

The expressions employed for the computation of $d\epsilon$, $\Delta\epsilon$, $d\psi$, $\Delta\psi$, $\delta\epsilon$, and $\delta\psi$ are those employed in Section 8.1.3. The values of $d\dot{\epsilon}$, $\Delta\dot{\epsilon}$, $\Delta\dot{\psi}$, $d\dot{\psi}$ are given by the expressions

$$d\dot{\epsilon} = \left[\dot{A}_1 X_1 + \dot{A}_3 X_3 + \sum_{i=1}^{8} A_i \dot{X}_i + (1-\lambda)\sum_{i=9}^{24} A_i \dot{X}_i\right] K, \tag{8.17}$$

$$\Delta\dot{\epsilon} = \left[\sum_{i=1}^{5} \dot{B}_i Y_i + \sum_{i=1}^{7} B_i \dot{Y}_i + (1-\lambda)\sum_{i=8}^{16} B_i \dot{Y}_i\right] K, \tag{8.18}$$

$$d\dot{\psi} = \left[\sum_{i=1}^{3} \dot{C}_i Z_i + \sum_{i=1}^{12} C_i \dot{Z}_i + (1-\lambda)\sum_{i=13}^{46} C_i \dot{Z}_i\right] K, \tag{8.19}$$

$$\Delta\dot{\psi} = \left[\sum_{i=1}^{7} \dot{D}_i W_i + (1-\lambda)\sum_{i=8}^{12} \dot{D}_i W_i + \sum_{i=1}^{9} D_i \dot{W}_i + (1-\lambda)\sum_{i=10}^{23} D_i \dot{W}_i\right] K, \tag{8.20}$$

where (8.21)

$$K = \frac{\pi}{180 \times 3600},$$

$$\begin{aligned}
\dot{T}_u &= (36525.0 \times 24.0 \times 60.0 \times 0.07436574)^{-1}, \\
\dot{g} &= \dot{L} - \dot{\Gamma}, \\
\dot{g}' &= \dot{\text{☾}} - \dot{\Gamma}', \\
\dot{\omega}' &= \dot{\Gamma} - \dot{\Omega};
\end{aligned}$$

$$\begin{aligned}
\dot{A}_1 &= -0''.00005\dot{T}_u, \\
\dot{A}_2 &= -0''.00001\dot{T}_u;
\end{aligned}$$

$$\begin{aligned}
\dot{X}_1 &= -Z_1(2\dot{\text{☾}}), \\
\dot{X}_2 &= -Z_3(2\dot{\text{☾}} - \dot{\Omega}), \\
\dot{X}_3 &= -Z_4(2\dot{\text{☾}} + \dot{g}'), \\
\dot{X}_4 &= -Z_6(\dot{\text{☾}} + \dot{\Gamma}'), \\
\dot{X}_5 &= -Z_8(\dot{\Omega} + \dot{g}'), \\
\dot{X}_6 &= Z_9(\dot{g}' - \dot{\Omega}), \\
\dot{X}_7 &= -Z_{11}(2\dot{\text{☾}} + \dot{g}' - \dot{\Omega}), \\
\dot{X}_8 &= -Z_{10}(4\dot{\text{☾}} - 2\dot{L} - \dot{g}'), \\
\dot{X}_9 &= -Z_{13}(4\dot{\text{☾}} - 2\dot{L}), \\
\dot{X}_{10} &= -Z_{15}(2\dot{\text{☾}} + 2\dot{g}'), \\
\dot{X}_{11} &= -Z_{14}(2\dot{L} + \dot{g}'), \\
\dot{X}_{12} &= -Z_{17}(\dot{\text{☾}} + \dot{\omega}'), \\
\dot{X}_{13} &= -Z_{18}(2\dot{L} - \dot{\text{☾}} - \dot{\omega}'), \\
\dot{X}_{14} &= [X_{13} \sin 2\Omega - Z_{18} \cos 2\Omega](2\dot{L} - \dot{\text{☾}} - \dot{\omega}' - 2\dot{\Omega}), \\
\dot{X}_{15} &= -Z_{20}(4\dot{\text{☾}} - 2\dot{L} - \dot{g}' - \dot{\Omega}), \\
\dot{X}_{16} &= -Z_{25}(2\dot{\text{☾}} - \dot{g}), \\
\dot{X}_{17} &= -Z_{26}(4\dot{\text{☾}} - 2\dot{L} + \dot{g}'), \\
\dot{X}_{18} &= -\sin(2\text{☾} - \Omega - 2L)[2\dot{\text{☾}} - \dot{\Omega} - 2\dot{L}], \\
\dot{X}_{19} &= -Z_{27}(2\dot{\text{☾}} - 2\dot{L} + \dot{\Omega}), \\
\dot{X}_{20} &= -Z_{30}(4\dot{\text{☾}} - 2\dot{L} - \dot{\Omega}), \\
\dot{X}_{21} &= -Z_{21}(2\dot{\text{☾}} + \dot{g}), \\
\dot{X}_{22} &= -Z_{28}(2\dot{L} + \dot{g}' - \dot{\Omega}), \\
\dot{X}_{23} &= -Z_{33}(2\dot{\text{☾}} - \dot{\Omega} + 2\dot{g}'), \\
\dot{X}_{24} &= -Z_{23}(2\dot{L} + 2\dot{g}');
\end{aligned}$$

$\dot{B}_1 = 0''.00091\dot{T}_u,$
$\dot{B}_2 = -0''.00029\dot{T}_u,$
$\dot{B}_3 = 0''.00004\dot{T}_u,$
$\dot{B}_4 = -0''.00006\dot{T}_u,$
$\dot{B}_5 = 0''.00003\dot{T}_u;$

$\dot{Y}_1 = -W_1(\dot{\Omega}),$
$\dot{Y}_2 = -W_2(2\dot{L}),$
$\dot{Y}_3 = -W_3(2\dot{\Omega}),$
$\dot{Y}_4 = -W_5(2\dot{L} + \dot{g}),$
$\dot{Y}_5 = -W_6(\dot{L} + \dot{\Gamma}),$
$\dot{Y}_6 = -W_7(2\dot{L} - \dot{\Omega}),$
$\dot{Y}_7 = -W_8(\dot{\Gamma}' + \dot{\omega}'),$
$\dot{Y}_8 = -W_{13}(\dot{\Omega} + \dot{g}),$
$\dot{Y}_9 = -W_{12}(2\dot{L} + 2\dot{g}),$
$\dot{Y}_{10} = -W_{15}(\dot{\Omega} - \dot{g}),$
$\dot{Y}_{11} = W_{16}(2\dot{L} - \dot{\Omega} - 2\dot{\Gamma}'),$
$\dot{Y}_{12} = -[W_{19}\cos g + Y_{13}W_4](2\dot{\Gamma} - \dot{\Omega} + \dot{g}),$
$\dot{Y}_{13} = -W_{19}(2\dot{\Gamma} - \dot{\Omega}),$
$\dot{Y}_{14} = -W_{21}(2\dot{\Gamma}'),$
$\dot{Y}_{15} = -W_{18}(2\dot{L} - \dot{\omega}' - \dot{\Gamma}),$
$\dot{Y}_{16} = -W_{20}(2\dot{L} - \dot{\Omega} + \dot{g});$

$\dot{C}_1 = -0''.00002\dot{T}_u,$
$\dot{C}_2 = 0''.00001\dot{T}_u,$
$\dot{C}_3 = -0''.00004\dot{T}_u;$

$\dot{Z}_1 = X_1(2\dot{☾}),$
$\dot{Z}_2 = \cos g'(\dot{g}'),$
$\dot{Z}_3 = X_2(2\dot{☾} - \dot{\Omega}),$
$\dot{Z}_4 = X_3(2\dot{☾} + \dot{g}'),$
$\dot{Z}_5 = [\cos(2L - ☾)\cos\Gamma' + \sin(2L - ☾)\sin\Gamma'](2\dot{L} - \dot{☾} - \dot{\Gamma}'),$
$\dot{Z}_6 = X_4(\dot{☾} + \dot{\Gamma}'),$
$\dot{Z}_7 = [X_1Y_2 + Z_1W_2](2\dot{☾} - 2\dot{L}),$
$\dot{Z}_8 = X_5(\dot{\Omega} + \dot{g}'),$

$$\dot{Z}_9 = X_6(\dot{\Omega} - \dot{g}'),$$
$$\dot{Z}_{10} = X_8(4\dot{☾} - 2\dot{L} - \dot{g}'),$$
$$\dot{Z}_{11} = X_7(2\dot{☾} - \dot{\Omega} + \dot{g}'),$$
$$\dot{Z}_{12} = [1 - 2\sin^2 g'](2\dot{g}'),$$
$$\dot{Z}_{13} = X_9(4\dot{☾} - 2\dot{L}),$$
$$\dot{Z}_{14} = X_{11}(2\dot{L} + \dot{g}'),$$
$$\dot{Z}_{15} = X_{10}(2\dot{☾} + 2\dot{g}'),$$
$$\dot{Z}_{16} = [X_2 \cos\Omega + Z_3 \sin\Omega](2\dot{☾} - 2\dot{\Omega}),$$
$$\dot{Z}_{17} = X_{12}(\dot{☾} + \dot{\omega}'),$$
$$\dot{Z}_{18} = X_{13}(2\dot{L} - \dot{☾} - \dot{\omega}'),$$
$$\dot{Z}_{19} = X_{14}(\dot{☾} - 2\dot{L} + \dot{\Omega} + \dot{\Gamma}'),$$
$$\dot{Z}_{20} = X_{15}(4\dot{☾} - 2\dot{L} - \dot{\Omega} - \dot{g}'),$$
$$\dot{Z}_{21} = X_{21}(2\dot{☾} + \dot{g}),$$
$$\dot{Z}_{22} = [X_4 Y_4 + Z_6 W_5](2\dot{L} + \dot{g} - \dot{☾} - \dot{\Gamma}'),$$
$$\dot{Z}_{23} = X_{24}(2\dot{L} + 2\dot{g}'),$$
$$\dot{Z}_{24} = [X_3 Y_2 + Z_4 W_2](2\dot{☾} - 2\dot{L} + \dot{g}'),$$
$$\dot{Z}_{25} = X_{16}(2\dot{☾} - \dot{g}),$$
$$\dot{Z}_{26} = X_{17}(4\dot{☾} - 2\dot{L} + \dot{g}'),$$
$$\dot{Z}_{27} = X_{19}(2\dot{☾} - 2\dot{L} + \dot{\Omega}),$$
$$\dot{Z}_{28} = X_{22}(2\dot{L} + \dot{g}' - \dot{\Omega}),$$
$$\dot{Z}_{29} = X_{18}(2\dot{L} - 2\dot{☾} + \dot{\Omega}),$$
$$\dot{Z}_{30} = X_{20}(4\dot{☾} - 2\dot{L} - \dot{\Omega}),$$
$$\dot{Z}_{31} = [\cos g' \cos g + Z_2 W_4](\dot{g}' - \dot{g}),$$
$$\dot{Z}_{32} = [Y_1 \cos(☾ + \omega') + Z_{17} W_1](\dot{\Omega} - \dot{☾} - \dot{\omega}'),$$
$$\dot{Z}_{33} = X_{23}(2\dot{☾} - \dot{\Omega} + 2\dot{g}'),$$
$$\dot{Z}_{34} = [X_1 Y_4 + Z_1 W_5](2\dot{L} + \dot{g} - 2\dot{☾}),$$
$$\dot{Z}_{35} = [\cos ☾ \cos L + \sin ☾ \sin L](\dot{☾} - \dot{L}),$$
$$\dot{Z}_{36} = [X_3 Y_3 + Z_4 W_3](2\dot{☾} + \dot{g}' - 2\dot{\Omega}),$$
$$\dot{Z}_{37} = [\cos g \cos g' - Z_2 W_4](\dot{g} + \dot{g}'),$$
$$\dot{Z}_{38} = [X_3 \cos g + Z_4 W_4](2\dot{☾} + \dot{g}' - \dot{g}),$$
$$\dot{Z}_{39} = [X_5 \cos g' - Z_2 Z_8](\dot{\Omega} + 2\dot{g}'),$$
$$\dot{Z}_{40} = [X_3 \cos g - Z_4 W_4](2\dot{☾} + \dot{g}' + \dot{g}),$$
$$\dot{Z}_{41} = [X_6 \cos g' + Z_2 Z_9](2\dot{g}' - \dot{\Omega}),$$
$$\dot{Z}_{42} = [Y_6 \cos g' + Z_2 W_7](2\dot{L} - \dot{\Omega} - \dot{g}'),$$

$$\dot{Z}_{43} = [X_8 \cos g + Z_{10} \sin g](4\dot{☾} - 2\dot{L} - \dot{g}' - \dot{g}),$$
$$\dot{Z}_{44} = [X_9 \cos g + Z_{13} W_4](4\dot{☾} - 2\dot{L} - \dot{g}),$$
$$\dot{Z}_{45} = [Y_3 \cos g' - Z_2 W_3](2\dot{\Omega} + \dot{g}'),$$
$$\dot{Z}_{46} = [X_{10} \cos g' - Z_2 Z_{15}](2\dot{☾} + 3\dot{g}');$$

$$\dot{D}_1 = -0''.01737\dot{T}_u,$$
$$\dot{D}_2 = -0''.00013\dot{T}_u,$$
$$\dot{D}_3 = 0''.00002\dot{T}_u,$$
$$\dot{D}_4 = -0''.00031\dot{T}_u,$$
$$\dot{D}_5 = 0''.00012\dot{T}_u,$$
$$\dot{D}_6 = -0''.00005\dot{T}_u,$$
$$\dot{D}_7 = 0''.00001\dot{T}_u, \ \dot{D}_8 = \dot{D}_9 = \dot{D}_{10} = 0,$$
$$\dot{D}_{11} = -0''.00001\dot{T}_u,$$
$$\dot{D}_{12} = 0''.00001\dot{T}_u;$$

$$\dot{W}_1 = Y_1(\dot{\Omega}),$$
$$\dot{W}_2 = Y_2(2\dot{L}),$$
$$\dot{W}_3 = Y_3(2\dot{\Omega}),$$
$$\dot{W}_4 = \cos g(\dot{g}),$$
$$\dot{W}_5 = Y_4(2\dot{L} + \dot{g}),$$
$$\dot{W}_6 = Y_5(\dot{L} + \dot{\Gamma}),$$
$$\dot{W}_7 = Y_6(2\dot{L} - \dot{\Omega}),$$
$$\dot{W}_8 = Y_7(\dot{\omega}' + \dot{\Gamma}'),$$
$$\dot{W}_9 = [Y_2 \cos 2\Gamma' + W_2 \sin 2\Gamma'](2\dot{L} - 2\dot{\Gamma}'),$$
$$\dot{W}_{10} = [Y_2 Y_3 + W_2 W_3](2\dot{L} - 2\dot{\Omega}),$$
$$\dot{W}_{11} = [1 - 2\sin^2 g](2\dot{g}),$$
$$\dot{W}_{12} = Y_9(2\dot{L} + 2\dot{g}),$$
$$\dot{W}_{13} = Y_8(\dot{g} + \dot{\Omega}),$$
$$\dot{W}_{14} = [1 - 2\sin^2 \omega'](2\dot{\omega}'),$$
$$\dot{W}_{15} = Y_{10}(\dot{\Omega} - \dot{g}),$$
$$\dot{W}_{16} = Y_{11}(2\dot{\Gamma}' - 2\dot{L} + \dot{\Omega}),$$
$$\dot{W}_{17} = [Y_1 Y_5 + W_1 W_6](\dot{L} + \dot{\Gamma} - \dot{\Omega}),$$
$$\dot{W}_{18} = Y_{15}(2\dot{L} + \dot{\Omega} - 2\dot{\Gamma}'),$$
$$\dot{W}_{19} = Y_{13}(2\dot{\Gamma} - \dot{\Omega}),$$

$$\dot{W}_{20} = Y_{16}(2\dot{L} - \dot{\Omega} + \dot{g}),$$
$$\dot{W}_{21} = Y_{14}(2\dot{\Gamma}'),$$
$$\dot{W}_{22} = [\cos L \cos \Gamma' + \sin L \sin \Gamma'](\dot{L} - \dot{\Gamma}'),$$
$$\dot{W}_{23} = [\cos \Gamma \cos \Gamma' + \sin \Gamma \sin \Gamma'](\dot{\Gamma} - \dot{\Gamma}'). \tag{8.21}$$

All the trigonometric functions in (8.21), for example, X_1, X_2, $\sin 2\Omega$, are obtained from the equations in Section 8.1.3. All the trigonometric functions taken from Section 8.1.3 are given as functions of the sines and cosines of the five mean longitudes. Therefore for nutation calculations it is only necessary to compute five sines and five cosines. All other trigonometric functions are derived from the ten basic sine, cosine combinations.

8.1.5 Determination of the Physical Librations of the Moon

The basic work in the determination of the physical librations of the Moon in longitude, inclination, and node was accomplished by Hayn over the period 1902 through 1920 ([17] to [22]). This was followed by Jönsson [23] whose efforts (1917, 1919) produced agreement with Hayn in regard to the librations in inclination and node, but a lack of agreement with respect to the libration in longitude. In 1948 the latter problem was attacked by Koziel [24], and his work yielded confirmation of the results achieved by Hayn in 1920 [22].

The present discussion is a brief interpretation of Koziel's recent chapter on librations [8], having as a primary objective the determination of a set of equations which will be useful for the rapid computation of the physical librations of the Moon as functions of time. I am indebted to D. G. Dethlefsen [32] for the contribution of this section.

The physical librations in longitude, inclination, and node are conventionally designated by τ_l, ρ_l, and σ_l, respectively, and are expressed as trigonometric series, a typical form for which is

$$\tau_l = \sum_i a_i \sin \alpha_i,$$
$$\alpha_i = c_1^{(i)} g + c_2^{(i)} g' + c_3^{(i)} \omega + c_4^{(i)} \omega',$$
$$c_1^{(i)}, c_2^{(i)}, c_3^{(i)}, c_4^{(i)} = 0, \pm 1, \pm 2, \ldots. \tag{8.22}$$

Like the oscillatory behavior of any physical system, the oscillation of the Moon is found to be characterized by free and forced librations. The free modes of oscillation, or free librations, arise from the general solution to the homogeneous differential equation describing the motion of the Moon, the frequencies of which are the resonant, or natural, frequencies of the body. When a forcing function, namely, an external driving force, is present, the equation of motion is not homogeneous and its general solution then contains the particular solution that gives rise to the forced liberations, the frequencies of which are those of the driving force. It was determined in the nineteenth century that the amplitudes of the free libration of the Moon are negligible in comparison with those for the forced libration. Thus the total physical libration is presently assumed to consist entirely of the latter components. With the Moon considered as a rigid body, the analysis is traditionally based on integration of Euler's equations of motion for such a body with one point fixed. Koziel includes the details of this analysis in his survey [8]. The results are presented in the form of tables listing numerical values of coefficients and the corresponding arguments for each component of the forced libration as functions of the dynamical or mechanical ellipticity f of the Moon, where

$$f = \frac{B(C - B)}{A(C - A)}, \qquad (8.23)$$

in which A, B, C denote the principal moments of inertia of the Moon.[3] The range of f covered by these tables is from 0.5 to 0.8, at intervals of 0.1. Koziel bases these results on Hayn's 1923 value of the mean inclination of the equatorial plane of the Moon with respect to the ecliptic plane [20], [25]:

$$I = 1°32'.20'' = 1°.53889 = 0.026859 \text{ radian}, \qquad (8.24)$$

and on Brown's value of the mean inclination of the orbit of the Moon with

[3] The principal moments of inertia, A, B, and C, of the Moon are those about the lunar axes a, b, and c, respectively. The c axis is the axis of rotation, or polar axis, of the Moon. The a axis is the primary radius of the Moon and is defined as the vector extending from the selenocenter to the mean center of the apparent lunar disk (see the discussion in Section 8.2). It thus intersects the lunar surface in Sinus Medii (Central Bay) in the equatorial plane of the Moon, and its direction is therefore nearly the direction of the geocenter. The b axis lies in the equatorial plane of the Moon to complete the right-handed a, b, c coordinate system. This selenocentric principal axis coordinate system is thus fixed in the Moon, and for that reason, in analogy to geographic coordinates on Earth, is sometimes referred to as the "selenographic" or, for the present analysis, the "true selenographic" coordinate system. At any given time, the position of the a axis differs from that of the line joining the selenocenter and the geocenter by the total geometrical and physical libration existing at that time.

The x, y, z true selenographic coordinate system defined in Section 8.2 is coincident with the a, b, c principal axis coordinate system.

respect to the ecliptic plane [26]:

$$i = 5°8'43'' = 5°.14528 = 0.089802 \text{ radian}. \tag{8.25}$$

For all terms whose coefficients have a magnitude of at least one-half second of arc within the range of f considered, the data in these tables has been plotted in Figures 8.3 and 8.4 for τ_l and in Figures 8.4, 8.5, and 8.6 for ρ_l and the product of sin I and σ_l.

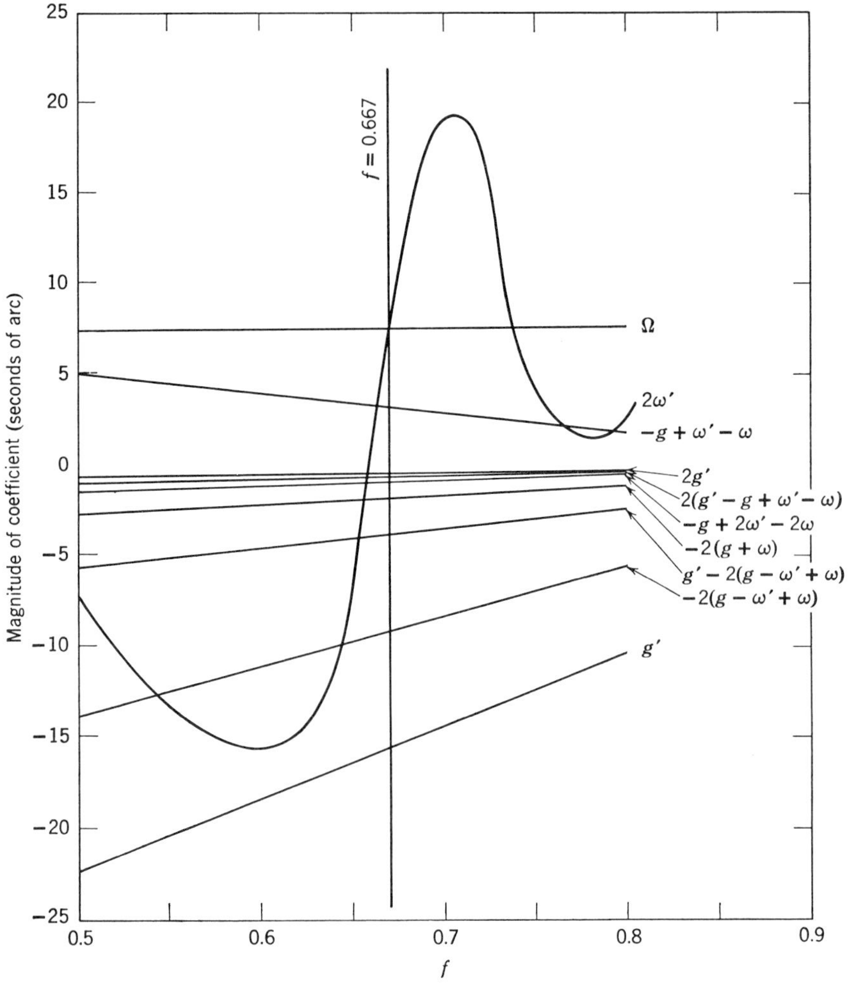

FIGURE 8.3 Coefficients for τ_l.

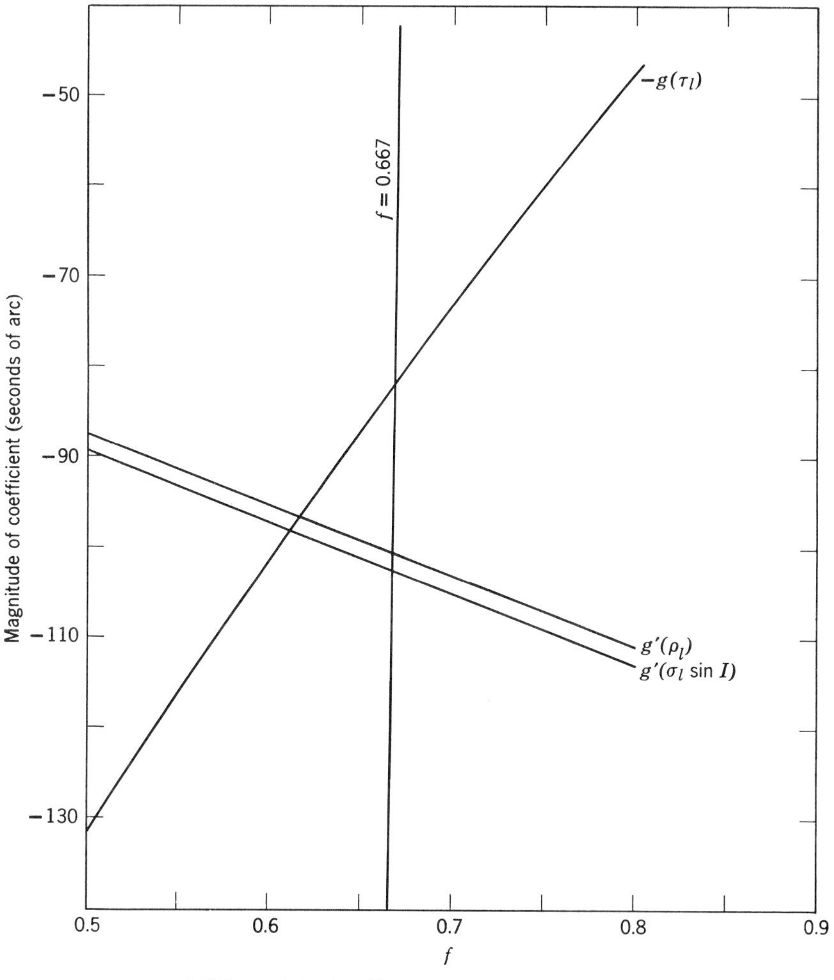

FIGURE 8.4 Coefficients for τ_l, ρ_l, and $\sigma_l \sin I$.

The arguments of the trigonometric functions illustrated in these figures are defined as follows:

$$
\begin{aligned}
g' &\equiv \text{☾} - \Gamma' = \text{mean anomaly of the Moon},\\
g &\equiv L - \Gamma = \text{mean anomaly of the Sun},\\
\omega' &\equiv \Gamma' - \Omega = \text{argument of perigee of the Moon},\\
\omega &\equiv \Gamma - \Omega = \text{angular distance of perigee of the Sun from the ascending node of the orbit of the Moon.}
\end{aligned}
$$

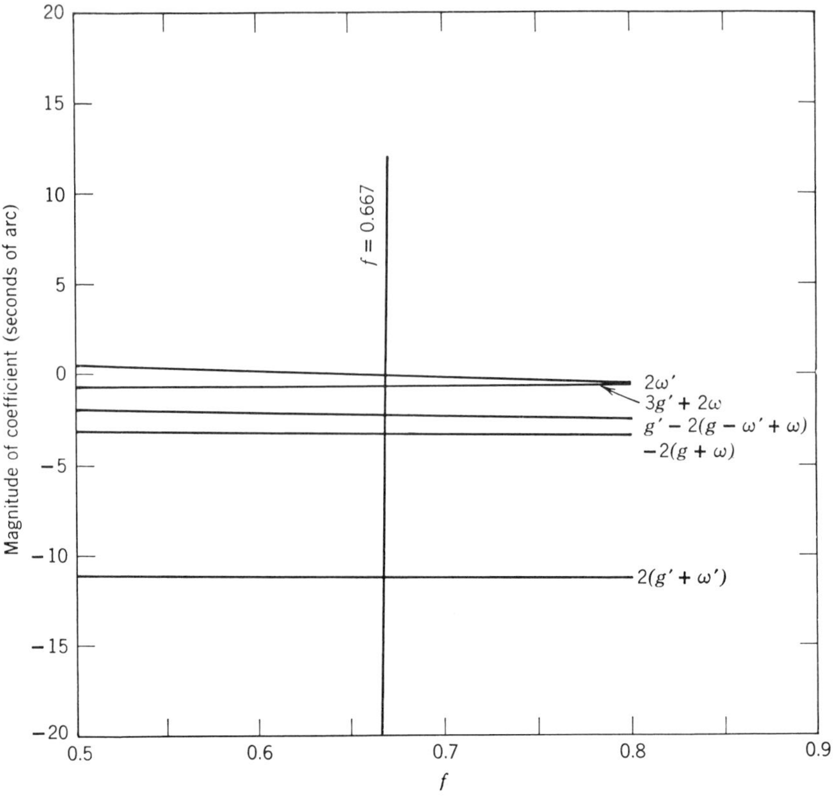

FIGURE 8.5 Coefficients for ρ_l and $\sigma_l \sin I$.

The mean longitudes, L, Γ, Γ', ☾, and Ω, are functions of time and can be found in Section 8.1.2.

The value of dynamical ellipticity employed here is computed from the values of the principal moments of inertia of the Moon. The principal moments of inertia for computing the dynamical ellipticity f are

$$\begin{aligned} A &= 0.88781798 \times 10^{35} \text{ kg m}^2, \\ B &= 0.88800195 \times 10^{35} \text{ kg m}^2, \\ C &= 0.88836978 \times 10^{35} \text{ kg m}^2. \end{aligned} \tag{8.26}$$

These values are current NASA standard constants and are derived from an improved value of the Earth-Moon mass ratio based on Mariner II flight data [27]. The value of dynamical ellipticity dictated by these values of the principal moments, (8.23), is

$$f = 0.667. \tag{8.27}$$

Entering the coefficient curves of Figures 8.3 through 8.6 with this value of f yields the following set of equations for the forced librations of the Moon, the terms of which are arranged in order of decreasing magnitude of the coefficients:

$$\begin{aligned}\tau_l = {} & 82''.4 \sin g - 15''.6 \sin g' \\ & + 9''.0 \sin 2(g - \omega' + \omega) + 7''.6 \sin \Omega \\ & + 7''.5 \sin 2\omega' - 3''.7 \sin (g' - 2g + 2\omega' - 2\omega) \\ & - 3''.2 \sin (g - \omega' + \omega) + 1''.7 \sin 2(g + \omega) \\ & + 0''.8 \sin (g - 2\omega' + 2\omega) \\ & - 0''.6 \sin 2(g' - g + \omega' - \omega) - 0''.4 \sin 2g',\end{aligned} \tag{8.28}$$

$$\begin{aligned}\rho_l = {} & -100''.8 \cos g' + 28''.2 \cos (g' + 2\omega') \\ & - 11''.1 \cos 2(g' + \omega') - 3''.3 \cos 2(g + \omega) \\ & - 2''.2 \cos (g' - 2g + 2\omega' - 2\omega) \\ & - 0''.6 \cos (3g' + 2\omega') - 0''.1 \cos 2\omega',\end{aligned} \tag{8.29}$$

$$\begin{aligned}\sigma_l \sin I = {} & -102''.8 \sin g' + 28''.2 \sin (g' + 2\omega') \\ & - 11''.1 \sin 2(g' + \omega') - 3''.3 \sin 2(g + \omega) \\ & - 2''.2 \sin (g' - 2g + 2\omega' - 2\omega) \\ & - 0''.6 \sin (3g' + 2\omega') - 0''.1 \sin 2\omega'.\end{aligned} \tag{8.30}$$

In terms of the mean longitudes these equations may be written as

$$\begin{aligned}\tau_l = {} & 82''.4 \sin (L - \Gamma) - 15''.6 \sin (☾ - \Gamma') \\ & + 9''.0 \sin 2(L - \Gamma') + 7''.6 \sin \Omega \\ & + 7''.5 \sin 2(\Gamma' - \Omega) - 3''.7 \sin (☾ + \Gamma' - 2L) \\ & - 3''.2 \sin (L - \Gamma') + 1''.7 \sin 2(L - \Omega) \\ & + 0''.8 \sin (L + \Gamma - 2\Gamma') - 0''.6 \sin 2(☾ - L) \\ & - 0''.4 \sin 2(☾ - \Gamma'),\end{aligned} \tag{8.31}$$

$$\begin{aligned}\rho_l = {} & -100''.8 \cos (☾ - \Gamma') + 28''.2 \cos (☾ + \Gamma' - 2\Omega) \\ & - 11''.1 \cos 2(☾ - \Omega) - 3''.3 \cos 2(L - \Omega) \\ & - 2''.2 \cos (☾ + \Gamma' - 2L) \\ & - 0''.6 \cos (3☾ - \Gamma' - 2\Omega) - 0''.1 \cos 2(\Gamma' - \Omega),\end{aligned} \tag{8.32}$$

$$\begin{aligned}\sigma_l \sin I = {} & -102''.8 \sin (☾ - \Gamma') + 28''.2 \sin (☾ + \Gamma' - 2\Omega) \\ & - 11''.1 \sin 2(☾ - \Omega) - 3''.3 \sin 2(L - \Omega) \\ & - 2''.2 \sin (☾ + \Gamma' - 2L) \\ & - 0''.6 \sin (3☾ - \Gamma' - 2\Omega) - 0''.1 \sin 2(\Gamma' - \Omega).\end{aligned} \tag{8.33}$$

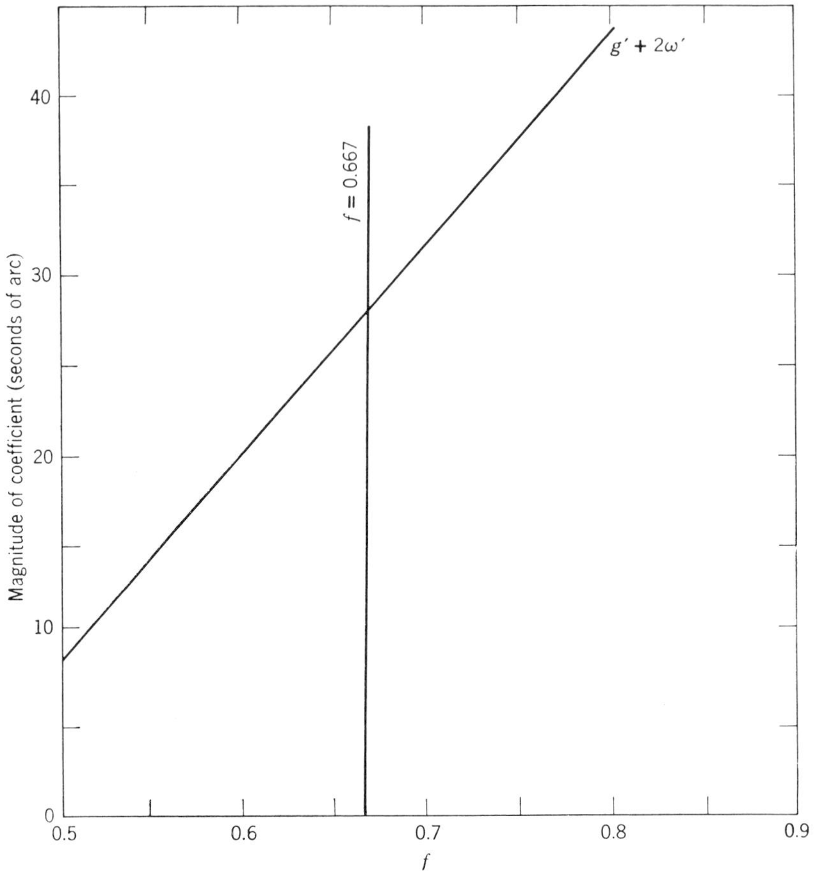

FIGURE 8.6 Coefficients for ρ_l and $\sigma_l \sin I$.

It is helpful to note that the value of f employed above is in exact agreement with the value suggested by Yakovkin in 1957 in a communication to Jeffreys [28]. At the same time Jeffreys' computation produced, for the forced motion, the neighboring value of

$$f = 0.673 \pm 0.0018.$$

As pointed out by Koziel, the function employed in the determination of the equation for the libration in longitude has a singularity at $f = 0.662$. Thus the use of the neighboring value, $f = 0.667$, necessitates the assumption that the curves representing the coefficients in Figures 8.3 and 8.4 are continuous in the critical neighborhood.

It should be noted that, in his description of Hayn's development of expressions for the librations in inclination and node, Koziel employs the tacit assumption

$$\begin{aligned} \sin(I + \rho_l) &\approx (I + \rho_l), \\ \sin(\tau_l - \sigma_l) &\approx (\tau_l - \sigma_l), \\ \cos(\tau_l - \sigma_l) &\approx 1, \end{aligned} \tag{8.34}$$

and that $\rho_l(\tau_l - \sigma_l)$ is of second order as compared with $I(\tau_l - \sigma_l)$. Considering the magnitudes involved—see (8.24), (8.28), (8.29), and (8.30)—these assumptions are adequately justified. Related to this is Koziel's use of the product $I\sigma_l$ rather than $\sigma_l \sin I$, which is the form utilized in (8.30) and (8.33).

Table 8.2 enumerates the results obtained for the value of inclination I by various investigators. The percentage deviation from the mean for the value employed in (8.24) (Hayn, 1923 [25]) is approximately 0.3.

Other sources for the libration equations include Kalensher [10] and Gabbard [5], each of whom records identical sets of relationships for the physical librations. However, when compared with the present analysis, certain inconsistencies are noted in the equations given by these authors. Examination of their relationship for τ_l indicates that a vertical line can be drawn on Figures 8.3 and 8.4 which defines their coefficients for sin g and sin g' exactly; the coefficient for sin $2\omega'$, however, is seen to be much closer to 3 seconds of arc than to their stated value of 18 seconds of arc. Additionally, it will be found from these curves that the value of f which corresponds to their coefficients for τ_l differs from that which corresponds to their coefficients for ρ_l and σ_l sin I, and also that each of these values is in the neighborhood of $f = 0.750$. It may be of assistance to note that, in regard to notation for the arguments of the trigonometric functions, Gabbard differs from Koziel and Kalensher in that Gabbard's primed quantities correspond to Koziel's or Kalensher's unprimed quantities, and vice versa.

In summary, these considerations, enumerated below, suggest the recommendation of (8.28), (8.29), and (8.30) or their equivalents, (8.31), (8.32), and (8.33), for adoption as the standard equations for computation of the physical librations of the Moon.

Table 8.2 Comparison of Values for the Mean Inclination of the Moon's Equator

Investigator	Year	Reference	I	sin I
Hayn	1907	[19]	$1°32'6'' = 1°.53500$	0.026788
Hayn	1923	[25]	$1°32'20'' = 1°.53889$	0.026855
Koziel	1948	[8]	$1°31'10'' = 1°.51944$	0.026516
Yakovkin	1950	[29]	$1°33'48'' = 1°.56333$	0.027282
Watts	1955	[30]	$1°33'50'' = 1°.56389$	0.027291
Mean			$1°32'39'' = 1°.54417$	0.026948

I. The value of f should be based on current NASA standard values of the principal moments of inertia of the Moon.
II. Use of Hayn's value of the mean equatorial inclination I and of Brown's value of mean orbital inclination i should be adopted.
III. Inclusion of all terms having coefficients of magnitude equal to at least one-half second of arc within the range of f should be considered, thus ensuring consideration of the smaller libration components.

8.1.6 Algorithmic Development of the Libration Equations

The equations for the physical librations in node, inclination, and longitude, respectively, are given by expressions of the form[4]

$$\sigma_l = \frac{1}{\sin I}\sum_{i=1}^{7} E_i S_i, \qquad \rho_l = \sum_{i=1}^{7} F_i T_i, \qquad \tau_l = \sum_{i=1}^{11} G_i U_i, \tag{8.35}$$

where I is the mean inclination of the Moon's equator, S_i, T_i, and U_i are trigonometric functions, and E_i, F_i, and G_i are constant coefficients.

Equations 8.35 contain a total of 25 sine and cosine trigonometric functions. As noted in the development of the nutation equations, Section 8.1.3, the machine computation of a large number of sine and cosine functions results in an unreasonably large computation time.

The number of machine computations of trigonometric functions can be reduced from 25 to 10 by expanding S_i, T_i, and U_i in terms of functions of the sines and cosines of the five mean longitudes. This procedure is adopted in the libration calculations. The resultant substitution of addition and multiplication steps for the corresponding steps of obtaining the sines and cosines produces a saving in overall computing time.

Table 8.3 Libration Constants

$I = 1°32'20'' = 0.026859$ radian	$F_6 = -3''.3$
$E_1 = -102''.8$	$F_7 = -2''.2$
$E_2 = -0''.1$	$G_1 = -15''.6$
$E_3 = 28''.2$	$G_2 = -0''.4$
$E_4 = -11''.1$	$G_3 = -82''.4$
$E_5 = -0''.6$	$G_4 = 7''.5$
$E_6 = 3''.3$	$G_5 = -1''.7$
$E_7 = -2''.2$	$G_6 = 3''.2$
$F_1 = -100''.8$	$G_7 = -0''.8$
$F_2 = -0''.1$	$G_8 = -9''.0$
$F_3 = 28''.2$	$G_9 = -3''.7$
$F_4 = -11''.1$	$G_{10} = -0''.6$
$F_5 = -0''.6$	$G_{11} = 7''.6$

[4] See (8.28), (8.29), and (8.30).

The inputs required to compute the librations are:

$$\Gamma,\ \Gamma',\ \Omega,\ \leftmoon,\ L,\ \dot{\Gamma},\ \dot{\Gamma}',\ \dot{\Omega},\ \dot{\leftmoon},\ \dot{L},$$

which can be obtained from Section 8.1.2 and the libration constants listed in Table 8.3.

The values of the trigonometric functions are obtained from (8.28), (8.29), (8.30), and the definitions of g', g, ω', and ω as (8.36):

$$\begin{aligned}
S_1 &= \sin g' = \sin \leftmoon \cos \Gamma' - \cos \leftmoon \sin \Gamma', \\
T_1 &= \cos g' = \cos \leftmoon \cos \Gamma' + \sin \leftmoon \sin \Gamma', \\
\sin \omega' &= \sin \Gamma' \cos \Omega - \cos \Gamma' \sin \Omega, \\
\cos \omega' &= \cos \Gamma' \cos \Omega + \sin \Gamma' \sin \Omega, \\
S_2 &= \sin 2\omega' = 2 \sin \omega' \cos \omega', \\
T_2 &= \cos 2\omega' = 1 - 2 \sin^2 \omega', \\
S_3 &= \sin (g' + 2\omega') = S_1 T_2 + T_1 S_2, \\
T_3 &= \cos (g' + 2\omega') = T_1 T_2 - S_1 S_2, \\
\sin 2g' &= 2 S_1 T_1, \\
\cos 2g' &= 1 - 2 S_1^2, \\
S_4 &= \sin (2g' + 2\omega') \\
&= T_2 \sin 2g' + S_2 \cos 2g', \\
T_4 &= \cos (2g' + 2\omega') \\
&= T_2 \cos 2g' - S_2 \sin 2g', \\
S_5 &= \sin (3g' + 2\omega') = S_1 T_4 + T_1 S_4, \\
T_5 &= \cos (3g' + 2\omega') = T_1 T_4 - S_1 S_4, \\
\sin g &= \sin L \cos \Gamma - \cos L \sin \Gamma, \\
\cos g &= \cos L \cos \Gamma + \sin L \sin \Gamma, \\
\sin \omega &= \sin \Gamma \cos \Omega - \cos \Gamma \sin \Omega, \\
\cos \omega &= \cos \Gamma \cos \Omega + \sin \Gamma \sin \Omega, \\
\sin 2g &= 2 \sin g \cos g, \\
\cos 2g &= 1 - 2 \sin^2 g, \\
\sin 2\omega &= 2 \sin \omega \cos \omega, \\
\cos 2\omega &= 1 - 2 \sin^2 \omega, \\
S_6 &= \sin (-2g - 2\omega) \\
&= -[\sin 2g \cos 2\omega + \cos 2g \sin 2\omega], \\
T_6 &= \cos (-2g - 2\omega) \\
&= \cos 2g \cos 2\omega - \sin 2g \sin 2\omega, \\
S_7 &= \sin (g' + 2\omega' - 2g - 2\omega) = S_3 T_6 + T_3 S_6, \\
T_7 &= \cos (g' + 2\omega' - 2g - 2\omega) = T_3 T_6 - S_3 S_6,
\end{aligned}$$

$$
\begin{aligned}
U_1 &= S_1,\\
U_2 &= \sin 2g',\\
U_3 &= -\sin g,\\
U_4 &= S_2,\\
U_5 &= S_6,\\
U_9 &= S_7,\\
U_{10} &= \sin (2g' + 2\omega' - 2g - 2\omega)\\
&= S_1T_7 + S_7T_1,\\
\cos (2g' + 2\omega' - 2g - 2\omega) &= T_1T_7 - S_1S_7,\\
U_{11} &= \sin \Omega,\\
U_8 &= \sin (2\omega' - 2g - 2\omega) = S_7T_1 - S_1T_7,\\
\cos (2\omega' - 2g - 2\omega) &= T_1T_7 + S_1S_7,\\
U_7 &= \sin (2\omega' - g - 2\omega)\\
&= U_8 \cos g + \cos (2\omega' - 2g - 2\omega) \sin g,\\
\cos (2\omega' - g - 2\omega) &= \cos (2\omega' - 2g - 2\omega) \cos g - U_8 \sin g,\\
\sin (\omega' - g) &= \sin \omega' \cos g - \cos \omega' \sin g,\\
\cos (\omega' - g) &= \cos \omega' \cos g + \sin \omega' \sin g,\\
U_6 &= \sin (\omega' - g - \omega)\\
&= \sin (\omega' - g) \cos \omega - \cos (\omega' - g) \sin \omega,\\
\cos (\omega' - g - \omega) &= \cos (\omega' - g) \cos \omega + \sin (\omega' - g) \sin \omega.
\end{aligned}
\tag{8.36}
$$

Application of (8.35) immediately yields σ_l, ρ_l, and τ_l. In passing, it should be noted that a number of the final trigonometric expressions in (8.36) are not used in obtaining the values of the librations, however, these terms become important in obtaining the derivatives of the libration expressions.

The time derivatives of the librations,

$$
\dot{\sigma}_l \equiv \frac{d}{d\tau}(\sigma_l), \qquad \dot{\rho}_l \equiv \frac{d}{d\tau}(\rho_l), \qquad \dot{\tau}_l \equiv \frac{d}{d\tau}(\tau_l), \tag{8.37}
$$

can be obtained in a similar fashion. More explicitly, from (8.31), (8.32), (8.33), and (8.35),

$$
\begin{aligned}
\dot{\sigma}_l &= \left[\frac{1}{\sin I}\sum_{i=1}^{7} E_i \dot{S}_i\right]\frac{\pi}{180 \times 3600},\\
\dot{\rho}_l &= \left[\sum_{i=1}^{7} F_i \dot{T}_i\right]\frac{\pi}{180 \times 3600},\\
\dot{\tau}_l &= \left[\sum_{i=1}^{11} G_i \dot{U}_i\right]\frac{\pi}{180 \times 3600},
\end{aligned}
\tag{8.38}
$$

where, as before, the dot indicates differentiation with respect to τ. Again, each expression in (8.38) must be multiplied by $\pi/(180 \times 3600)$ since the constants E_i, F_i, and G_i are given in seconds of arc rather than radians.

The algorithm is completed by the computation of the values of $\dot{S}_i$, $\dot{T}_i$, and $\dot{U}_i$ as follows:

$$
\begin{aligned}
\dot{g}' &= \dot{☾} - \dot{\Gamma}, & \dot{S}_6 &= -T_6(2\dot{g} + 2\dot{\omega}),\\
\dot{g} &= \dot{L} - \dot{\Gamma}, & \dot{T}_6 &= S_6(2\dot{g} + 2\dot{\omega}),\\
\dot{\omega}' &= \dot{\Gamma}' - \dot{\Omega}, & \dot{S}_7 &= T_7(\dot{g}' + 2\dot{\omega}' - 2\dot{g} - 2\dot{\omega}),\\
\dot{\omega} &= \dot{\Gamma} - \dot{\Omega}, & \dot{T}_7 &= -S_7(\dot{g}' + 2\dot{\omega}' - 2\dot{g} - 2\dot{\omega}),\\
\dot{S}_1 &= T_1(\dot{g}'), & \dot{U}_1 &= \dot{S}_1,\\
\dot{T}_1 &= -S_1(\dot{g}'), & \dot{U}_2 &= [\cos 2g'](2\dot{g}'),\\
\dot{S}_2 &= T_2(2\dot{\omega}'), & \dot{U}_3 &= -[\cos g](\dot{g}),\\
\dot{T}_2 &= -S_2(2\dot{\omega}'), & \dot{U}_4 &= \dot{S}_2,\\
\dot{S}_3 &= T_3(\dot{g}' + 2\dot{\omega}'), & \dot{U}_5 &= \dot{S}_6,\\
\dot{T}_3 &= -S_3(\dot{g}' + 2\dot{\omega}'), & \dot{U}_6 &= [\cos(\omega' - g - \omega)](\dot{\omega}' - \dot{g} - \dot{\omega}),\\
\dot{S}_4 &= T_4(2\dot{g}' + 2\dot{\omega}'), & \dot{U}_7 &= [\cos(2\omega' - g - 2\omega)](2\dot{\omega}' - \dot{g} - 2\dot{\omega}),\\
\dot{T}_4 &= -S_4(2\dot{g}' + 2\dot{\omega}'), & \dot{U}_8 &= [\cos(2\omega' - 2g - 2\omega)](2\dot{\omega}' - 2\dot{g} - 2\dot{\omega}),\\
\dot{S}_5 &= T_5(3\dot{g}' + 2\dot{\omega}'), & \dot{U}_9 &= \dot{S}_7,\\
\dot{T}_5 &= -S_5(3\dot{g}' + 2\dot{\omega}'), & \dot{U}_{10} &= [\cos(2g' + 2\omega' - 2g - 2\omega)]\\
& & &\quad \times (2\dot{g}' + 2\dot{\omega}' - 2\dot{g} - 2\dot{\omega}),\\
& & \dot{U}_{11} &= [\cos \Omega](\dot{\Omega}).
\end{aligned}
\tag{8.39}
$$

8.1.7 Evaluation of the Precession of the Equinox

The *Explanatory Supplement* [6] presents expressions for the determination of the three terms ζ_0, z_p, and θ_p, illustrated in Figure 8.7. These three terms determine the general precession of the equinox between two dates. However, the equations for ζ_0, z_p, and θ_p listed in [6] neglect certain second-order terms. A complete set of equations, including these second-order terms, are given in *Connaissance des Temps* [3]. These equations are based on the work of H. Andoyer [31].

The more complete expressions in [3] are adopted in this chapter. Each of the equations for the components of the total precession is expressed in terms of the quantities

$T_0 \equiv$ time from 1900.0 to the selected reference epoch in tropical centuries,
$T \equiv$ elapsed time since the reference epoch in tropical centuries.

The quantity T_0 is a constant determined by the particular reference

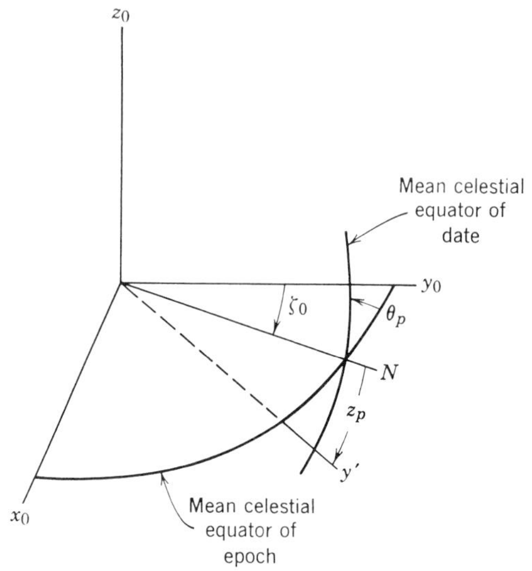

FIGURE 8.7 Coordinate rotations induced by the precession of the equinox.

epoch chosen. The reference epoch chosen in [1], [3], [6], [9], [10] because of the ready accessibility of data relative to this epoch, is the reference epoch $\equiv$ 1950.0, for which

$$T_0 = 0.5. \tag{8.40}$$

Gabbard, Escobal, and Birkholz [9] give an equation for T in terms of the Julian date (J.D.) and the Julian date of the reference epoch $(\text{J.D.})_{1950.0}$. A change in reference epoch therefore entails a change of the constant in (8.40) and a corresponding change in the expression for T.

A more convenient form for T is the following expression which is a function of the constant in (8.40):

$$T = \frac{\text{J.D.} - [(\text{J.D.})_{1900.0} + 36524.220T_0]}{36524.219879}, \tag{8.41}$$

where the denominator is an 11 significant figure representation of the number of ephemeris days in a tropical century (neglecting the very small secular variation). The reference Julian date $(\text{J.D.})_{1900.0}$ is

$$(\text{J.D.})_{1900.0} = 2415020.313. \tag{8.42}$$

One additional constant is required. It is the derivative $\dot{T}$ of T with respect to the modified time variable τ. From (8.6) and (8.41),

$$\dot{T} = \frac{d}{d\tau}(T) = (36524.219879 \times 24 \times 60 \times 0.07436574)^{-1}. \tag{8.43}$$

With the previous understandings it is possible to consider three general precession terms which are defined as follows:

$$\zeta_0 + z_p \equiv \text{general precession in right ascension},$$

$$\theta_p \equiv \text{precession in declination},$$

$$\frac{\pi}{2} - \zeta_0 \equiv \text{angle from mean equinox of epoch to the ascending node of mean equator of date.}$$

The precession equations to be developed require the values of z_p, θ_p, ζ_0, and the corresponding derivatives

$$\dot{z}_p \equiv \frac{dz_p}{d\tau}, \qquad \dot{\theta}_p \equiv \frac{d\theta_p}{d\tau}, \qquad \dot{\zeta}_0 \equiv \frac{d\zeta_0}{d\tau}.$$

Values of the six variables above are determined from the following equations [3].

$$C_1 = 2304''.253 + 13''.973 \times 10^{-1}T_0 + 0''.006 \times 10^{-2}T_0^2,$$

$$C_2 = (3''.023 - 0''.027 \times 10^{-1}T_0) \times 10^{-1},$$

$$D_1 = 2004''.685 - 8''.533 \times 10^{-1}T_0 - 0''.037 \times 10^{-2}T_0^2,$$

$$D_2 = (4''.267 + 0''.037 \times 10^{-1}T_0) \times 10^{-1},$$

$$E_1 = C_1 - 0''.001 \times 10^{-1}T_0,$$

$$E_2 = (10''.950 + 0''.039 \times 10^{-1}T_0) \times 10^{-1},$$

$$\zeta_0 = C_1T + C_2T^2 + 1''.800 \times 10^{-2}T^3,$$

$$\theta_p = D_1T - D_2T^2 - 4''.180 \times 10^{-2}T^3,$$

$$z_p = E_1T + E_2T^2 + 1''.832 \times 10^{-2}T^3,$$

$$\dot{\zeta}_0 = [C_1 + 2C_2T + 5''.400 \times 10^{-2}T^2]\dot{T}\frac{\pi}{3600 \times 180},$$

$$\dot{\theta}_p = [D_1 - 2D_2T - 12''.540 \times 10^{-2}T^2]\dot{T}\frac{\pi}{3600 \times 180},$$

$$\dot{z}_p = [E_1 + 2E_2T + 5''.496 \times 10^{-2}T^2]\dot{T}\frac{\pi}{3600 \times 180}. \tag{8.44}$$

As before, the multiplication by $\pi/(3600 \times 180)$ is for the conversion from seconds of arc to radians. Equations 8.44 complete the algorithm of the precession calculations.

Sections 8.1.1 through 8.1.7 have developed all the auxiliary equations which are required in order to perform the selenographic transformations. These transformations are explicitly discussed in Section 8.2.

8.2 SELENOGRAPHIC COORDINATE TRANSFORMATIONS

8.2.1 Basic Coordinate Systems

The study of selenographic coordinate transformations is concerned with three fundamental planes: *the ecliptic*, which is the plane of the Earth's orbit around the Sun; *the celestial equator*, which is perpendicular to the Earth's axis of rotation; and *the lunar equator*. It is convenient to place the center of the Moon at the origin of the celestial sphere and to translate the ecliptic and celestial equatorial planes parallel to themselves, such that each of the three fundamental planes contains the origin.

The three fundamental planes, as defined above, are not uniquely determined, as none of the planes is fixed in space. It is therefore assumed that each of the planes is defined at the same instant of time as that corresponding to the velocity and position vectors to be transformed. It is customary to refer to the instantaneous position of the reference planes as the *position of date*. Unless otherwise specified, the position of all reference planes is the position of date.

The x axis for the ecliptic and the celestial equator is in the direction of the vernal equinox, which is the ascending node on the celestial equator of the Sun's apparent orbit around the Earth.[5] The z axis corresponding to the celestial equator is the north celestial pole. The directed line perpendicular to the ecliptic which makes an angle less than 180° with the north celestial pole is the z axis corresponding to the ecliptic. In each case the y axis is chosen to complete a right-handed coordinate system.

It is conventional, in the definition of a mean and true equator and (vernal) equinox of date, for the term *mean* to indicate that the periodic nutation terms have been neglected. The actual location at a given instant therefore corresponds to the true position of date.

The x, y, and z axes corresponding to the lunar equator are fixed in the Moon and correspond to the principal axes of inertia. The z axis is aligned with the Moon's axis of rotation. The x axis is in the Moon's equatorial plane pointing in a direction which is approximately along a line to the center of the Earth, and the y axis is chosen to complete a right-handed coordinate system. The intersection of the x, z plane with the surface of the

[5] It is thus seen that the x axis lies along the intersection of the ecliptic and the celestial equator. See ([1], [2], and [4], Chapter 4).

Moon defines the lunar prime meridian. This coordinate system, which is related to the lunar equator, is called a *selenographic coordinate system.* A more precise definition of the selenographic coordinate system requires an examination of Cassini's laws.

8.2.2 The Laws of Cassini

Before considering Cassini's laws it is necessary to consider one additional plane. This is the plane of the Moon's motion about the Earth, or the *lunar orbit plane*, which is identical with the plane of the Earth's apparent orbit about the Moon as viewed from the selenocenter. Since the ascending node, with respect to the ecliptic, is independent of the view taken, this plane is referred to as the lunar orbit plane.

The rotation of the Moon can be described as small periodic oscillations from a mean motion described by the empirical laws postulated by J. Cassini in 1721. These laws may be stated as follows.

I. The mean axis of rotation is fixed in the Moon, perpendicular to the mean lunar equator. The mean period of rotation is equal to the mean sidereal period of revolution of the Moon around the Earth.
II. The mean lunar equator intersects the ecliptic (of date) at a constant inclination equal to I.[6]
III. The mean lunar equator, the ecliptic, and the lunar orbit plane meet in a line. The angle i between the lunar orbit plane and the ecliptic is constant, as is the angle $i + I$ between the mean lunar equator and the lunar orbit plane. The ecliptic is seen to lie always between the mean lunar equator and the lunar orbit plane.

Cassini's third law has led many authors [5], [6], [7], [8] to label the nodal position of the lunar equator at the descending node of the lunar orbit as the "ascending node of the lunar equator." This terminology is accepted reluctantly in the present analysis.

Cassini's laws define the mean selenographic coordinate system with the z axis coincident with the mean axis of rotation. The x axis is the line in the lunar equatorial plane from the center of the Moon to the center of the Earth at the time when $☾ = \Omega = \lambda$, where λ is the geocentric longitude of the Moon [6], [8], and [1, Chapter 4].

As previously noted, the true selenographic coordinate system, which is fixed in the Moon, oscillates about the mean selenographic coordinate system. The term mean thus indicates that the physical librations are neglected.

[6] The value of I is discussed in Section 8.1.5.

8.2.3 Listing of Coordinate Transformations

The following six coordinate systems have been defined in Sections 8.2.1 and 8.2.2:

1. Mean selenographic,
2. True selenographic,
3. Ecliptic, mean equinox (of date),
4. Ecliptic, true equinox (of date),
5. Mean (celestial) equator (of date),
6. True equator (of date).

This section is concerned with the transformation of a pair of position and velocity vectors from one of the first two coordinate systems to one of the last four coordinate systems, or from one of the last four coordinate systems to either the first or second coordinate system. If C_1 denotes the initial, and C_2 the final, coordinate system, then C_1 can be chosen as any of the six stated coordinate systems but C_2 is of course limited as follows. If C_1 is equal to any number between 3 and 6, then $C_2 = 1$ or 2. If $C_1 = 1$ or 2, then C_2 is a number between 3 and 6.

It is noted that none of the coordinate systems above is inertial. This entails difficulties if any one of these coordinate systems is employed for the integration of the equations of motion. The difficulty is due not only to the necessity of employing additional terms in the equations of motion but additionally in evolving a scheme for locating various celestial bodies, with respect to the chosen system.

A convenient fixed reference system is a system corresponding to the mean celestial equator of epoch, C_7. The locations of lunar, solar, and planetary coordinates are readily available in this particular coordinate system, especially for the epoch of 1950.0. This coordinate system has therefore been appended to the set previously obtained. A discussion of the transformation to the mean equatorial coordinate system of epoch can be found in [4], [5], [6], [7], [9], [10]. The selenographic coordinate transformation transforms position and velocity vectors from a member of the set A:

3. Ecliptic, mean equinox,
4. Ecliptic, true equinox,
5. Mean equator,
6. True equator,
7. Mean equator of epoch,

to a member of the set B, namely the following:

1. Mean selenographic,
2. True selenographic.

Before the particular transformations are examined, consideration is directed to Section 8.2.4, which develops the equations for the transformation of position and velocity vectors from an arbitrary reference system A to an arbitrary reference system B.

8.2.4 Basic Rotation Matrices

Consider the transformation of coordinates from system A with origin O and axes Ox_0, Oy_0, and Oz_0 to system B with origin O and axes Ox, Oy, and Oz, where system A can be rotated into system B as described below [11], and in ([1], Chapter 4).

A rotation is made about the axis Oz_0 rotating the axis Ox_0 through an angle ϕ from its initial position to ON, where ON is the intersection of the x_0, y_0, and the x, y planes, that is, the equatorial planes in systems A and B, respectively. The angle ϕ increases in a counterclockwise direction about Oz_0.

A second rotation is made about the axis ON rotating the axis Oz_0 through an angle θ to Oz, where θ is the angle between the equatorial x_0, y_0 and x, y planes. The angle θ increases in a counterclockwise direction about ON.

The first two rotations align the axes Oz_0 and Oz and the equatorial x_0, y_0 and x, y planes. The transformation is, therefore, completed by rotating the ON axis to Ox.

The third rotation is about the Oz axis, rotating the axis ON through an angle ψ to its final position Ox which completes the transformation. The angle ψ increases in a counterclockwise direction about Oz.

The angles ϕ, θ, and ψ are known as Euler's angles. They are, respectively, the angle between the Ox_0 axis and the intersection of the equatorial planes, the angle between the equatorial planes, and the angle between the intersection of the equatorial planes and the Ox axis. Figure 8.8 illustrates the relation between the Eulerian angles and the relative orientation of the initial and final planes. A similar discussion of Euler's angles can be found in Smart [12].

For the present problem, system A corresponds to a choice of C_1 between 3 and 7, and system B corresponds to a choice of C_2 equal to 1 or 2. If $\mathbf{r}$ and $\dot{\mathbf{r}}$ are position and velocity vectors, respectively, in system A where

$$\mathbf{r} \equiv \begin{bmatrix} x \\ y \\ z \end{bmatrix}, \qquad \dot{\mathbf{r}} \equiv \begin{bmatrix} \dot{x} \\ \dot{y} \\ \dot{z} \end{bmatrix}, \tag{8.45}$$

and $\mathbf{r}'$ and $\dot{\mathbf{r}}'$ are the same position and velocity vectors in system B, then from ([1], Section 3.4) it follows that

$$\mathbf{r}' = M\mathbf{r}, \qquad \dot{\mathbf{r}}' = M\dot{\mathbf{r}} + \dot{M}\mathbf{r}, \tag{8.46}$$

where

$$\dot{M} \equiv \frac{d}{d\tau}(M), \tag{8.47}$$

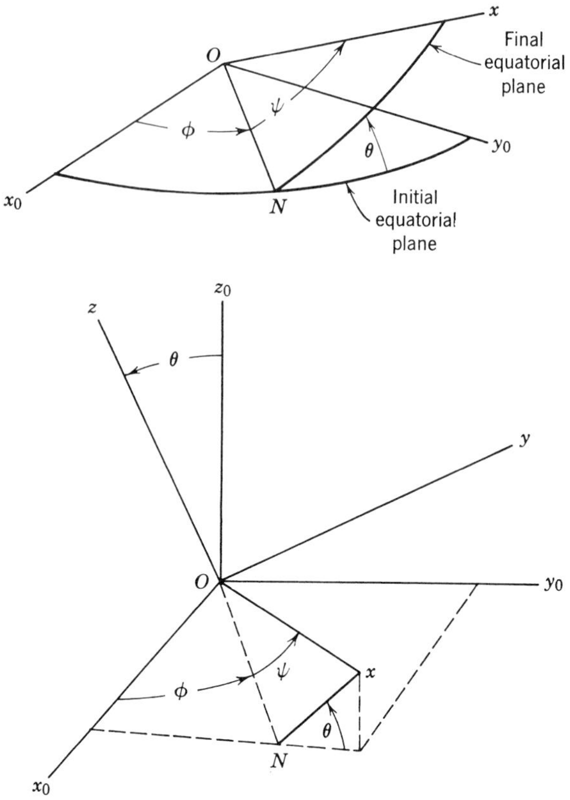

FIGURE 8.8 Euler's angles ϕ, θ, and ψ.

with the M and $\dot{M}$ matrices equal to

$$M = \begin{bmatrix} M_{11} & M_{12} & M_{13} \\ M_{21} & M_{22} & M_{23} \\ M_{31} & M_{32} & M_{33} \end{bmatrix},$$

$$\dot{M} = \begin{bmatrix} \dot{M}_{11} & \dot{M}_{12} & \dot{M}_{13} \\ \dot{M}_{21} & \dot{M}_{22} & \dot{M}_{23} \\ \dot{M}_{31} & \dot{M}_{32} & \dot{M}_{33} \end{bmatrix}, \tag{8.48}$$

and

$$\mathbf{r}' \equiv \begin{bmatrix} x' \\ y' \\ z' \end{bmatrix}, \qquad \dot{\mathbf{r}}' \equiv \begin{bmatrix} \dot{x}' \\ \dot{y}' \\ \dot{z}' \end{bmatrix}.$$

The dot indicates the differentiation with respect to the modified time variable τ. It can be shown that the matrix elements are given by the following expressions derived in ([1], Section 3.4.1).

$$
\begin{aligned}
M_{11} &= \cos\phi\cos\psi - \cos\theta\sin\phi\sin\psi, \\
M_{12} &= \sin\phi\cos\psi + \cos\theta\cos\phi\sin\psi, \\
M_{13} &= \sin\psi\sin\theta, \\
M_{21} &= -\sin\psi\cos\phi - \cos\theta\sin\phi\cos\psi, \\
M_{22} &= -\sin\phi\sin\psi + \cos\theta\cos\phi\cos\psi, \\
M_{23} &= \sin\theta\cos\psi, \\
M_{31} &= \sin\phi\sin\theta, \\
M_{32} &= -\cos\phi\sin\theta, \\
M_{33} &= \cos\theta.
\end{aligned} \tag{8.49}
$$

By formal differentiation the time rate of change of these elements is given by

$$
\begin{aligned}
\dot{M}_{11} = {} & +[\cos\psi]\dot{C}(\phi) + [\cos\phi]\dot{C}(\psi) \\
& - [\sin\phi\sin\psi]\dot{C}(\theta) - [\cos\theta\sin\psi]\dot{S}(\phi) \\
& - [\cos\theta\sin\phi]\dot{S}(\psi), \\
\dot{M}_{12} = {} & +[\cos\psi]\dot{S}(\phi) + [\sin\phi]\dot{C}(\psi) \\
& + [\cos\phi\sin\psi]\dot{C}(\theta) + [\cos\theta\sin\psi]\dot{C}(\phi) \\
& + [\cos\theta\cos\phi]\dot{S}(\psi), \\
\dot{M}_{13} = {} & +[\sin\theta]\dot{S}(\psi) + [\sin\psi]\dot{S}(\theta), \\
\dot{M}_{21} = {} & -[\cos\phi]\dot{S}(\psi) - [\sin\psi]\dot{C}(\phi) \\
& - [\sin\phi\cos\psi]\dot{C}(\theta) - [\cos\theta\cos\psi]\dot{S}(\phi) \\
& - [\cos\theta\sin\phi]\dot{C}(\psi), \\
\dot{M}_{22} = {} & -[\sin\psi]\dot{S}(\phi) - [\sin\phi]\dot{S}(\psi) \\
& + [\cos\phi\cos\psi]\dot{C}(\theta) + [\cos\theta\cos\psi]\dot{C}(\phi) \\
& + [\cos\theta\cos\phi]\dot{C}(\psi), \\
\dot{M}_{23} = {} & +[\cos\psi]\dot{S}(\theta) + [\sin\theta]\dot{C}(\psi), \\
\dot{M}_{31} = {} & +[\sin\theta]\dot{S}(\phi) + [\sin\phi]\dot{S}(\theta), \\
\dot{M}_{32} = {} & -[\sin\theta]\dot{C}(\phi) - [\cos\phi]\dot{S}(\theta), \\
\dot{M}_{33} = {} & +\dot{C}(\theta),
\end{aligned} \tag{8.50}
$$

where, for the sake of compactness,

$$
\begin{aligned}
\dot{C}(\phi) &\equiv \frac{d}{d\tau}(\cos\phi), & \dot{S}(\theta) &\equiv \frac{d}{d\tau}(\sin\theta), \\
\dot{S}(\phi) &\equiv \frac{d}{d\tau}(\sin\phi), & \dot{C}(\psi) &\equiv \frac{d}{d\tau}(\cos\psi), \\
\dot{C}(\theta) &\equiv \frac{d}{d\tau}(\cos\theta), & \dot{S}(\psi) &\equiv \frac{d}{d\tau}(\sin\psi).
\end{aligned}
$$

This development treats the transformation of position and velocity vectors from system A to system B or transformations into the selenographic coordinate system. It is also necessary to obtain the transformation from system B to system A or out of the selenographic coordinate system. Evidently this can be accomplished by obtaining Euler's angles for the inverse transformation and utilizing the transformation above. A somewhat more direct technique is described below.

Let the initial position and velocity vectors defined in (8.45) be specified in system B corresponding to a choice of C_1 equal to 1 or 2. Similarly, let the transformed position and velocity vectors defined in (8.48) be specified in system A corresponding to a choice of C_2 as some value between 3 and 7. It can then be shown that

$$\mathbf{r}' = M^T\mathbf{r}, \qquad \dot{\mathbf{r}}' = M^T\dot{\mathbf{r}} + \dot{M}^T\mathbf{r}, \tag{8.51}$$

since the inverse M^{-1} of M is equal to its transpose M^T. The matrices M^T and $\dot{M}^T$ are, therefore, defined as follows:

$$M^T = \begin{bmatrix} M_{11} & M_{21} & M_{31} \\ M_{12} & M_{22} & M_{32} \\ M_{13} & M_{23} & M_{33} \end{bmatrix}, \qquad \dot{M}^T = \begin{bmatrix} \dot{M}_{11} & \dot{M}_{21} & \dot{M}_{31} \\ \dot{M}_{12} & \dot{M}_{22} & \dot{M}_{32} \\ \dot{M}_{13} & \dot{M}_{23} & \dot{M}_{33} \end{bmatrix}, \tag{8.52}$$

where the matrix components are given by (8.49) and (8.50).

The preceding analysis transforms position and velocity vectors into and out of the selenographic coordinate system. The necessary Eulerian angles are supplied by the analyst in accordance with Section 8.2.5 below. However, it should be emphasized again that the analysis of this section can be utilized with any general coordinate transformation excluding a translation of origin[7] as, for example, the transformations into and out of the areocentric coordinate system discussed in ([1], Chapter 4).

A similar discussion of Eulerian angles and the resultant transformation matrices can be found in [5].

8.2.5 Transformation Selection

The most important portion of the selenographic transformation process is discussed in this section. The required data are the Julian date, J.D., the selection parameter λ defined in Section 8.1.3, the components x, y, z of the position vector, the components $\dot{x}$, $\dot{y}$, $\dot{z}$ of the velocity vector, the code C_1 of Section 8.2.3 which signifies in which coordinate system x, y, z, $\dot{x}$, $\dot{y}$, and $\dot{z}$ are specified, and the code C_2 which specifies in which coordinate system the position and velocity vectors are desired. The parameters obtained are the components x', y', z' of the position vector and the components $\dot{x}'$, $\dot{y}'$, $\dot{z}'$ of the velocity vector in the coordinate system indicated by the parameter C_2. Figure 8.9 shows the relationship of all the parameters discussed in the preceding sections.

[7] The translation of origin can be performed after the rotation.

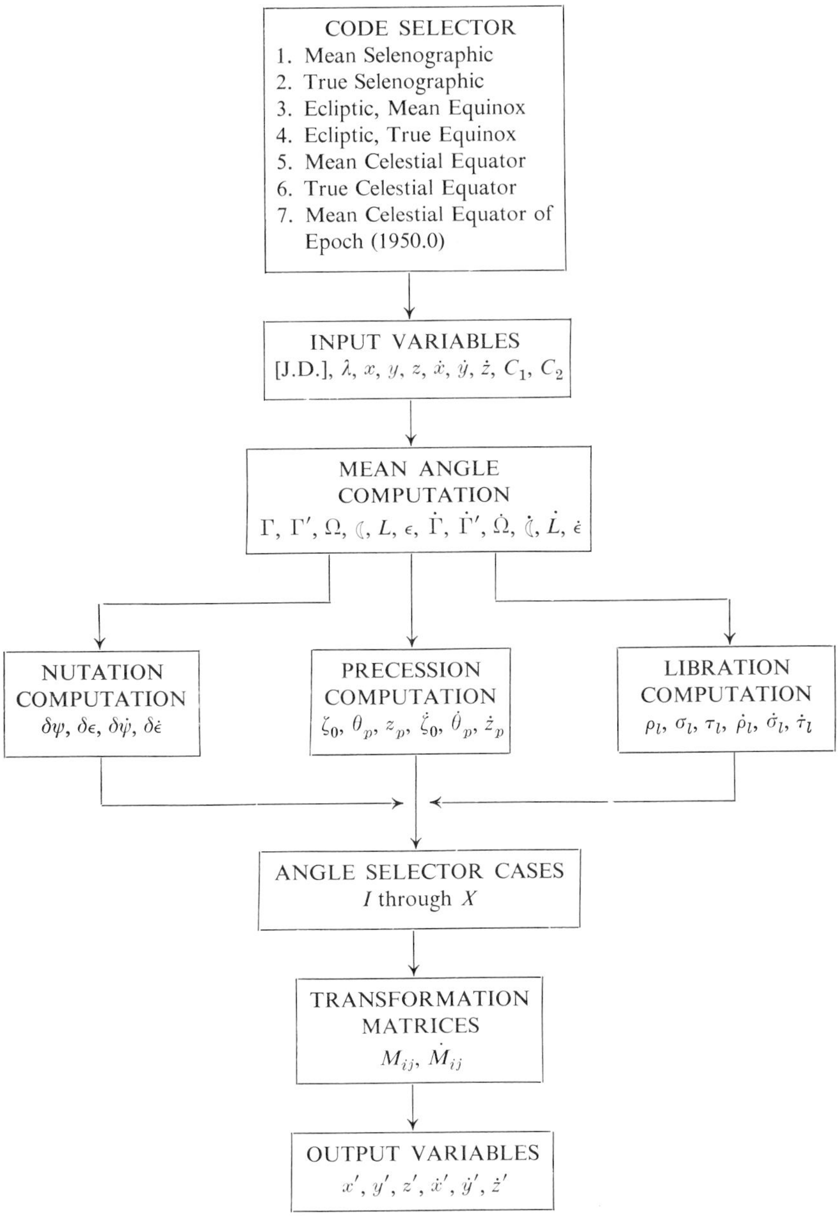

FIGURE 8.9 Selenographic coordinate transformation philosophy.

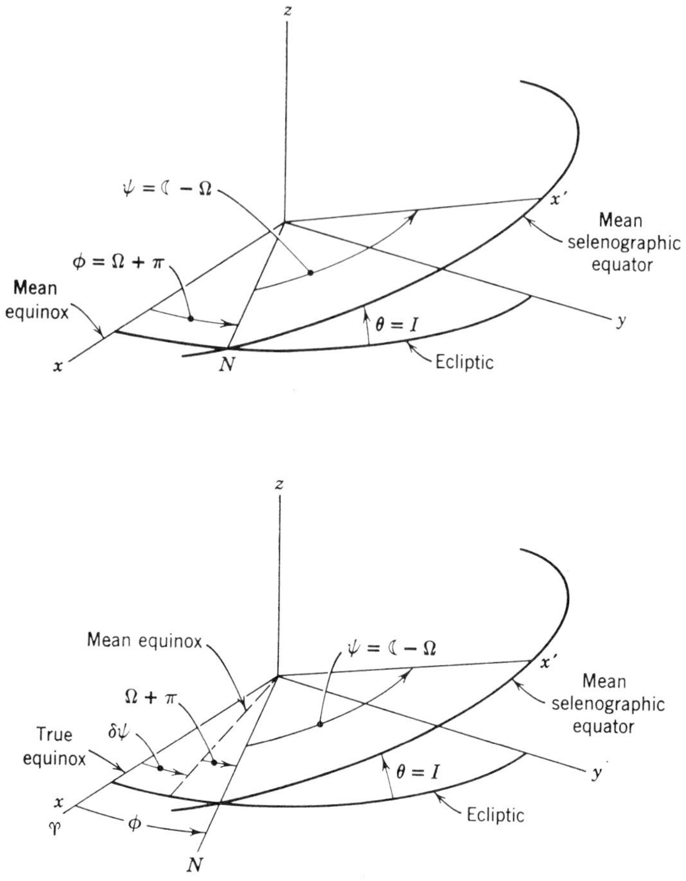

FIGURE 8.10 Euler's angles for the transformation to the mean selenographic coordinate system from the ecliptic, mean equinox and the ecliptic, true equinox coordinate systems.

The Eulerian angles obtained below are functions of the outputs of the previous sections, that is, the mean angles, the nutations, the librations and the precession.

This section, in essence, generates the necessary Eulerian angles which are used in Section 8.2.4 to obtain the desired transformed position and velocity vectors.

Figure 8.10 shows the Eulerian angles for the transformations from the ecliptic, mean equinox to the mean selenographic and from the ecliptic, true equinox to the mean selenographic coordinate system. The appropriate values of ϕ, θ, ψ, $\dot{C}(\phi)$, etc., required for (8.49) and (8.50) are as follows (see Figure 8.10).

I. *Ecliptic, Mean Equinox to Mean Selenographic or Inverse*

$$
\begin{aligned}
&(3 \to 1 \quad \text{or} \quad 1 \to 3),\\
\theta &= I \text{ (see footnote}^{8}\text{)},\\
\cos\phi &= -\cos\Omega,\\
\sin\phi &= -\sin\Omega,\\
\psi &= ☾ - \Omega, \pmod{2\pi},\\
\dot{C}(\theta) &= 0,\\
\dot{C}(\phi) &= (-\sin\phi)\dot{\Omega},\\
\dot{C}(\psi) &= (-\sin\psi)(\dot{☾} - \dot{\Omega}),\\
\dot{S}(\theta) &= 0,\\
\dot{S}(\phi) &= (\cos\phi)\dot{\Omega},\\
\dot{S}(\psi) &= (\cos\psi)(\dot{☾} - \dot{\Omega}).
\end{aligned} \tag{8.53}
$$

II. *Ecliptic, True Equinox to Mean Selenographic or Inverse*

$$
\begin{aligned}
&(4 \to 1 \quad \text{or} \quad 1 \to 4),\\
\theta &= I,\\
\cos\phi &= -\cos(\Omega + \delta\psi),\\
\sin\phi &= -\sin(\Omega + \delta\psi,)\\
\psi &= ☾ - \Omega, \pmod{2\pi},\\
\dot{C}(\theta) &= 0, \qquad \dot{S}(\theta) = 0,\\
\dot{C}(\phi) &= (-\sin\phi)(\dot{\Omega} + \delta\dot{\psi}),\\
\dot{S}(\phi) &= (\cos\phi)(\dot{\Omega} + \delta\dot{\psi}),\\
\dot{C}(\psi) &= (-\sin\psi)(\dot{☾} - \dot{\Omega}),\\
\dot{S}(\psi) &= (\cos\psi)(\dot{☾} - \dot{\Omega}).
\end{aligned} \tag{8.54}
$$

The values of Ω, ☾, $\dot{\Omega}$, and $\dot{☾}$ are obtained from Section 8.1.2. Section 8.1.3 yields the values of $\delta\psi$ and $\delta\dot{\psi}$.

Figure 8.11 shows the Eulerian angles for the transformations from the ecliptic, mean equinox to the true selenographic, and from the ecliptic, true equinox to the true selenographic coordinate system.

III. *Ecliptic, Mean Equinox to True Selenographic or Inverse*

$$
\begin{aligned}
&(3 \to 2 \quad \text{or} \quad 2 \to 3),\\
\theta &= I + \rho_l,\\
\cos\phi &= -\cos(\Omega + \sigma_l),\\
\sin\phi &= -\sin(\Omega + \sigma_l),\\
\psi &= (☾ + \tau_l) - (\Omega + \sigma_l), \pmod{2\pi},\\
\dot{C}(\theta) &= (-\sin\theta)\dot{\rho}_l,\\
\dot{S}(\theta) &= (\cos\theta)\dot{\rho}_l,\\
\dot{C}(\phi) &= (-\sin\phi)(\dot{\Omega} + \dot{\sigma}_l),\\
\dot{S}(\phi) &= (\cos\phi)(\dot{\Omega} + \dot{\sigma}_l),\\
\dot{C}(\psi) &= (-\sin\psi)(\dot{☾} + \dot{\tau}_l - \dot{\Omega} - \dot{\sigma}_l),\\
\dot{S}(\psi) &= (\cos\psi)(\dot{☾} + \dot{\tau}_l - \dot{\Omega} - \dot{\sigma}_l).
\end{aligned} \tag{8.55}
$$

[8] The value of I for these transformations is defined by (8.24).

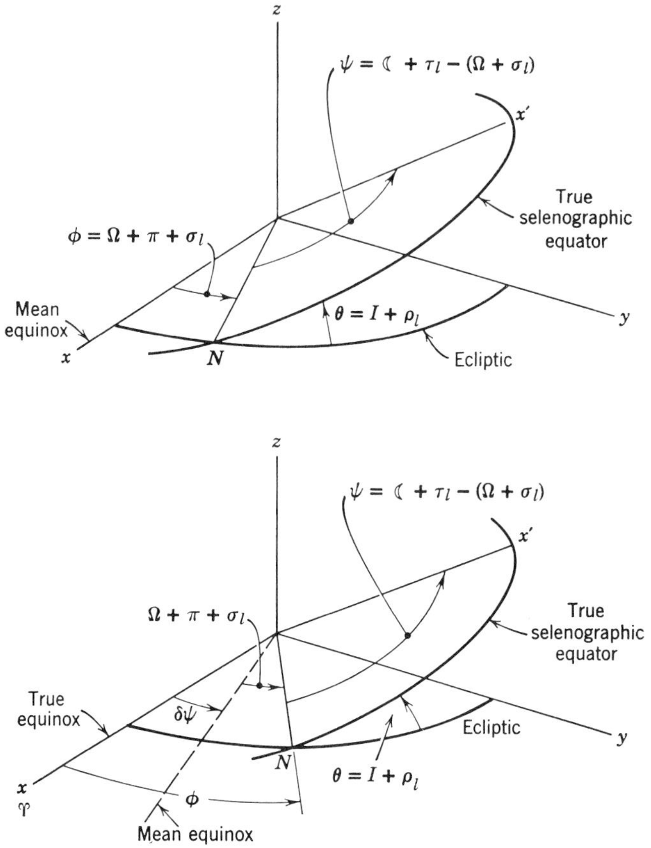

FIGURE 8.11 Euler's angles for the transformation to the true selenographic coordinate system from the ecliptic, mean equinox and the ecliptic, true equinox coordinate systems.

IV. *Ecliptic, True Equinox to True Selenographic or Inverse*

$$(4 \rightarrow 2 \quad \text{or} \quad 2 \rightarrow 4),$$

$$\begin{aligned}
\theta &= I + \rho_l, \\
\cos\phi &= -\cos(\Omega + \delta\psi + \sigma_l), \\
\sin\phi &= -\sin(\Omega + \delta\psi + \sigma_l), \\
\psi &= (☾ + \tau_l) - (\Omega + \sigma_l), \ (\text{mod } 2\pi), \\
\dot{C}(\theta) &= (-\sin\theta)\dot{\rho}_l, \\
\dot{S}(\theta) &= (\cos\theta)\dot{\rho}_l, \\
\dot{C}(\phi) &= (-\sin\phi)(\dot{\Omega} + \delta\dot{\psi} + \dot{\sigma}_l), \\
\dot{S}(\phi) &= (\cos\phi)(\dot{\Omega} + \delta\dot{\psi} + \dot{\sigma}_l), \\
\dot{C}(\psi) &= (-\sin\psi)(\dot{☾} + \dot{\tau}_l - \dot{\Omega} - \dot{\sigma}_l), \\
\dot{S}(\psi) &= (\cos\psi)(\dot{☾} + \dot{\tau}_l - \dot{\Omega} - \dot{\sigma}_l).
\end{aligned} \tag{8.56}$$

The values of ρ_l, σ_l, τ_l, $\dot{\rho}_l$, $\dot{\sigma}_l$, and $\dot{\tau}_l$ are obtained from Section 8.1.6.

The previously obtained knowledge relative to Euler's angles for the ecliptic to selenographic transformation can be utilized to obtain Euler's angles for the celestial equator to selenographic coordinate transformation. Figure 8.12 illustrates the relative orientation of the celestial equator, the ecliptic, and the selenographic equator.

The angles $c + \pi$, a, and $\psi - \Delta$ are the respective values of ϕ, θ, and ψ for the transformation from the ecliptic to the selenographic coordinate system. The third angle, b, is the angle between the celestial equator and the ecliptic. It therefore follows that b is equal to the mean obliquity ϵ for the mean and the sum $\epsilon + \delta\epsilon$ of the mean obliquity and the nutation in obliquity $\delta\epsilon$ for the true celestial equator.

The mean and the true equinox are, respectively, the intersection of the ecliptic with the mean and the true celestial equator. Figure 8.13 illustrates the Eulerian angles corresponding to transformations from the mean and true celestial equator coordinate system. In each case the Ox axis is the appropriate equinox.

At this point it is convenient to define other auxiliary angles. The angle ϕ_1 is the angle measured in the same sense as ϕ between the equinox or Ox axis and the intersection of the initial and intermediate planes. In the present case the initial plane is the celestial equator, the intermediate plane is the

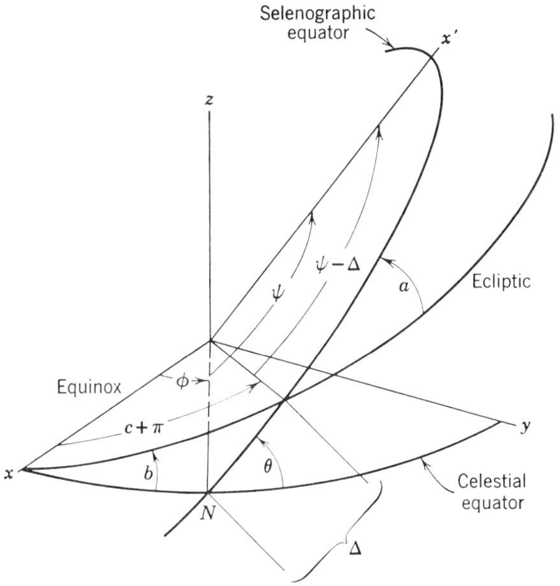

FIGURE 8.12 Euler's angles for transformations from celestial equator to selenographic coordinate system.

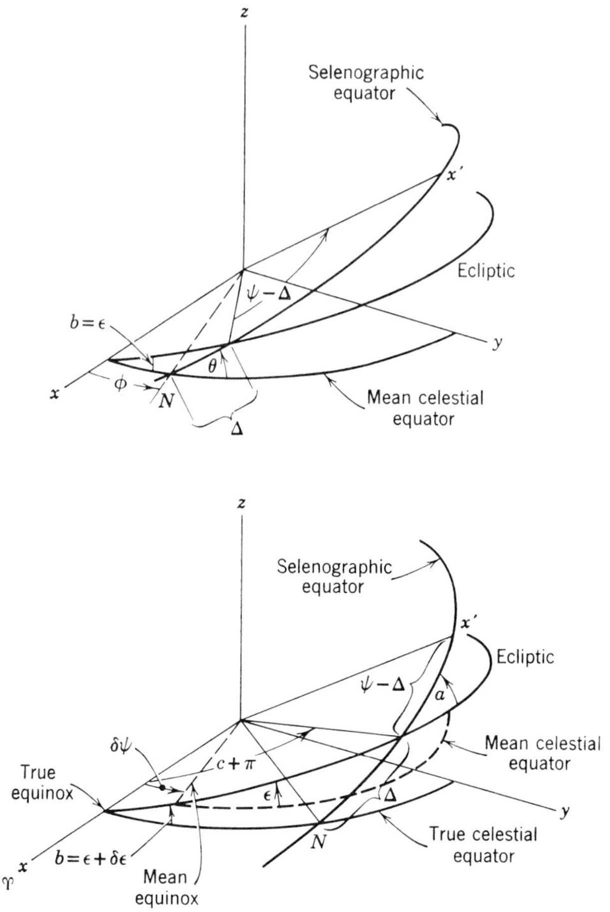

FIGURE 8.13 Euler's angles for the mean and true celestial equator coordinate systems.

ecliptic, and therefore $\phi_1 \equiv 0$. The angle $\phi_2 \equiv \phi - \phi_1$ is the angle measured, in the sense of ϕ, from the above intersection to the line ON of intersection of the initial and final planes. The final plane is always the selenographic equator. Another important angle d is defined as

$$d \equiv \psi - \Delta. \tag{8.57}$$

In essence, the introduction of the auxiliary angles, a, b, c, etc., is only for convenience. By using these angles and redefining them for each transformation it is possible to develop a single set of equations which can then be used to obtain the necessary Eulerian angles. From a, b, c, etc., the corresponding values of ϕ, θ, ψ are then obtained by (8.69); this equation will be introduced presently.

The values of a, b, c, d, ϕ_1, $\dot{a}$, $\dot{b}$, $\dot{c}$, and $\dot{d}$ for the transformation from the equatorial to the selenographic coordinate systems are given as follows (see Figures 8.12 and 8.13).

V. *Mean Equator to Mean Selenographic or Inverse*

$$(5 \rightarrow 1 \quad \text{or} \quad 1 \rightarrow 5),$$

$$\begin{aligned} a &= I, \quad \phi_1 = 0, & \dot{a} &= 0, \\ b &= \epsilon \quad (\text{see footnote}^9), & \dot{b} &= \dot{\epsilon}, \\ c &= \Omega, & \dot{c} &= \dot{\Omega}, \\ d &= ☾ - \Omega, & \dot{d} &= \dot{☾} - \dot{\Omega}. \end{aligned} \tag{8.58}$$

VI. *Mean Equator to True Selenographic or Inverse*

$$(5 \rightarrow 2 \quad \text{or} \quad 2 \rightarrow 5),$$

$$\begin{aligned} a &= I + \rho_l, \quad \phi_1 = 0, & \dot{a} &= \dot{\rho}_l, \\ b &= \epsilon, & \dot{b} &= \dot{\epsilon}, \\ c &= \Omega + \sigma_l, & \dot{c} &= \dot{\Omega} + \dot{\sigma}_l, \\ d &= ☾ + \tau_l - (\Omega + \sigma_l), & \dot{d} &= \dot{☾} + \dot{\tau}_l - (\dot{\Omega} + \dot{\sigma}_l). \end{aligned} \tag{8.59}$$

VII. *True Equator to Mean Selenographic or Inverse*

$$(6 \rightarrow 1 \quad \text{or} \quad 1 \rightarrow 6),$$

$$\begin{aligned} a &= I, \quad \phi_1 = 0, & \dot{a} &= 0 \\ b &= \epsilon + \delta\epsilon, & \dot{b} &= \dot{\epsilon} + \delta\dot{\epsilon}, \\ c &= \Omega + \delta\psi, & \dot{c} &= \dot{\Omega} + \delta\dot{\psi}, \\ d &= ☾ - \Omega, & \dot{d} &= \dot{☾} - \dot{\Omega}. \end{aligned} \tag{8.60}$$

VIII. *True Equator to True Selenographic or Inverse*

$$(6 \rightarrow 2 \quad \text{or} \quad 2 \rightarrow 6),$$

$$\begin{aligned} a &= I + \rho_l, \quad \phi_1 = 0, & \dot{a} &= \dot{\rho}_l, \\ b &= \epsilon + \delta\epsilon, & \dot{b} &= \dot{\epsilon} + \delta\dot{\epsilon}, \\ c &= \Omega + \delta\psi + \sigma_l, & \dot{c} &= \dot{\Omega} + \delta\dot{\psi} + \dot{\sigma}_l, \\ d &= ☾ + \tau_l - \Omega - \sigma_l, & \dot{d} &= \dot{☾} + \dot{\tau}_l - \dot{\Omega} - \dot{\sigma}_l. \end{aligned} \tag{8.61}$$

[9] The values of ϵ and $\dot{\epsilon}$ are obtained from Section 8.1.2.

It is convenient to introduce the following set of auxiliary equations for each of the transformations in V through VIII:

$$\dot{C}(a) \equiv \frac{d}{d\tau}(\cos a) = (-\sin a)\dot{a},$$

$$\dot{S}(a) \equiv \frac{d}{d\tau}(\sin a) = (\cos a)\dot{a},$$

$$\dot{C}(b) \equiv \frac{d}{d\tau}(\cos b) = (-\sin b)\dot{b},$$

$$\dot{S}(b) \equiv \frac{d}{d\tau}(\sin b) = (\cos b)\dot{b},$$

$$\dot{C}(c) \equiv \frac{d}{d\tau}(\cos c) = (-\sin c)\dot{c},$$

$$\dot{S}(c) \equiv \frac{d}{d\tau}(\sin c) = (\cos c)\dot{c},$$

$$\dot{C}(d) \equiv \frac{d}{d\tau}(\cos d) = (-\sin d)\dot{d},$$

$$\dot{S}(d) \equiv \frac{d}{d\tau}(\sin d) = (\cos d)\dot{d},$$

$$\dot{C}(\phi_1) \equiv \frac{d}{d\tau}(\cos \phi_1) = 0,$$

$$\dot{S}(\phi_1) \equiv \frac{d}{d\tau}(\sin \phi_1) = 0. \tag{8.62}$$

Figure 8.14 shows the relation between the mean celestial equator of epoch and the selenographic coordinate systems. The mean celestial equator of date is utilized as the intermediate plane rather than the ecliptic. The angle a is equal to the Eulerian angle θ' for the transformation from mean celestial equator of date to the selenographic coordinate system. The angle b is equal to θ_p, the angle between the mean celestial equator of epoch and the mean celestial equator of date.

The angles ζ_0, z_p, and θ_p are obtained from Section 8.1.7. The angles $c + \pi$, d, and ϕ_1 are given by the equations

$$\phi_1 = \frac{\pi}{2} - \zeta_0, \tag{8.63}$$

$$c + \pi = \frac{3\pi}{2} - z_p + \phi', \tag{8.64}$$

$$d = \psi - \Delta = \psi', \tag{8.65}$$

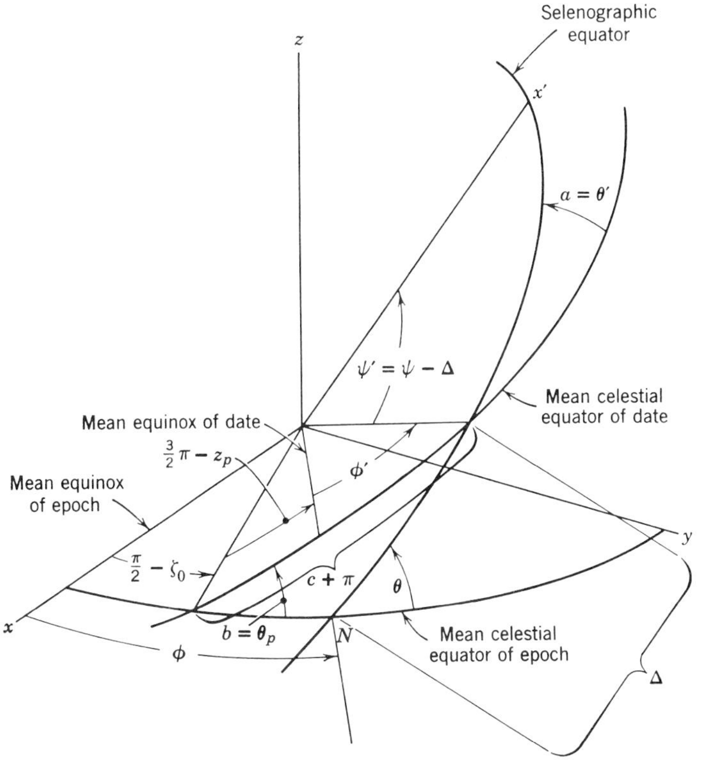

FIGURE 8.14 Euler's angles for the transformation from mean celestial equator of epoch to selenographic coordinate system.

where ϕ', θ', and ψ' are Euler's angles for the transformation from the mean celestial equator of date to the selenographic coordinate system. Similarly, a', b', c', and d' are used to denote the values of a, b, c, and d for the transformation from the mean celestial equator of date to the selenographic coordinate system. To reiterate, the primed angles a', b', c', etc., are used to determine values of θ', ϕ', and ψ', which in turn are used to determine the angles a, b, c, etc. The angles a, b, c, etc., by means of (8.69) yield the sought for rotation angles.

The values of a', b', c', d', $\dot{a}'$, $\dot{b}'$, $\dot{c}'$, and $\dot{d}'$, corresponding, respectively, to the transformation from mean celestial equator of epoch to mean selenographic, and from mean celestial equator of epoch to true selenographic are as follows.

IX. *Mean Equator of Epoch (1950.0) to Mean Selenographic or Inverse*

$$(7 \rightarrow 1 \quad \text{or} \quad 1 \rightarrow 7),$$

$$\begin{aligned} a' &= I, & \dot{a}' &= 0, \\ b' &= \epsilon, & \dot{b}' &= \dot{\epsilon}, \\ c' &= \Omega, & \dot{c}' &= \dot{\Omega}, \\ d' &= \text{☾} - \Omega, & \dot{d}' &= \dot{\text{☾}} - \dot{\Omega}. \end{aligned} \tag{8.66}$$

X. *Mean Equator of Epoch (1950.0) to True Selenographic or Inverse*

$$(7 \rightarrow 2 \quad \text{or} \quad 2 \rightarrow 7),$$

$$\begin{aligned} a' &= I + \rho_l, & \dot{a}' &= \dot{\rho}_l, \\ b' &= \epsilon & \dot{b}' &= \dot{\epsilon}, \\ c' &= \Omega + \sigma_l, & \dot{c}' &= \dot{\Omega} + \dot{\sigma}_l, \\ d' &= \text{☾} + \tau_l - (\Omega + \sigma_l), & \dot{d}' &= \dot{\text{☾}} + \dot{\tau}_l - (\dot{\Omega} + \dot{\sigma}_l). \end{aligned} \tag{8.67}$$

Use of simple spherical trigonometry yields the values of ϕ', θ', ψ', a, b, etc., for the transformations in IX and X as (8.68):

$$\cos\theta' = \cos a' \cos b' + \sin a' \sin b' \cos c',$$

$$\sin\theta' = +\sqrt{1 - \cos^2\theta'},$$

$$\cos\phi' = \frac{\cos a' \sin b' - \sin a' \cos b' \cos c'}{\sin\theta'},$$

$$\sin\phi' = \frac{-\sin a' \sin c'}{\sin\theta'},$$

$$\cos\Delta' = \frac{\sin a' \cos b' - \cos a' \sin b' \cos c'}{\sin\theta'},$$

$$\sin\Delta' = \frac{-\sin b' \sin c'}{\sin\theta'},$$

$$\cos\psi' = \cos\Delta' \cos d' - \sin\Delta' \sin d',$$

$$\sin\psi' = \sin\Delta' \cos d' + \cos\Delta' \sin d',$$

$$\cos a = \cos\theta',$$

$$\sin a = \sin\theta',$$

$$\cos b = \cos\theta_p,$$

$$\sin b = \sin\theta_p,$$

$$\cos c = \cos\phi' \sin z_p - \sin\phi' \cos z_p,$$

$$\sin c = \sin \phi' \sin z_p + \cos \phi' \cos z_p,$$

$$\cos d = \cos \psi',$$

$$\sin d = \sin \psi',$$

$$\cos \phi_1 = \sin \zeta_0,$$

$$\sin \phi_1 = \cos \zeta_0,$$

$$\dot{C}(\theta') \equiv \frac{d}{d\tau}(\cos \theta')$$
$$= -\sin \theta'[(\cos \Delta')\dot{a}' + (\cos \phi')\dot{b}'] - [\sin a' \sin b' \sin c']\dot{c}',$$

$$\dot{S}(\theta') \equiv \frac{d}{d\tau}(\sin \theta') = (\cos \theta')\dot{\theta}' = -[\text{ctn}\, \theta']\dot{C}(\theta') = -\frac{\cos \theta'}{\sin \theta'}\dot{C}(\theta'),$$

$$\dot{C}(\phi') \equiv \frac{d}{d\tau}(\cos \phi')$$
$$= -[(\sin a' \sin b' + \cos a' \cos b' \cos c')/\sin \theta']\dot{a}'$$
$$+ [\cos \theta'/\sin \theta']\dot{b}' + [\sin a' \cos b' \sin c'/\sin \theta']\dot{c}' - \frac{\cos \phi'}{\sin \theta'}\dot{S}(\theta'),$$

$$\dot{S}(\phi') \equiv \frac{d}{d\tau}(\sin \phi')$$
$$= -[(\cos a' \sin c')/\sin \theta']\dot{a}' - [\sin a' \cos c'/\sin \theta']\dot{c}' - \frac{\sin \phi'}{\sin \theta'}\dot{S}(\theta'),$$

$$\dot{C}(\Delta') \equiv \frac{d}{d\tau}(\cos \Delta')$$
$$= [\cos \theta'/\sin \theta']\dot{a}' - [(\sin a' \sin b' + \cos a' \cos b' \cos c')/\sin \theta']\dot{b}'$$
$$+ [\cos a' \sin b' \sin c'/\sin \theta']\dot{c}' - \frac{\cos \Delta'}{\sin \theta'}\dot{S}(\theta'),$$

$$\dot{S}(\Delta') \equiv \frac{d}{d\tau}(\sin \Delta')$$
$$= -[(\cos b' \sin c')/\sin \theta']\dot{b}' - [(\sin b' \cos c')/\sin \theta']\dot{c}' - \frac{\sin \Delta'}{\sin \theta'}\dot{S}(\theta'),$$

$$\dot{C}(\psi') \equiv \frac{d}{d\tau}(\cos \psi') = -[\sin \psi']\dot{d}' + [\cos d']\dot{C}(\Delta') - [\sin d']\dot{S}(\Delta'),$$

$$\dot{S}(\psi') \equiv \frac{d}{d\tau}(\sin \psi') = [\cos \psi']\dot{d}' + [\cos d']\dot{S}(\Delta') + [\sin d']\dot{C}(\Delta'),$$

$$\dot{C}(a) = \dot{C}(\theta'),$$

$$\dot{S}(a) = \dot{S}(\theta'),$$

$$\dot{C}(b) = -[\sin \theta_p]\dot{\theta}_p,$$

$$\dot{S}(b) = [\cos \theta_p]\dot{\theta}_p,$$

$$\dot{C}(c) = [\sin z_p]\dot{C}(\phi') - [\cos z_p]\dot{S}(\phi') + [\sin c]\dot{z}_p,$$

$$\dot{S}(c) = [\sin z_p]\dot{S}(\phi') + [\cos z_p]\dot{C}(\phi') - [\cos c]\dot{z}_p,$$

$$\dot{C}(d) = \dot{C}(\psi'),$$

$$\dot{S}(d) = \dot{S}(\psi'),$$

$$\dot{C}(\phi_1) = [\cos \zeta_0]\dot{\zeta}_0,$$

$$\dot{S}(\phi_1) = -[\sin \zeta_0]\dot{\zeta}_0. \tag{8.68}$$

In summary, the Eulerian angles and the first time derivative of the sines and cosines of these angles have been obtained for transformations I through IV. For transformations V through X the auxiliary angles a, b, c, d, ϕ_1 and the first time derivative of the sines and cosines of these angles have been obtained.

The angle d is defined by (8.57) and the angle ϕ_1 is defined in the paragraph immediately preceding that equation. The angles a, b, and c are depicted in Figure 8.12. The algorithm of computation is completed by the following set of equations obtained by simple spherical trigonometry (8.69):

$$\cos \theta = \cos a \cos b + \sin a \sin b \cos c,$$

$$\sin \theta = +\sqrt{1 - \cos^2 \theta},$$

$$\cos \phi_2 = [\cos a \sin b - \sin a \cos b \cos c]/\sin \theta,$$

$$\sin \phi_2 = [-\sin a \sin c]/\sin \theta,$$

$$\cos \Delta = [\sin a \cos b - \cos a \sin b \cos c]/\sin \theta,$$

$$\sin \Delta = [-\sin b \sin c]/\sin \theta,$$

$$\cos \phi = \cos \phi_1 \cos \phi_2 - \sin \phi_1 \sin \phi_2,$$

$$\sin \phi = \sin \phi_1 \cos \phi_2 + \cos \phi_1 \sin \phi_2,$$

$$\cos \psi = \cos \Delta \cos d - \sin \Delta \sin d,$$

$$\sin \psi = \sin \Delta \cos d + \cos \Delta \sin d,$$

$$\dot{C}(\theta) \equiv \frac{d}{d\tau}(\cos\theta)$$
$$= [\cos b]\dot{C}(a) + [\cos a]\dot{C}(b) + [\sin b \cos c]\dot{S}(a)$$
$$+ [\sin a \cos c]\dot{S}(b) + [\sin a \sin b]\dot{C}(c),$$

$$\dot{S}(\theta) \equiv \frac{d}{d\tau}(\sin\theta) = -\frac{\cos\theta}{\sin\theta}\dot{C}(\theta),$$

$$\dot{C}(\phi_2) \equiv \frac{d}{d\tau}(\cos\phi_2)$$
$$= \{[\sin b]\dot{C}(a) + [\cos a]\dot{S}(b) - [\cos b \cos c]\dot{S}(a)$$
$$- [\sin a \cos c]\dot{C}(b) - [\sin a \cos b]\dot{C}(c)\}/\sin\theta - (\cos\phi_2/\sin\theta)\dot{S}(\theta),$$

$$\dot{S}(\phi_2) \equiv \frac{d}{d\tau}(\sin\phi_2)$$
$$= -\{[\sin c]\dot{S}(a) + [\sin a]\dot{S}(c)\}/\sin\theta - (\sin\phi_2/\sin\theta)\dot{S}(\theta),$$

$$\dot{C}(\Delta) \equiv \frac{d}{d\tau}(\cos\Delta)$$
$$= \{[\cos b]\dot{S}(a) + [\sin a]\dot{C}(b) - [\sin b \cos c]\dot{C}(a) - [\cos a \cos c]\dot{S}(b)$$
$$- [\cos a \sin b]\dot{C}(c)\}/\sin\theta - (\cos\Delta/\sin\theta)\dot{S}(\theta),$$

$$\dot{S}(\Delta) \equiv \frac{d}{d\tau}(\sin\Delta)$$
$$= -\{[\sin c]\dot{S}(b) + [\sin b]\dot{S}(c)\}/\sin\theta - (\sin\Delta/\sin\theta)\dot{S}(\theta),$$

$$\dot{C}(\phi) \equiv \frac{d}{d\tau}(\cos\phi)$$
$$= [\cos\phi_2]\dot{C}(\phi_1) + [\cos\phi_1]\dot{C}(\phi_2) - [\sin\phi_2]\dot{S}(\phi_1) - [\sin\phi_1]\dot{S}(\phi_2),$$

$$\dot{S}(\phi) \equiv \frac{d}{d\tau}(\sin\phi)$$
$$= [\cos\phi_2]\dot{S}(\phi_1) + [\sin\phi_1]\dot{C}(\phi_2) + [\sin\phi_2]\dot{C}(\phi_1) + [\cos\phi_1]\dot{S}(\phi_2),$$

$$\dot{C}(\psi) \equiv \frac{d}{d\tau}(\cos\psi)$$
$$= [\cos d]\dot{C}(\Delta) + [\cos\Delta]\dot{C}(d) - [\sin d]\dot{S}(\Delta) - [\sin\Delta]\dot{S}(d),$$

$$\dot{S}(\psi) = [\cos d]\dot{S}(\Delta) + [\sin\Delta]\dot{C}(d) + [\sin d]\dot{C}(\Delta) + [\cos\Delta]\dot{S}(d). \qquad (8.69)$$

8.2.6 Computational Algorithm for Selenographic Transformations

The computational algorithm describing selenographic transformations proceeds as follows. The value of I is first obtained from (8.24).

A choice is made, on the basis of the values of C_1 and C_2, between (8.53), (8.54), (8.55), (8.56), (8.58), (8.59), (8.60), (8.61), (8.66), and (8.67); Table 8.4 shows the variables that must be supplied for each of the possible transformations. The first time derivative of each listed variable is required if and only if the variable is required. The transformation code and the location of the necessary variables are shown in Figure 8.9.

Table 8.4 Transformation Code and Associated Variables

$1 \rightleftarrows 3$	Ω, ☾
$1 \rightleftarrows 4$	Ω, ☾, $\delta\psi$
$3 \rightleftarrows 2$	Ω, ☾, ρ_l, σ_l, τ_l
$4 \rightleftarrows 2$	Ω, ☾, $\rho_l, \sigma_l, \tau_l, \delta\psi$
$5 \rightleftarrows 1$	Ω, ☾, ϵ
$5 \rightleftarrows 2$	Ω, ☾, $\epsilon, \rho_l, \sigma_l, \tau_l$
$6 \rightleftarrows 1$	Ω, ☾, $\epsilon, \delta\epsilon, \delta\psi$
$6 \rightleftarrows 2$	Ω, ☾, $\epsilon, \rho_l, \sigma_l, \tau_l, \delta\epsilon, \delta\psi$
$7 \rightleftarrows 1$	Ω, ☾, $\epsilon, \zeta_0, \theta_p, z_p$
$7 \rightleftarrows 2$	Ω, ☾, $\epsilon, \rho_l, \sigma_l, \tau_l, \zeta_0, \theta_p, z_p$

For transformations involving (8.53), (8.54), (8.55), and (8.56) the necessary variables for the rotation matrices defined by (8.48), (8.49), and (8.50) have been computed. These transformations can therefore be directly determined.

For transformations involving (8.58) through (8.61) proceed to (8.69). After (8.69) proceed to (8.49) and (8.50), which supply the necessary inputs to the rotation matrices.

For transformations involving (8.66) and (8.67) there is a subsequent instruction to proceed to (8.68). After (8.68) proceed to (8.69) and finally to (8.49) and (8.50), which supply the inputs to the rotation matrices.

Treatments of the coordinate transformations performed above can be found in [1], [5], [6], [7], [9], [10], [14], [15].

8.3 SIMPLIFIED COORDINATE TRANSFORMATIONS

8.3.1 Formulation of Velocity Transformations

As noted previously, (8.45), the transformation of a position vector from one of the ecliptic or celestial equator coordinate systems into either the mean or true selenographic coordinate system can be represented as follows:

$$\mathbf{r}' = M\mathbf{r}, \tag{8.70}$$

where **r** is the position vector in the original and **r**′ is the same position vector in the final coordinate system (in this case the selenographic coordinate system). The matrix M is the transformation matrix and has been represented (8.48) as

$$M = \begin{bmatrix} M_{11} & M_{12} & M_{13} \\ M_{21} & M_{22} & M_{23} \\ M_{31} & M_{32} & M_{33} \end{bmatrix}. \tag{8.71}$$

If, as before, **r** and **r**′ represent the position vector in the initial and final coordinate systems, respectively, then for transformations from the selenographic coordinate system to one of the ecliptic or equatorial systems, it follows that

$$\mathbf{r}' = M^T\mathbf{r}, \tag{8.72}$$

where M^T is the transpose of M.

The velocity transformation is made in two steps. One step is a transformation from the ecliptic or equatorial system to an inertial selenographic coordinate system which is fixed at the location of the actual selenographic system at the instantaneous Julian date. The second transformation is from the inertial coordinate system to the noninertial actual selenographic coordinate system. If $\dot{\mathbf{r}}_{s0}$ and $\dot{\mathbf{r}}_s$ are the velocity vectors in the inertial and the real selenographic systems, respectively, then from [1, Section 4.1.1] it follows that

$$\dot{\mathbf{r}}_{s0} = \dot{\mathbf{r}}_s + \boldsymbol{\omega} \times \mathbf{r}_s, \tag{8.73}$$

where $\mathbf{r}_s$ is the position vector[10] and $\boldsymbol{\omega}$ is the angular velocity vector.

8.3.2 Simplification of Velocity Transformations

The angular velocity vector is a strong function of certain mean angles and a very weak function of the physical librations for the true selenographic system. For the mean selenographic system, $\boldsymbol{\omega}$ is a function of the mean longitudes alone.

It can be shown that the libration dependence of $\boldsymbol{\omega}$ is always completely negligible. Therefore, within the required accuracy, (8.73) is independent of whether the transformation involves the mean or the true selenographic coordinate system.

The vector $\boldsymbol{\omega}$ can be broken into two components (see Figures 8.10, 8.11, and 8.12), a component of magnitude $\dot{\Omega}$ along the pole to the ecliptic and a component $\dot{\zeta} - \dot{\Omega}$ along the selenographic north pole. If **i**, **j**, and **k** denote

[10] Evidently $\mathbf{r}_s$ is the same in either selenographic system.

unit vectors along selenographic x, y, z axes and if $\mathbf{r}_s \equiv x_s\mathbf{i} + y_s\mathbf{j} + z_s\mathbf{k}$, then

$$\boldsymbol{\omega}\times\mathbf{r}_s = [(\dot{\Omega}\sin\{\text{☾} - \Omega\}\sin I)\mathbf{i} + (\dot{\Omega}\cos\{\text{☾} - \Omega\}\sin I)\mathbf{j} + (\dot{\text{☾}} - \{1 - \cos I\}\dot{\Omega})\mathbf{k}]\times[x_s\mathbf{i} + y_s\mathbf{j} + z_s\mathbf{k}], \quad (8.74)$$

where the value of I is obtained from (8.24).

Examination of the rate of change of the mean longitude in (8.11) shows that the term $\{1 - \cos I\}\dot{\Omega}$ is negligible in comparison to $\dot{\text{☾}}$. Furthermore, both $\dot{\text{☾}}$ and $\dot{\Omega}$ can be set equal to the constant lead terms of their expansion within the required accuracy.

Substitution of (8.74) into (8.73) with the two preceding simplifications yields the following equation

$$\dot{\mathbf{r}}_{s0} = \dot{\mathbf{r}}_s + \begin{bmatrix} (\alpha\cos\{\text{☾} - \Omega\}z_s - \beta y_s) \\ (\beta x_s - \alpha\sin\{\text{☾} - \Omega\}z_s) \\ (\alpha\sin\{\text{☾} - \Omega\}y_s - \alpha\cos\{\text{☾} - \Omega\}x_s) \end{bmatrix}, \quad (8.75)$$

where, from (8.9) and (8.11),

$$\alpha = -\frac{1934.1420\pi\sin I}{180\times 36{,}525\times 24\times 60\times 0.07436574} \cong \dot{\Omega}\sin I$$

and

$$\beta = \frac{481{,}267.88\pi}{180\times 36{,}525\times 24\times 60\times 0.07436574} \cong \dot{\text{☾}}.$$

The final step in the velocity transformation is the transformation between any of the ecliptic or equatorial systems and the "inertial selenographic system." For these transformations the $\dot{M}\mathbf{r}$ velocity correction is completely negligible. It therefore follows that, for transformations into the "inertial selenographic coordinate system,"

$$\dot{\mathbf{r}}_{s0} = M\dot{\mathbf{r}}, \quad (8.77)$$

where $\dot{\mathbf{r}}$ is the velocity vector in the initial system, namely, the ecliptic or equatorial coordinate system. For transformations out of the "inertial selenographic coordinate system,"

$$\dot{\mathbf{r}}' = M^T\dot{\mathbf{r}}_{s0}, \quad (8.78)$$

where $\dot{\mathbf{r}}'$ is the velocity vector in the final system, namely, the ecliptic or equatorial coordinate system.

From (8.70), (8.71), (8.72), (8.75), (8.77), and (8.78) it follows that for transformations into the selenographic coordinate system, that is, $3\to 1$, $3\to 2$, $4\to 1$, $4\to 2$, $5\to 1$, $5\to 2$, $6\to 1$, $6\to 2$, $7\to 1$, and $7\to 2$ with the code defined in Section 8.2.3,

$$\mathbf{r}' \equiv \begin{bmatrix} x' \\ y' \\ z' \end{bmatrix} = M\mathbf{r}, \tag{8.79}$$

$$\dot{\mathbf{r}}' = M\dot{\mathbf{r}} - \begin{bmatrix} (\alpha \cos\{☾ - \Omega\}z' - \beta y') \\ (\beta x' - \alpha \sin\{☾ - \Omega\}z') \\ (\alpha \sin\{☾ - \Omega\}y' - \alpha \cos\{☾ - \Omega\}x') \end{bmatrix}. \tag{8.80}$$

For the transformations out of the selenographic coordinate system, that is, $1 \to 3$, $2 \to 3$, $1 \to 4$, $2 \to 4$, $1 \to 5$, $2 \to 5$, $1 \to 6$, $2 \to 6$, $1 \to 7$, and $2 \to 7$,

$$\mathbf{r}' = M^T\mathbf{r} = M^T \begin{bmatrix} x \\ y \\ z \end{bmatrix}, \tag{8.81}$$

$$\begin{aligned} \dot{\mathbf{r}}' &= M^T\dot{\mathbf{r}} + M^T \begin{bmatrix} (\alpha \cos\{☾ - \Omega\}z - \beta y) \\ (\beta x - \alpha \sin\{☾ - \Omega\}z) \\ (\alpha \sin\{☾ - \Omega\}y - \alpha \cos\{☾ - \Omega\}x) \end{bmatrix} \\ &= M^T \begin{bmatrix} (\dot{x} + \alpha \cos\{☾ - \Omega\}z - \beta y) \\ (\dot{y} + \beta x - \alpha \sin\{☾ - \Omega\}z) \\ (\dot{z} + \alpha \sin\{☾ - \Omega\}y - \alpha \cos\{☾ - \Omega\}x) \end{bmatrix}. \end{aligned} \tag{8.82}$$

The constants α and β are obtained from (8.75).

8.4 SUMMARY

This chapter has discussed and formulated the very complex selenographic transformation. Accuracy and speed of computation have been stressed. The first portion of the chapter described the causes of precession and nutation. By means of a combination of theoretical and empirical analysis it was shown how the mean angles of the lunar longitudes and obliquity could be computed. Very accurate expressions for the nutations in obliquity and longitude were also displayed. An examination of the physical lunar librations or wobblings was undertaken and an algorithmic development of the librations of the Moon presented. Having obtained the required input, namely, the mean angles, librations, and nutations along with the precession, if desired, the accurate lunar coordinate rotations could then be considered.

It was shown how, by means of the basic laws of Cassini, the required rotations could be performed as variations from a basic set of rotation or mean Eulerian angles. By adding small excursions to the fundamental rotation angles the effects of nutation and libration were incorporated easily. Various transformations of the selenographic class were developed.

Finally, the very rigorous velocity transformations developed earlier were simplified by performing some justified approximations. The results of these approximations yielded a much simplified velocity transformation which has found substantial use in industry.

EXERCISES

1. What is precession? What is the cause of precession? Discuss generally.
2. What is nutation? What is the cause of nutation? Discuss generally.
3. What is lunisolar precession, planetary precession, general precession?
4. Show that, if storage space is at a premium, in modern computing machinery the total nutation in obliquity and longitude (8.14) can be written

$$\delta\epsilon = d\epsilon + \Delta\epsilon,$$
$$\delta\psi = d\psi + \Delta\psi,$$

where

$$\begin{aligned}
\Delta\epsilon = {} & +(9''.21 + 0''.00091T_u)\cos\Omega - (0''.0904 - 0''.00004T_u)\cos 2\Omega \\
& + (0''.5522 - 0''.00029T_u)\cos 2L \\
& + (0''.0216 - 0''.00006T_u)\cos(3L - \Gamma) \\
& - (0''.0093 - 0''.00003T_u)\cos(L + \Gamma) - 0''.0066\cos(2L - \Omega) \\
& + 0''.0007\cos(4L - 2\Gamma) - 0''.0024\cos(2☾ - 2g' - \Omega) \\
& + 0''.0008\cos(L + \Omega - \Gamma) + 0''.0005\cos(L - \Gamma - \Omega) \\
& + 0''.0003\cos(2L + 2g' - 2☾ - \Omega) + 0''.0003\cos(L + \Gamma - \Omega) \\
& + 0''.0002\cos(2\Gamma - \Omega) - 0''.0002\cos(2L + 2g' + \Omega - 2☾) \\
& - 0''.0002\cos(3L - \Gamma - \Omega) + 0''.0002\cos(2☾ - 2g'), \\
d\epsilon = {} & +(0''.0884 - 0''.00005T_u)\cos 2☾ + 0''.0183\cos(2☾ - \Omega) \\
& + (0''.0113 - 0''.00001T_u)\cos(2☾ + g') - 0''.0050\cos(2☾ - g') \\
& - 0''.0031\cos(\Omega + g') + 0''.0030\cos(\Omega - g') \\
& + 0''.0022\cos(4☾ - 2L - g') + 0''.0023\cos(2☾ + g' - \Omega) \\
& + 0''.0014\cos(4☾ - 2L) - 0''.0011\cos(2L + g') \\
& + 0''.0011\cos(2☾ + 2g') - 0''.0010\cos(2☾ - \Omega - g') \\
& - 0''.0007\cos(2L + g' - 2☾ - \Omega) \\
& + 0''.0007\cos(2L + \Omega + g' - 2☾) \\
& + 0''.0005\cos(4☾ - 2L - \Omega - g') - 0''.0003\cos(2☾ + L - \Gamma) \\
& + 0''.0003\cos(2☾ + \Gamma - L) + 0''.0003\cos(4☾ + g' - 2L) \\
& - 0''.0002\cos(2L + 2g') + 0''.0003\cos(2L - 2☾ - \Omega) \\
& - 0''.0003\cos(2L + g' - \Omega) + 0''.0003\cos(2L + \Omega - 2☾) \\
& + 0''.0003\cos(4☾ - 2L - \Omega) + 0''.0002\cos(2☾ + 2g' - \Omega),
\end{aligned}$$

$$\begin{aligned}
\Delta\psi = {} & -(17''.2327 + 0''.01737T_u)\sin\Omega + (0''.2088 + 0''.00002T_u)\sin 2\Omega \\
& - (1''.2729 + 0''.00013T_u)\sin 2L \\
& + (0''.1261 - 0''.00031T_u)\sin(L - \Gamma) \\
& + (0''.0214 - 0''.00005T_u)\sin(L + \Gamma) \\
& - (0''.0497 - 0''.00012T_u)\sin(3L - \Gamma) \\
& + (0''.0124 + 0''.00001T_u)\sin(2L - \Omega) \\
& + (0''.0016 - 0''.00001T_u)\sin(2L - 2\Gamma) \\
& - (0''.0015 - 0''.00001T_u)\sin(4L - 2\Gamma) \\
& + 0''.0045\sin(2☾ - 2g' - \Omega) \\
& - 0''.0015\sin(L + \Omega - \Gamma) + 0''.0010\sin(L - \Gamma - \Omega) \\
& + 0''.0005\sin(2L + 2g' - 2☾ - \Omega) - 0''.0005\sin(L + \Gamma - \Omega) \\
& - 0''.0004\sin(2\Gamma - \Omega) + 0''.0004\sin(2L + 2g' + \Omega - 2☾) \\
& + 0''.0003\sin(3L - \Gamma - \Omega) - 0''.0003\sin(2☾ - 2g') \\
& + 0''.0045\sin(2L + 2g' - 2☾) + 0''.0021\sin(2\Omega - 2L) \\
& + 0''.0010\sin(2\Omega + 2g' - 2☾) - 0''.0003\sin(L + g' - ☾) \\
& - 0''.0002\sin(g' + \Gamma - ☾), \\
d\psi = {} & -(0''.2037 + 0''.00002T_u)\sin 2☾ + (0''.0675 + 0''.00001T_u)\sin g' \\
& - (0''.0342 + 0''.00004T_u)\sin(2☾ - \Omega) - 0''.0261\sin(2☾ + g') \\
& + 0''.0114\sin(2☾ - g') + 0''.0060\sin(2☾ - 2L) \\
& - 0''.0149\sin(2L + g' - 2☾) + 0''.0058\sin(\Omega + g') \\
& - 0''.0057\sin(\Omega - g') - 0''.0052\sin(4☾ - 2L - g') \\
& - 0''.0044\sin(2☾ + g' - \Omega) - 0''.0032\sin(4☾ - 2L) \\
& + 0''.0026\sin(2L + g') - 0''.0026\sin(2☾ + 2g') \\
& + 0''.0019\sin(2☾ - \Omega - g') - 0''.0014\sin(2L + g' - 2☾ - \Omega) \\
& - 0''.0013\sin(2L + \Omega + g' - 2☾) \\
& - 0''.0009\sin(4☾ - 2L - \Omega - g') \\
& + 0''.0007\sin(2☾ + L - \Gamma) - 0''.0006\sin(2☾ + \Gamma - L) \\
& - 0''.0006\sin(4☾ + g' - 2L) + 0''.0006\sin(2L + 2g') \\
& + 0''.0006\sin(2L - 2☾ - \Omega) + 0''.0005\sin(2L + g' - \Omega) \\
& - 0''.0005\sin(2L + \Omega - 2☾) - 0''.0005\sin(4☾ - 2L - \Omega) \\
& - 0''.0004\sin(2☾ + 2g' - \Omega) + 0''.0028\sin 2g'.
\end{aligned}$$

5. A mission from Cape Kennedy to the crater Kepler on the Moon is in the phases of preliminary planning. What coordinate systems will be involved in each phase of the mission? Make certain that an inertial system is used to perform the integration of the equations of motion.

6. By use of Figure 8.14 and standard spherical trigonometry verify that

$$\cos \theta = \cos a \cos b + \sin a \sin b \cos c,$$

$$\sin \theta = \pm\sqrt{1 - \cos^2 \theta}.$$

Why can the minus sign be eliminated in the $\sin \theta$ equation?

7. Under what circumstances are equations 8.14 to be preferred to the corresponding equations in Problem 4?

8. Why does the Moon librate? Explain generally.

REFERENCES

[1] P. R. Escobal, *Methods of Orbit Determination*, John Wiley and Sons, New York, 1965.

[2] S. Newcomb, *A Compendium of Spherical Astronomy*, The Macmillan Company, 1906.

[3] *Connaissance des Temps*, Bureau des Longitudes, France, 1965.

[4] J. M. A. Danby, *Fundamentals of Celestial Mechanics*, The Macmillan Company, New York, 1962.

[5] T. P. Gabbard, *Lunar Rotation and Selenographic Coordinates*, Lockheed Astrodynamics Research Center Technical Memo, LTM 50197, August 1962.

[6] *Explanatory Supplement to the Astronomical Ephemeris and the American Ephemeris and Nautical Almanac*, H.M.S. Nautical Almanac Office, London, 1961.

[7] R. C. Hutchinson, *Inertial Orientation of the Moon*, MIT Instrumentation Laboratory Report R-385, October 1962.

[8] Z. Kopal, *Physics and Astronomy of the Moon*, Academic Press, New York, 1962, Chapter 2.

[9] T. P. Gabbard, P. R. Escobal, and M. L. Birkholz, *A Cowel Special Perturbations Computer Program for Geocentric, Earth-Satellite, and Lunar/Cislunar Trajectories*, Lockheed Astrodynamics Research Center, LTM 50259, November 1962.

[10] B. E. Kalensher, *Selenographic Coordinates*, JPL Technical Report 32–41, February 1961.

[11] R. A. Becker, *Introduction to Theoretical Mechanics*, McGraw-Hill Book Company. New York, 1954.

[12] W. M. Smart, *Celestial Mechanics*, Longmans, London, 1953.

[13] H. L. Roth, *Description of Nutation Subroutine*, TRW Space Technology Laboratories Interim Report, 3400-6037-KU000, May 1965.

[14] T. P. Gabbard, *Nutation Transformation*, Lockheed Astrodynamics Research Center, Technical Memo, LTM 50292, December 1962.

[15] C. E. Rodert, M. L. Wheelon, and T. P. Gabbard, *Time Derivatives of the Precession, Nutation, and Selenographic Transformation Matrices*, Lockheed Astrodynamics Research Center Technical Memo, LTM 50294, January 1963.

[16] E. W. Woolard, "A Redevelopment of the Theory of Nutation," *The Astronomical Journal*, Vol. 58, No. 1, February 1953.

[17] F. Hayn, "Selenographische Koordinaten," *Abhandlung der Mathematischen-Physikalischen Klasse der Königlichen Sächsischen Gesellschaft der Wissenschaften*, Vol. 27, 1902.

[18] F. Hayn [17], Vol. 29, 1906.

[19] F. Hayn [17], Vol. 30, 1907.

[20] F. Hayn [17], Vol. 33, 1914.

[21] F. Hayn, *Astronomische Nachrichten*, Vol. 199, 1914.

[22] F. Hayn [21], Vol. 211, 1920.

[23] A. Jönsson, "Über die Rotation des Mondes," *Meddelandenfran Lunds Astronomiska Observatorium*, Serie II, No. 15, 1917.

[24] K. Koziel, *Acta Astronomica*, Series a, Vol. 4, 1948.

[25] F. Hayn, *Enzyklopädie der Mathematischen Wissenschaften*, Vol. 6, Part 2, 1923, Chapter 20.

[26] E. W. Brown, *Tables of the Motion of the Moon*, New Haven, 1919.

[27] V. C. Clarke, *Constants and Related Data for Use in Trajectory Calculations*, JPL Technical Report 32-604, March 1964.

[28] H. Jeffreys, "The Moon's Libration in Longitude," *MNRAS*, Vol. 117, No. 5, 1957.

[29] A. Yakovkin, *Transactions of the International Astronomical Union*, Vol. 8, 1950.

[30] C. B. Watts, "A New Method of Measuring the Inclination of the Moon's Equator," *Astronomical Journal*, Vol. 60, December 1955.

[31] H. Andoyer, *Bulletin Astronomique*, Vol. 28, 1911.

[32] D. G. Dethlefsen, "Physical Librations of the Moon," Contribution to [33].

[33] H. L. Roth and P. R. Escobal, *Transformations Involving the Selenographic Coordinate System*, TRW Space Technology Laboratories, Summary Technical Report, 3400-6045-RU000, June 1965.

[34] H. L. Roth and G. A. Crichton, *On the Selection of a Selenographic Coordinate Transformation Package for a Trajectory Integration Program*, TRW Systems, Program Technical Report, Task MSC/TRW A-42, 3838-6002-RU000, October 1965.

[35] J. C. Maxwell, "On A Dynamical Top," *Transactions of the Royal Society of Edinburgh*, Vol. 21, 1857.

APPENDIX 1

Numerical Generation of State Transition Matrices

Special perturbations, as discussed in Chapter 7 of this book, is a procedure in which the initial state of a trajectory can be propagated to a final state via numerical techniques. Hence, given an initial position and velocity vector, the numerical integration is started and the space object under investigation proceeds to follow a path subject to the forces of gravitation, drag, lift, etc. Nowhere in the special perturbations procedures previously discussed in this work has an attempt been made to answer the question of how the space object will arrive at a particular point in space such as a target planet. It is known that the initial conditions of the trajectory procedure, if varied correctly, will yield the desired end conditions—the question is how should the initial conditions, the position and velocity vectors for example, be varied to achieve the desired goal. From a mission planning and guidance overview it is therefore desirable to establish the relationship between the initial and final conditions. These relationships, which work in conjunction with the methods discussed in Chapter 7, are discussed in this Appendix.

STATE TRANSITION MATRICES

As illustrated in Figure A1.1, when a space vehicle is undergoing transition from trajectory state 1 to 2, it tends to drift from a given nominal path, with error $\Delta\mathbf{r}$ owing to perturbative influences. This error probably increases with time because of cumulative perturbations on the vehicle. If it is assumed that the nominal path is a given reference route that has been determined by some given process, patched conic theory for example, then a deviation vector, $\Delta\mathbf{r}$, can be defined as

$$\Delta\mathbf{r} \equiv (\mathbf{r}_A - \mathbf{r}_N),$$

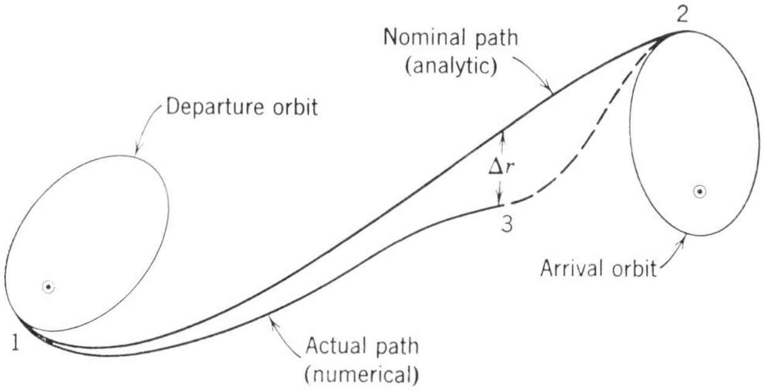

FIGURE A1.1 Trajectory deviation.

where $\mathbf{r}_A$ is the actual trajectory path radius vector, and $\mathbf{r}_N$ is the adopted nominal transfer trajectory radius vector which will accomplish the mission.

The question of correction comes to mind in a natural manner when the deviation $\Delta\mathbf{r}$ reaches a critical error magnitude. Before discussion of orbital correction schemes such that arc 32 of Figure A1.1 is executed, a little thought should reveal that a given correction scheme will utilize the rates of change of final position and velocity vectors $\mathbf{r}$, $\dot{\mathbf{r}}$ with respect to initial position and velocity vectors $\mathbf{r}_0$, $\dot{\mathbf{r}}_0$. Perhaps the inverse rates will also be useful. The immediate objective in point therefore is the generation of the matrix $[\Lambda]$, that is,

$$[\Lambda] \equiv \begin{bmatrix} [M] & [N] \\ [M^*] & [N^*] \end{bmatrix}, \tag{A1.1}$$

where

$$[M] \equiv \begin{bmatrix} \dfrac{\partial x}{\partial x_0} & \dfrac{\partial x}{\partial y_0} & \dfrac{\partial x}{\partial z_0} \\ \dfrac{\partial y}{\partial x_0} & \dfrac{\partial y}{\partial y_0} & \dfrac{\partial y}{\partial z_0} \\ \dfrac{\partial z}{\partial x_0} & \dfrac{\partial z}{\partial y_0} & \dfrac{\partial z}{\partial z_0} \end{bmatrix}, \qquad [N] \equiv \begin{bmatrix} \dfrac{\partial x}{\partial \dot{x}_0} & \dfrac{\partial x}{\partial \dot{y}_0} & \dfrac{\partial x}{\partial \dot{z}_0} \\ \dfrac{\partial y}{\partial \dot{x}_0} & \dfrac{\partial y}{\partial \dot{y}_0} & \dfrac{\partial y}{\partial \dot{z}_0} \\ \dfrac{\partial z}{\partial \dot{x}_0} & \dfrac{\partial z}{\partial \dot{y}_0} & \dfrac{\partial z}{\partial \dot{z}_0} \end{bmatrix},$$

$$[M^*] \equiv \begin{bmatrix} \dfrac{\partial \dot{x}}{\partial x_0} & \dfrac{\partial \dot{x}}{\partial y_0} & \dfrac{\partial \dot{x}}{\partial z_0} \\ \dfrac{\partial \dot{y}}{\partial x_0} & \dfrac{\partial \dot{y}}{\partial y_0} & \dfrac{\partial \dot{y}}{\partial z_0} \\ \dfrac{\partial \dot{z}}{\partial x_0} & \dfrac{\partial \dot{z}}{\partial y_0} & \dfrac{\partial \dot{z}}{\partial z_0} \end{bmatrix}, \qquad [N^*] \equiv \begin{bmatrix} \dfrac{\partial \dot{x}}{\partial \dot{x}_0} & \dfrac{\partial \dot{x}}{\partial \dot{y}_0} & \dfrac{\partial \dot{x}}{\partial \dot{z}_0} \\ \dfrac{\partial \dot{y}}{\partial \dot{x}_0} & \dfrac{\partial \dot{y}}{\partial \dot{y}_0} & \dfrac{\partial \dot{y}}{\partial \dot{z}_0} \\ \dfrac{\partial \dot{z}}{\partial \dot{x}_0} & \dfrac{\partial \dot{z}}{\partial \dot{y}_0} & \dfrac{\partial \dot{z}}{\partial \dot{z}_0} \end{bmatrix}.$$

It should be noted that $\mathbf{r}$ and $\dot{\mathbf{r}}$, the position and velocity components, are referred to a given coordinate system, say, for purposes herein, a heliocentric inertial system. The four preceding matrices are the so-called sensitivity matrices which find much use in guidance analysis. The same matrices are developed from a two-body point of view in ([1], Chapter 9).

DERIVATION OF THE VARIATIONAL SYSTEM

In this section the objective is to derive a system of differential equations whose integration will generate the guidance matrix of the previous section, that is, $[\Lambda]$. To accomplish this objective the equations of motion of a given particle in space can be written in the functional form

$$\ddot{\mathbf{r}} = \ddot{\mathbf{r}}(\mathbf{r}, \dot{\mathbf{r}}), \tag{A1.2}$$

which implies that the acceleration of a given vehicle is functionally dependent on both the position and the velocity vectors of the vehicle. For the sake of compactness let the following column vector be introduced:

$$\mathbf{q} \equiv \begin{bmatrix} x_1 \\ x_2 \\ x_3 \\ x_4 \\ x_5 \\ x_6 \end{bmatrix} \equiv \begin{bmatrix} x \\ y \\ z \\ \dot{x} \\ \dot{y} \\ \dot{z} \end{bmatrix}. \tag{A1.3}$$

It follows at once that

$$\dot{\mathbf{q}} = \begin{bmatrix} \dot{x}_1 \\ \dot{x}_2 \\ \dot{x}_3 \\ \dot{x}_4 \\ \dot{x}_5 \\ \dot{x}_6 \end{bmatrix} = \begin{bmatrix} \dot{x} \\ \dot{y} \\ \dot{z} \\ \ddot{x} \\ \ddot{y} \\ \ddot{z} \end{bmatrix} = \begin{bmatrix} x_4 \\ x_5 \\ x_6 \\ \ddot{x}(x_1, \ldots, x_6) \\ \ddot{y}(x_1, \ldots, x_6) \\ \ddot{z}(x_1, \ldots, x_6) \end{bmatrix}. \tag{A1.4}$$

In the language of finite differences, the differential of (A1.4) is given directly by

$$\Delta\dot{\mathbf{q}} = \begin{bmatrix} \Delta\dot{x}_1 \\ \Delta\dot{x}_2 \\ \Delta\dot{x}_3 \\ \Delta\dot{x}_4 \\ \Delta\dot{x}_5 \\ \Delta\dot{x}_6 \end{bmatrix} = \begin{bmatrix} \Delta x_4 \\ \Delta x_5 \\ \Delta x_6 \\ \sum_{i=1}^{6} \frac{\partial \ddot{x}}{\partial x_i} \Delta x_i \\ \sum_{i=1}^{6} \frac{\partial \ddot{y}}{\partial x_i} \Delta x_i \\ \sum_{i=1}^{6} \frac{\partial \ddot{z}}{\partial x_i} \Delta x_i \end{bmatrix}. \tag{A1.5}$$

Hence in matrix notation the following relation between $\Delta\dot{\mathbf{q}}$ and $\Delta\mathbf{q}$ can be verified by direct expansion:

$$\Delta\dot{\mathbf{q}} = [A(t)]\,\Delta\mathbf{q}, \tag{A1.6}$$

where the 6×6 matrix $[A(t)]$ is defined by

$$[A(t)] = \begin{bmatrix} [0_3] & [U_3] \\ [F_p] & [F_v] \end{bmatrix}, \tag{A1.7}$$

with $[0_3]$ defined as the 3×3 null matrix, $[U_3]$ defined as the 3×3 diagonal unity matrix, and

$$[F_p] \equiv \begin{bmatrix} \frac{\partial \ddot{x}}{\partial x} & \frac{\partial \ddot{x}}{\partial y} & \frac{\partial \ddot{x}}{\partial z} \\ \frac{\partial \ddot{y}}{\partial x} & \frac{\partial \ddot{y}}{\partial y} & \frac{\partial \ddot{y}}{\partial z} \\ \frac{\partial \ddot{z}}{\partial x} & \frac{\partial \ddot{z}}{\partial y} & \frac{\partial \ddot{z}}{\partial z} \end{bmatrix}, \qquad [F_v] \equiv \begin{bmatrix} \frac{\partial \ddot{x}}{\partial \dot{x}} & \frac{\partial \ddot{x}}{\partial \dot{y}} & \frac{\partial \ddot{x}}{\partial \dot{z}} \\ \frac{\partial \ddot{y}}{\partial \dot{x}} & \frac{\partial \ddot{y}}{\partial \dot{y}} & \frac{\partial \ddot{y}}{\partial \dot{z}} \\ \frac{\partial \ddot{z}}{\partial \dot{x}} & \frac{\partial \ddot{z}}{\partial \dot{y}} & \frac{\partial \ddot{z}}{\partial \dot{z}} \end{bmatrix}.$$

In passing, it is well to note that $[A(t)]$ can be determined analytically from the equations of motion derived in ([1], Chapter 2), that is, from the partial differentiation of the symbolic relation

$$\ddot{\mathbf{r}} = \begin{bmatrix} \ddot{x} \\ \ddot{y} \\ \ddot{z} \end{bmatrix} = \nabla\Phi + \Sigma_{\mathbf{P}} + \Sigma_{\mathbf{T}} + \Sigma_{\mathbf{L}} + \Sigma_{\mathbf{D}}, \tag{A1.8}$$

where Φ is the general potential of the central planet[1] dominating the motion,

[1] See Appendix 2.

Σ_P are the planetary perturbations, Σ_T are the thrust perturbations, Σ_L are the lift perturbations, and Σ_D are the drag perturbations. Hence the 3×3 matrix of position partials, $[F_p]$, is evaluated directly from (A1.8), and the 3×3 matrix of velocity partials, $[F_v]$, is likewise obtained from (A1.8).

Consider next (A1.6), that is

$$\Delta \frac{d}{dt} \mathbf{q} = \frac{d}{dt}(\Delta \mathbf{q}) = [A(t)]\, \Delta \mathbf{q}, \tag{A1.9}$$

and divide through by Δx_{j0}, an initial set of constant increments, so that

$$\frac{1}{\Delta x_{j0}} \frac{d}{dt}(\Delta \mathbf{q}) = [A(t)] \frac{\Delta \mathbf{q}}{\Delta x_{j0}}, \tag{A1.10}$$

where, since Δx_{j0} is constant,

$$\frac{d}{dt}\left(\frac{\Delta \mathbf{q}}{\Delta x_{j0}}\right) = [A(t)] \frac{\Delta \mathbf{q}}{\Delta x_{j0}}. \tag{A1.11}$$

Because $\Delta \dot{\mathbf{q}}$ is a column vector defined by the differential of (A1.3), it follows that the finite differences approximate the actual partials as $\Delta x_{j0} \to 0$. Thus $\Delta \mathbf{q}/\Delta x_{j0} \to \partial \mathbf{q}/\partial \mathbf{q}_0$ or in component form to $\partial x_i/x_{j0}$, for all appropriate i and j. Hence it is possible to write (A1.11) as

$$\frac{d}{dt}\left(\frac{\partial x_i}{\partial x_{j0}}\right) = [A(t)] \frac{\partial x_i}{\partial x_{j0}}. \tag{A1.12}$$

Notice that, if $j = 1$, (A1.12) is a column vector equation which yields the rate of change of $x_1, \ldots, x_6$ with respect to $x_{10} = x_0$, that is $\partial x/\partial x_0$, $\partial y/\partial x_0$, $\partial z/\partial x_0$, $\partial \dot{x}/\partial x_0$, $\partial \dot{y}/\partial x_0$, $\partial \dot{z}/\partial x_0$. Similarly, if $j = 2$, (A1.12) is another column vector equation which yields the rate of change of $x_1, \ldots, x_6$ with respect to $x_{20} = y_0$, etc. Hence $j = 1$ corresponds to the first column of $[\Lambda]$, $j = 2$ corresponds to the second column of $[\Lambda]$, etc., where $[\Lambda]$ is the matrix defined by (A1.1). Equation A1.12 may be written concisely as

$$[\dot{\Lambda}] = [A(t)][\Lambda] \tag{A1.13}$$

and is usually referred to as the variational equation of system (A1.8) because of its similarity with the Euler-Lagrange equation of the calculus of variations. Initial conditions for $[\Lambda]$ may be obtained from inspection of (A1.1) at $t = t_0$ or $x = x_0$, $y = y_0$, $z = z_0$, etc. Since $\partial x_0/\partial x_0 = 1$, $\partial y_0/\partial y_0 = 1$, or, in general,

$$[\Lambda(t_0)] = \begin{bmatrix} 1 & 0 & 0 & 0 & 0 & 0 \\ 0 & 1 & 0 & 0 & 0 & 0 \\ 0 & 0 & 1 & 0 & 0 & 0 \\ 0 & 0 & 0 & 1 & 0 & 0 \\ 0 & 0 & 0 & 0 & 1 & 0 \\ 0 & 0 & 0 & 0 & 0 & 1 \end{bmatrix} = [U_6], \tag{A1.14}$$

the initial condition for numerical integration of (A1.13) is thus the 6×6 unity diagonal matrix $[U_6]$. It may be well to reiterate what is happening when (A1.13) is integrated forward in time with initial conditions (A1.14). Essentially, for any given starting time t_0, the partials of which the elements of (A1.1) are comprised are generated as a function of time for all $t > t_0$.

THE ADJOINT SYSTEM OF EQUATIONS

Another approach to the integration of (A1.13) will perhaps be of value to the reader and shed further light on the generation of the guidance matrix.

Solution of nth-order linear differential equations with or without constant coefficients makes use of adjoint differential equations as a means of obtaining so-called integrating factors [2]. In brief, given a system of equations of the form

$$\dot{x} = [M(t)]x, \tag{A1.15}$$

where $\dot{x}$ and x are column vectors, the adjoint of this system is defined as

$$\dot{y} = -[M(t)]^T y, \tag{A1.16}$$

that is, the system obtained by taking the negative of the transpose of the original matrix of the system. A useful result to be employed presently can be obtained by noting that

$$\frac{d}{dt}(y^T x) = \dot{y}^T x + y^T \dot{x} \tag{A1.17}$$

and, substituting from (A1.16), $\dot{y}^T = -y^T[M]$. Hence, eliminating $\dot{x}$ via (A1.15),

$$\frac{d}{dt}(y^T x) = -y^T[M]x + y^T[M]x = 0 \tag{A1.18}$$

and therefore

$$y^T x = \text{constant}. \tag{A1.19}$$

Proceeding more formally, let the adjoint of (A1.6) be denoted by

$$\dot{\boldsymbol{\lambda}} = -[A(t)]^T \boldsymbol{\lambda}, \tag{A1.20}$$

where $\boldsymbol{\lambda}$ is a 6×1 column matrix. From the column matrix $\boldsymbol{\lambda}$, as was done with $\Delta \mathbf{q}$, it is possible to form the matrix equation

$$[\dot{\Gamma}] = -[A(t)]^T[\Gamma], \tag{A1.21}$$

where $[\Gamma]$ is a 6×6 square matrix. The physical significance of $\boldsymbol{\lambda}$ and $[\Gamma]$ are not known at present.

Consider the transpose of (A1.21); that is,

$$[\dot{\Gamma}]^T = -([A]^T[\Gamma])^T = -[\Gamma]^T[A]. \tag{A1.22}$$

Consider using (A1.22) in the expression for the time derivative of the product of $[\Gamma]^T$ and $\Delta\mathbf{q}$, namely,

$$\frac{d}{dt}([\Gamma]^T\,\Delta\mathbf{q}) = [\Gamma]^T\,\Delta\dot{\mathbf{q}} + [\dot{\Gamma}]^T\,\Delta\mathbf{q},$$

so that

$$\frac{d}{dt}([\Gamma]^T\,\Delta\mathbf{q}) = [\Gamma]^T\,\Delta\dot{\mathbf{q}} - [\Gamma]^T[A]\,\Delta\mathbf{q},$$

and substitute (A1.6) to obtain

$$\frac{d}{dt}([\Gamma]^T\,\Delta\mathbf{q}) = [\Gamma]^T[A]\,\Delta\mathbf{q} - [\Gamma]^T[A]\,\Delta\mathbf{q} = 0. \tag{A1.23}$$

The very important conclusion to be drawn from (A1.23) is that the time derivative of the product $[\Gamma]^T\,\Delta\mathbf{q}$ is invariant with time! Hence the following relations hold:

$$[\Gamma(t)]^T\,\Delta\mathbf{q}(t) = [\Gamma(t_0)]^T\,\Delta\mathbf{q}(t_0) = \text{constant}, \tag{A1.24}$$

$$[\Gamma(t)]^T\,\Delta\mathbf{q}(t) = [\Gamma(t_f)]^T\,\Delta\mathbf{q}(t_f) = \text{constant}, \tag{A1.25}$$

where the subscripts 0 and f refer to initial and final conditions on a given integration interval of (A1.21).

In (A1.24), consider that $[U_6(t_0)]$ is chosen for $[\Gamma(t_0)]^T$ as the initial condition for forward integration of (A1.21). Then post multiplication by $[\Delta\mathbf{q}(t)]^{-1}$ yields

$$[\Gamma(t)]^T = \frac{\Delta\mathbf{q}(t_0)}{\Delta\mathbf{q}(t)}, \tag{A1.26}$$

while, if (A1.21) is integrated backward from t_f with $[U_6(t_f)]$ as initial condition for $[\Gamma(t_f)]^T$,

$$[\Gamma(t)]^T = \frac{\Delta\mathbf{q}(t_f)}{\Delta\mathbf{q}(t)}.$$

In the limiting cases, with forward integration and initial condition $U_6(t_0)$,

$$[\Gamma(t)]^T = \frac{\partial\mathbf{q}(t_0)}{\partial\mathbf{q}(t)}; \tag{A1.27}$$

and with backward integration and initial condition $U_6(t_f)$,

$$[\Gamma(t)]^T = \frac{\partial\mathbf{q}(t_f)}{\partial\mathbf{q}(t)}. \tag{A1.28}$$

It follows at once that $\partial \mathbf{q}(t_f)/\partial \mathbf{q}(t)$ is nothing more than a symbolic representation of the desired partial derivatives, that is, the physical significance of $\Gamma(t)$ becomes clear, since by (A1.27),

$$\{[\Gamma(t)]^T\}^{-1} = \frac{\partial \mathbf{q}(t)}{\partial \mathbf{q}(t_0)} = [\Lambda(t)]. \tag{A1.29}$$

As should be noted, integration of system (A1.13) provides the desired information or sensitivity partials without the necessity of matrix inversion. However, if the epoch t_0 is moved, the integration must be started from the new initial epoch. But, if (A1.22), that is,

$$[\dot{\Gamma}]^T = -[\Gamma]^T[A(t)], \tag{A1.30}$$

is integrated backward from t_f in a continuous process of six simultaneous integrations, a history of $[\Lambda(t)]$ may be obtained by simple inversion which is valid for all variable t_0. Both approaches have their relative merits and distinct applications.

Illustrative Example

Assuming that only gravitational influences are to be considered, derive the $[A(t)]$ matrix defined by (A1.7).

By definition,

$$[0_3] \equiv \begin{bmatrix} 0 & 0 & 0 \\ 0 & 0 & 0 \\ 0 & 0 & 0 \end{bmatrix}, \qquad [U_3] \equiv \begin{bmatrix} 1 & 0 & 0 \\ 0 & 1 & 0 \\ 0 & 0 & 1 \end{bmatrix}.$$

Since gravitational influences are only position dependent,

$$[F_v] \equiv \begin{bmatrix} 0 & 0 & 0 \\ 0 & 0 & 0 \\ 0 & 0 & 0 \end{bmatrix}$$

and the only other required matrix is $[F_p]$. Remembering that $\ddot{\mathbf{r}} = \nabla\Phi$, where Φ is the gravitational potential, it is possible to write

$$\ddot{\mathbf{r}} = \begin{bmatrix} \dfrac{\partial \Phi}{\partial x} \\[2ex] \dfrac{\partial \Phi}{\partial y} \\[2ex] \dfrac{\partial \Phi}{\partial y} \end{bmatrix},$$

which yields

$$[F_p] = \begin{bmatrix} \dfrac{\partial \ddot{x}}{\partial x} & \dfrac{\partial \ddot{x}}{\partial y} & \dfrac{\partial \ddot{x}}{\partial z} \\[2ex] \dfrac{\partial \ddot{y}}{\partial x} & \dfrac{\partial \ddot{y}}{\partial y} & \dfrac{\partial \ddot{y}}{\partial z} \\[2ex] \dfrac{\partial \ddot{z}}{\partial x} & \dfrac{\partial \ddot{z}}{\partial y} & \dfrac{\partial \ddot{z}}{\partial z} \end{bmatrix} = \begin{bmatrix} \dfrac{\partial^2 \Phi}{\partial x^2} & \dfrac{\partial^2 \Phi}{\partial x\,\partial y} & \dfrac{\partial^2 \Phi}{\partial x\,\partial z} \\[2ex] \dfrac{\partial^2 \Phi}{\partial y\,\partial x} & \dfrac{\partial^2 \Phi}{\partial y^2} & \dfrac{\partial^2 \Phi}{\partial y\,\partial z} \\[2ex] \dfrac{\partial^2 \Phi}{\partial z\,\partial x} & \dfrac{\partial^2 \Phi}{\partial z\,\partial y} & \dfrac{\partial^2 \Phi}{\partial z^2} \end{bmatrix}$$

THE CORRECTION PROCESS

Once the appropriate state transition matrices are known, numerous correction processes can be developed. A simple approach follows. Assume

$$\mathbf{r} = \mathbf{r}(\mathbf{r}_0, \dot{\mathbf{r}}_0),$$
$$\dot{\mathbf{r}} = \dot{\mathbf{r}}(\mathbf{r}_0, \dot{\mathbf{r}}_0), \qquad \text{(A1.31)}$$

so that by formal differentiation

$$\Delta\mathbf{r} = \frac{\partial \mathbf{r}}{\partial \mathbf{r}_0}\Delta\mathbf{r}_0 + \frac{\partial \mathbf{r}}{\partial \dot{\mathbf{r}}_0}\Delta\dot{\mathbf{r}}_0 = [M]\,\Delta\mathbf{r}_0 + [N]\,\Delta\dot{\mathbf{r}}_0, \qquad \text{(A1.32)}$$

$$\Delta\dot{\mathbf{r}} = \frac{\partial \dot{\mathbf{r}}}{\partial \mathbf{r}_0}\Delta\mathbf{r}_0 + \frac{\partial \dot{\mathbf{r}}}{\partial \dot{\mathbf{r}}_0}\Delta\dot{\mathbf{r}}_0 = [M^*]\,\Delta\mathbf{r}_0 + [N^*]\,\Delta\dot{\mathbf{r}}_0. \qquad \text{(A1.33)}$$

Equations (A1.32) and (A1.33) represent a linear system of equations which relate the errors $(\Delta\mathbf{r}, \Delta\dot{\mathbf{r}})$ at a final state to the errors at the initial state $(\Delta\mathbf{r}_0, \Delta\dot{\mathbf{r}}_0)$. Solving the system yields

$$\Delta\mathbf{r}_0 = [[M][N^*] - [M^*][N]]^{-1}[\Delta\mathbf{r}[N^*] - \Delta\dot{\mathbf{r}}[N]],$$
$$\Delta\dot{\mathbf{r}}_0 = [[M][N^*] - [M^*][N]]^{-1}[[M]\,\Delta\dot{\mathbf{r}} - [M^*]\,\Delta\mathbf{r}]. \qquad \text{(A1.34)}$$

Therefore, once the miss errors $\Delta\mathbf{r}$ and $\Delta\dot{\mathbf{r}}$ have been obtained from the procedures of special perturbations, the change in the initial conditions,

$$(\mathbf{r}_0)_{n+1} = (\mathbf{r}_0)_n + \Delta\mathbf{r}_0, \qquad (\dot{\mathbf{r}}_0)_{n+1} = (\dot{\mathbf{r}}_0)_n + \Delta\dot{\mathbf{r}}_0, \qquad \text{(A1.35)}$$

can be determined. Many variations of these equations are possible [3].

REFERENCES

[1] P. R. Escobal, *Methods of Orbit Determination*, John Wiley and Sons, New York, 1965.

[2] E. A. Coddington and N. Levinson, *Theory of Ordinary Differential Equations*, McGraw-Hill Book Company, New York, 1965.

[3] R. H. Battin, *Astronautical Guidance*, McGraw-Hill Book Company, New York, 1964.

APPENDIX 2

Generalized Potential Function

The purpose of this appendix is to state the formulas for the generalized potential function and to list the appropriate numerical coefficients of the Earth's zonal and tesseral harmonics. This function is widely used in general and special perturbations techniques (Chapter 7).

A generalized form of the potential function can be written as

$$\Phi = \frac{k^2 m}{r} + \frac{k^2 m}{r}\left[\sum_{n=1}^{\infty}\left(\frac{a_e}{r}\right)^n \sum_{m=1}^{n}(C_{nm}\cos m\lambda + S_{nm}\sin m\lambda)P_{nm}(\sin\phi)\right], \tag{A2.1}$$

where a_e = radius of Earth $\cong$ 6,378,165 m, k = gravitational constant, m = the mass of the Earth, C_{nm} and S_{nm} = coefficients of the Earth's potential, ϕ = geocentric latitude of a point in space, r = radial distance from the dynamical center to a point in space, λ = longitude measured on the equator between the Greenwich meridian and the meridian of the point in space, and $P_{nm}(\sin\phi)$ = associated Legendre functions defined by

$$P_{nm}(\nu) = \frac{(1-\nu^2)^{m/2}}{2^n n!}\cdot\frac{d^{n+m}}{d\nu^{n+m}}(\nu^2-1)^n$$

with $\nu \equiv \sin\phi$.

Equation A2.1 can be separated to account for the zonal and tesseral effects as follows:

$$\Phi = \frac{k^2 m}{r} + \frac{k^2 m}{r}\left\{\sum_{n=2}^{\infty}\left(\frac{a_e}{r}\right)^n C_{n0}P_{n0}(\sin\phi)\right.$$

$$\left.+\sum_{n=2}^{\infty}\sum_{m=1}^{n}\left(\frac{a_e}{r}\right)^n(C_{nm}\cos m\lambda + S_{nm}\sin m\lambda)P_{nm}(\sin\phi)\right\}, \tag{A2.2}$$

where, from (A2.1),

$$P_{n0} = \frac{1}{2^n n!}\cdot\frac{d^n}{d\nu^n}(\nu^2-1)^n.$$

Table A2.1 Tesseral Harmonic Coefficients

n	m	$\bar{C}_{nm} \times 10^6$	$\bar{S}_{nm} \times 10^6$
2	2	+2.379	−1.351
3	1	+1.936	+0.266
3	2	+0.734	−0.538
3	3	+0.561	+1.620

Smithsonian Reports [1] list numerical data for the normalized coefficients $\bar{C}_{nm}$ and $\bar{S}_{nm}$ which are related to C_{nm} and S_{nm} by

$$C_{nm} = N_{nm}\bar{C}_{nm}, \qquad S_{nm} = N_{nm}\bar{S}_{nm} \tag{A2.3}$$

with

$$N_{nm} = \left[\frac{2 \cdot (n-m)!\,(2n+1)}{(n+m)!}\right]^{1/2}$$

as displayed in Tables A2.1 and A2.2.

A convenient form of (A2.2) which is used frequently in analytic investigations can be obtained by the following transformation

$$C_{nm} \cos m\lambda + S_{nm} \sin m\lambda = J_{nm} \cos (m[\lambda - \lambda_{nm}])$$

which implies that

$$C_{nm} = J_{nm} \cos m\lambda_{nm}, \qquad S_{nm} = J_{nm} \sin m\lambda_{nm}.$$

Under this transformation (A2.2) becomes

$$\Phi = \frac{k^2m}{r} + \frac{k^2m}{r}\left\{-\sum_{n=2}^{\infty}\left(\frac{a_e}{r}\right)^n J_n P_{n0}(\sin\phi) + \sum_{n=2}^{\infty}\sum_{m=1}^{n}\left(\frac{a_e}{r}\right)^n J_{nm}\cos(m[\lambda - \lambda_{nm}])P_{nm}(\sin\phi)\right\}, \tag{A2.4}$$

where the associated values of J_{nm} and λ_{nm} are given in Table A2.3. Notice that $C_{n0} = -J_n$.

Table A2.2 Zonal Harmonic Coefficients

$C_{20} = -1082.645 \times 10^{-6}$	$C_{80} = +0.270 \times 10^{-6}$
$C_{30} = +2.546 \times 10^{-6}$	$C_{90} = +0.053 \times 10^{-6}$
$C_{40} = +1.649 \times 10^{-6}$	$C_{100} = +0.054 \times 10^{-6}$
$C_{50} = +0.210 \times 10^{-6}$	$C_{110} = -0.302 \times 10^{-6}$
$C_{60} = -0.646 \times 10^{-6}$	$C_{120} = +0.357 \times 10^{-6}$
$C_{70} = +0.333 \times 10^{-6}$	$C_{130} = +0.114 \times 10^{-6}$
$C_{140} = -0.179 \times 10^{-6}$	

Table A2.3 Smithsonian Coefficients J_{nm}, λ_{nm} of the Geopotential

n	m	J_{nm}	λ_{nm} (Degrees)
2	2	1.766×10^{-6}	−14.79
3	1	2.111×10^{-6}	+7.818
3	2	0.311×10^{-6}	−18.12
3	3	0.239×10^{-6}	+23.6
4	1	0.702×10^{-6}	−140.7
4	2	0.165×10^{-6}	+31.7
4	3	5.21×10^{-8}	−4.19
4	4	4.99×10^{-9}	−19.2
5	1	0.111×10^{-6}	−127.5
5	2	0.109×10^{-6}	−10.09
5	3	1.72×10^{-8}	−0.257
5	4	2.12×10^{-9}	−3.39
5	5	1.51×10^{-9}	−15.04
6	1	4.26×10^{-8}	−150.2
6	2	4.63×10^{-8}	−39.63
6	3	1.29×10^{-9}	−9.9
6	4	1.97×10^{-9}	−23.7
6	5	4.48×10^{-10}	−24.9
6	6	3.73×10^{-11}	+12.6

REFERENCE

[1] C. A. Lundquist and G. Veis, *Geodetic Parameters for A 1966 Smithsonian Institution Standard Earth*, Special Report 200, Volumes 1, 2 and 3, 1966.

Nomenclature of Methods of Astrodynamics

Though a definite and repetitive effort has been made to define symbols in the text as they appear, the nomenclature of this field is so extensive that a tabulation of symbols is deemed necessary.

SUPERSCRIPTS

$\cdot$	Relating to modified time differentiation
$'$	Relating to general differentiation
	Relating to geocentric latitude
$''$	Seconds of arc
$*$	Particular parameter or special form of an analytical expression
$\smile$	Particular parameter or special form of an analytical expression
$-$	Used to denote average or special form of an analytical expression or parameter
$\circ$	Degrees
d	Days
hr	Hours

SPECIAL SYMBOLS

$\equiv$	Identically equal to
	Equal to by definition
$\doteq$	Replace left side of equation with right side of equation
$\simeq$	Approximately equal to
♈	Vernal equinox (sign of the ram's horns)
∞	Infinity
$\angle x, y$	Angle between x and y
$\rightarrow$	Yields
$\lvert x \rvert$	Absolute value of x

SUBSCRIPTS

a	Relating to apofocus
	Relating to after

ap	Relating to arrival conditions
A	Denoting primary body
$\mathbf{A}$	Coefficient vector in the hybrid technique
b	Relating to a specific body
	Relating to before
B	Barycenter of a system
	Denoting secondary body
$\mathbf{B}$	Coefficient vector in the hybrid technique
$\mathbf{C}$	Coefficient vector in the hybrid technique
c	Indicating a computed quantity
	Indicating a reference quantity
	Relating to central
dp	Relating to departure conditions
D	Relating to drag
	Determinant of a system of equations
E	East
F	Relating to final conditions
e	Relating to planet Earth
	Relating to elliptic orbits
g	Relating to Greenwich or Greenwich meridian
h	Relating to the azimuth-elevation coordinate system
	Relating to hyperbolic orbits
i	General index
I	Relating to initial conditions
j	General index
ij	With ij any integer values relate to the distance between body i and body j
k	General index
l	Relating to limiting
	Relating to lunar
	Relating to launch
L	Relating to lift
m	Relating to molecular
	Relating to maximum
0	Relating to observations
	Relating to reference
	Relating to some epoch time
p	Relating to perifocus
	Relating to planet
	Relating to penumbra
r	Relating to relative
	Relating to rotating

s Relating to Sun
Relating to subvehicle point
Relating to satellite
sp Relating to specific impulse
T Relating to transfer parameters
t Relating to topocentric conditions
u Relating to time measured in Julian centuries
v Relating to vehicle-centered coordinate system
Relating to orbital axes in the orbit plane coordinate system
x, y, z Relating to components along x, y, z orthogonal axes
δ Relating to hour-angle declination coordinate system
ϵ Relating to ecliptic
Ω Relating to the ascending node
ω Relating to orbit plane coordinate system
∞ Relating to exospheric

SPECIAL SYMBOLS

☉ Sun
☿ Mercury
♀ Venus
⊕ Earth
☾ Moon
Mean longitude of the Moon
♂ Mars
♃ Jupiter
♄ Saturn
♅ Uranus
♆ Neptune
♇ Pluto

ENGLISH SYMBOLS

A Azimuth angle
Miscellaneous constants
Area
Moment of inertia
a Semimajor axis of a conic section
Matrix coefficient
a_e Equational radius of Earth
B Miscellaneous constants
C Jacobi's constant
Miscellaneous constants
Moment of inertia

C_D	Drag coefficient
C_e	Element ($= e \cos E_0$)
C_h	Element ($= e \cosh F_0$)
c	Speed of light
D	Drag force magnitude
	Dynamical parameter $= r\dot{r}/\sqrt{\mu}$
	Mean elongation of the Moon from the Sun
E	Eccentric anomaly
	Miscellaneous constants
e	Orbital eccentricity
	Mathematical constant
F	Force
	Hyperbolic anomaly (in correspondence with E)
	Miscellaneous constants
	Radius or semimajor axis ratio
	Solar flux
f	Geometrical flattening of reference spheroid adopted for central planet
	Dynamical flattening of Moon
	Functional notation
	Coefficient of f and g series
	Series expansion in Encke's method
G	Station location and shape coefficients
	Universal gravitational constant
	Miscellaneous constants
g	Coefficient of f and g series
	Gravitational acceleration
	Mean anomaly of the Sun
g'	Mean anomaly of the Moon
H	Bielliptic coefficients
	Altitude of station measured normal to adopted ellipsoid
h	Elevation angle
	Height
I	Specific impulse
	Mean inclination of the Moon's equator
$\mathbf{I}$	Unit vector along the principal axis of a given coordinate system
i	Orbital inclination
	The imaginary ($= \sqrt{-1}$)
J	Harmonic coefficients of the Earth's potential function
J.D.	Julian date

J	Unit vector advanced to **I** by a right angle in the fundamental plane
K	A constant
	Degrees Kelvin
K	Unit vector defined by $\mathbf{I} \times \mathbf{J} = \mathbf{K}$
k	Gravitational constant
L	Libration point
	Lift force magnitude
	Mean longitude $(= \Omega + \omega + M)$
L	Unit vector from observational station to satellite
M	Mean anomaly $[= n(t - T)]$
	Mean molecular weight
N	Number of revolutions
m	General symbol for mass
n	Mean motion $(= k\sqrt{\mu}/a^{3/2})$
	Number of revolutions
P	Orbital period (time from perigee crossing to perigee crossing)
	Weight ratio function
P_h	Perifocus
P	Unit vector which points from the dynamical center toward perifocus
p	Orbital semiparameter $[= a(1 - e^2)]$
	Gas pressure
Q	Unit vector advanced to **P** by a right angle in the direction and plane of motion
q	Generalized element
	Perifocal distance $[= a(1 - e)]$
	Parameter in the method of Encke
q	Generalized state vector
R	Perturbative function $(= \Phi - V)$
	Magnitude of station coordinate vector
	Universal gas constant
R	Station coordinate vector
r	Magnitude of satellite radius vector
r	Satellite radius vector
S	Satellite symbol
S	Vector used in three-body problem
S_e	Element $(= e \sin E_0)$
S_h	Element $(= e \sinh F_0)$
s	A parameter taking the value of 1 or -1
	Square root of semiparameter

T	Time measured in centuries
	Time of perifocal passage
	Thrust force magnitude
	Absolute temperature
$\mathbf{T}$	Thrust vector
t	Universal or ephemeris time
$\mathbf{U}$	Unit vector from the dynamical center pointing toward a given satellite
u	Argument of latitude
V	General symbol for velocity vector magnitude
	Spherical potential of planets
$\mathbf{V}$	Unit vector advanced to **U** by a right angle in the direction and plane of motion
v	True anomaly
w	Weight
$\mathbf{W}$	Unit vector perpendicular to orbit plane
X, Y, Z	Rectangular coordinates of station coordinate vector
	Coefficients utilized for lunar coordinate transformations
x, y, z	Rectangular coordinates of an object
z	Precession variable
$\mathbf{Z}$	Unit vector in the zenith direction

GREEK SYMBOLS

α	Right ascension
'	Miscellaneous constants or coefficients
β	Celestial latitude
	Miscellaneous constants or coefficients
Γ	Sun's mean longitude of perigee
Γ'	Mean longitude of lunar perigee
γ	Flight path angle
Δ	Increment or difference
∇	Gradient operation
	$\left[\nabla(\cdot) = \dfrac{\partial(\cdot)}{\partial x}\mathbf{I} + \dfrac{\partial(\cdot)}{\partial y}\mathbf{J} + \dfrac{\partial(\cdot)}{\partial z}\mathbf{K}\right]$
δ	Declination
	Variation
ζ	Coefficient of oblateness
	Precession variable
ϵ	Obliquity of the ecliptic
	Optimization parameter
	Specified tolerance
	Nutation in obliquity

$\boldsymbol{\epsilon}$	Deviation vector $(= \mathbf{r}_2 - \mathbf{r}_1)$
θ	Sidereal time
	Plane change angle
	Precession variable
λ	Longitude
	Atmospheric lag angle
	Lagrange multiplier
	Constant multiplier
μ	Sum of masses or mass
	Lagrange multiplier
ν	Transfer coefficients
ξ	Optimization parameter
ρ	Mass density
	Libration in inclination
$\boldsymbol{\rho}$	Slant range vector
σ	Libration in node
	Stage structure factor
τ	Modified time $[= k(t - t_0)]$
	Libration in longitude
Φ	Gravitational potential
ϕ	Geodetic latitude
	Transfer coefficients
ψ	Nutation in longitude
Ω	Longitude of ascending node
ω	Argument of perigee
	Weight coefficients
	Angular distance of the perigee of the Sun
ω'	Argument of perigee of the Moon
$\tilde{\omega}$	Longitude of perihelion $(= \Omega + \omega)$
π	Sine parallax
	Mathematical constant

ABBREVIATIONS

a.u.	astronomical units
C	centigrade
cm	centimeters
c.m.	central masses
c.u.	characteristic units
c.s.u	circular satellite units
deg.	degrees
e.m.	earth masses
e.r.	earth radii

F	Fahrenheit
ft	feet
gm	grams
hr	hours
K	Kelvin
km	kilometers
m	meters
	minutes
MSI	mean sphere of influence
R	Rankine
sec	seconds
s.m.	solar masses
STP	standard temperature and pressure

Answers and Hints

CHAPTER 1

1. Greater.
2. Obtain the Julian Date (J.D.) ([1], Chapter 1). Use (1.2) to compute T_u. Use (1.5) and (1.6) to compute L, $\tilde{\omega}$, Ω, e, i, a. Use the standard formulas to determine **P** and **Q** ([1], Chapter 3). Solve Kepler's equation, $M_i = E_i - e \sin E_i$, to obtain E_i, $i = ♂, \oplus$, and obtain $\mathbf{r}_i = x_{\omega i}\mathbf{P}_i + y_{\omega i}\mathbf{Q}_i$, where

$$x_\omega = a(\cos E - e), \qquad y_\omega = a\sqrt{1 - e^2} \sin E.$$

Finally, the angle is obtained as $\cos (< ♂, \oplus) = \mathbf{r}_{♂} \cdot \mathbf{r}_{\oplus}$.
3. See Footnote 2, Chapter 1.
4. Replace (1.25) with the set

$$\alpha = \alpha(a_1, a_2, \ldots, a_q),$$
$$\delta = \delta(a_1, a_2, \ldots, a_q),$$
$$\pi = \pi(a_1, a_2, \ldots, a_q),$$

and differentiate to obtain the residuals $\Delta\alpha$, $\Delta\delta$, $\Delta\pi$; namely, $\Delta\alpha \equiv \alpha_0 - \alpha_c$, $\Delta\delta \equiv \delta_0 - \delta_c$, $\Delta\pi \equiv \pi_\oplus - \pi_c$, where the subscripts denote observed minus computed variables. In this case the computed right ascension, declination and parallax can be obtained from the precision integrated $\mathbf{r}_☾$.
5. Use (1.32) to obtain $M_♃$ where $M_{CH_4} = 16.04$ and $M_{H_2} = 2.016$. Obtain the density of the Jovian atmosphere, ρ, from (1.31). Finally, compute $D = \frac{1}{2}C_D A \rho V^2$.
6. Using the same approach as in Exercise 2, determine $\mathbf{r}_\oplus$ and $\mathbf{r}_♄$. Obtain the J.D. from [1], Chapter 1. Finally, compute $|\mathbf{r}_♄ - \mathbf{r}_\oplus| \equiv \rho_{\oplus ♄}$ and obtain the time required from $\rho_{\oplus ♄}/c$.
7. Assume that the molecular weight of air will not change in 10,000 ft, so that $T = T_m$; then use (1.39). The speed of sound is obtained from (1.44), where the corresponding parameters are obtained from Table 1.5.
8. Use the technique of Exercise 2 to obtain the position vector of the Earth on August 24, 1970 (obtain the J.D. from [1], Chapter 1). Then use (1.47) and (1.48) to obtain the corresponding right ascension and declination of the Sun. Compute the auxiliary function $f(\psi')$ from (1.49) ($n = 6$) and obtain the density from (1.50).
9. Substitute the data given into (1.51).
10. Assume that the highest planetary eccentricity is 0.25. Then $(0.25)^n = 0.00000001$, where n is the number of terms.

CHAPTER 2

1. $\tan\theta = (1 - e^2)^2/e$.
2. $t = -(ae + 2e)/5a$.
3. Minimize the function

$$r^2 = (x_{\omega e} - x_{\omega p})^2 + (y_{\omega e} - y_{\omega p})^2$$

by parameterizing the equations as

$$x_{\omega e} = a(\cos E - e), \qquad x_{\omega p} = q - \tfrac{1}{2}D^2,$$

$$y_{\omega e} = a\sqrt{1 - e^2}\sin E, \qquad y_{\omega p} = \sqrt{2q}\, D,$$

with a, e, and q taken as constants. Because the time phasing is correct Kepler's equation need not be considered. Therefore equate $\partial r^2/\partial E$ and $\partial r^2/\partial D$ to zero and solve the resulting system. Note that the solution by this method will be the difference in perihelion distances.

4. See Section 2.4 and omit all eccentricities that are not linear.
5. $x = \dfrac{ad}{a^2 + b^2 + c^2}, \qquad y = \dfrac{bd}{a^2 + b^2 + c^2}, \qquad z = \dfrac{cd}{a^2 + b^2 + c^2}.$
6. Whenever all the $\partial g_i/\partial x_i = 0$ at the extreme point that is sought.

CHAPTER 3

1. Two deboost maneuvers whose total is 0.136 c.s.u. must be executed.
2. The Syncom must be advanced to the probe by 96°.5.
3. The most optimum maneuver is a Hohmann transfer ($R = 11.33$).
4. For the triple impulse maneuver $\Delta V = 0.675$ c.s.u. The change of plane is not optimum, as can be seen by solving (3.86) and (3.87) simultaneously.
5. Use (3.80) with condition (3.84).
6. Repeat the derivation of (3.80) using correctly oriented impulses.
7. Equation 3.107 tells all.
8. Equation 3.119 tells all.

CHAPTER 4

1. 3.4 light years.
2. The true anomaly of intersection is 155°. Use the hyperbolic form of Kepler's equation to obtain M. Finally, $t = \dfrac{M}{n} + T$.
3. Consider the general case (n – stage vehicle) with different specific impulses for each stage. Let j denote the jth stage, W_e denote the empty weight, W_{fj} denote the weight of fuel at the end of the jth burn, and W_{ft} denote the total mass of fuel. Then for any impulsive maneuver

$$\Delta V_j = I_{spj} \log\left(\frac{W_e + W_{f(j-1)}}{W_e + W_{fj}}\right),$$

or

$$\frac{\bar{I}}{I_{spj}}\Delta V_j = \bar{I}\log\left(\frac{W_e + W_{f(j-1)}}{W_e + W_{fj}}\right),$$

where $\bar{I}/I_{spj} \equiv \lambda_j$ is an arbitrary constant; that is, it is the weight assigned to the jth velocity increment. It follows that

$$\sum_{j=1}^{n} \lambda_j \,\Delta V_j = \bar{I}\left[\log\left(\frac{W_e + W_{ft}}{W_e + W_{f1}}\right) + \log\left(\frac{W_e + W_{f1}}{W_e + W_{f2}}\right) + \cdots + \log\left(\frac{W_e + W_{f(n-1)}}{W_e}\right)\right]$$

$$= \bar{I}\log\left(\frac{W_e + W_{ft}}{W_e}\right),$$

since $M_{fn} = 0$ (because the total fuel available is assumed to have been utilized). If desired, a fixed amount of unused fuel can be assumed to be contained in the empty weight M_e.

The constant $\bar{I}$ can be chosen to adjust the sum $\sum_{j=1}^{n} \lambda_j$. In fact it is convenient to choose $\bar{I}$ such that $\sum_{j=1}^{n} \lambda_j = n$. This will assure direct correlation with the unweighted or constant specific impulse case. Because $\lambda_j \equiv \bar{I}/I_{spj}$, it follows that

$$\bar{I} = n\left(\sum_{j=1}^{n} \frac{I}{I_{spj}}\right)^{-1}.$$

Because W_e is fixed and is determined by the selected vehicle, payload, and reserve fuel mass, the previous relationship, namely

$$S \equiv \sum_{j=1}^{n} \lambda_j \,\Delta V_j = \bar{I}\log\left(\frac{W_e + W_{ft}}{W_e}\right)$$

shows that minimization of the total vehicle mass or weight corresponds to minimization of the total weighted velocity increment S.

4. No. The mission is not possible ($\Delta V = 28{,}800$ ft/sec for the transfer maneuver from the Earth's orbit to the vicinity of Jupiter). This does not even account for the escape impulse from Earth (see Equation 4.6).

5. The broken plane transfer is of interest to manned missions because at certain periods of time a rescue or abort mission could not be performed because of the excessive velocity increment required by planar three-dimensional transfers.

7. The speed of escape is approximately equal to 73.4 mps.

9. Draw a picture of the mission and place all the weights on the trajectory segments. Then, starting backwards from Earth (end of mission), sum the ratios.

10. The result is the same as the circular case.

CHAPTER 5

1. Positions and velocities account for six conditions, the accelerations would require another three conditions, which are not available.

2. $A_0 = 1$; $A_1 = 2\beta$; $A_2 = -2\zeta + \beta$; $A_3 = -2\beta\zeta$; $A_4 = \zeta^2$.

3. Perform the translations $x_r = u + L_2$, $y_r = v$.

4. Substitute the trial solution into the differential equation.

5. $y_r \simeq 0.3$.

6. Invert (5.34) to obtain $\mathbf{r}_r = f(\mathbf{r})$, and substitute $\mathbf{r} = x_\omega \mathbf{P} + y_\omega \mathbf{Q}$ so that

$$\mathbf{r}_r = g(x_\omega, y_\omega, \mathbf{P}, \mathbf{Q}).$$

Then, for $y_r = [\alpha x_r + \beta]^{1/2}$, where $\alpha \equiv 2(4\gamma_1 - 1)/3\gamma_2$ and $\beta \equiv \alpha L_2$, substitute $\mathbf{r}_r = f(\mathbf{r})$ to obtain

$$g(x_\omega, y_\omega, P_y, Q_y) = [\alpha g(x_\omega, y_\omega, P_x, Q_x) + \beta]^{1/2}.$$

Finally, simultaneously solve the previous relationship with (5.7).

7. The gravity force for the Earth is equal to $-(1 - \mu)/r_{12}^2$ and the potential is equal to $(1 - \mu)/r_{12}$. The gravity potential of the Moon is equal to $-\mu/r_{32}^2$ and the potential is equal to μ/r_{32}. The centrifugal force is nr; the potential is given by $\frac{1}{2}n^2r^2$. Equate the forces and solve the resulting equation.

CHAPTER 6

1. The assumptions are (a) adoption of a realistic model for the shape of the Earth can be made; (b) the positions or ephemerides of the Moon and planets are accurately known; and (c) the state of the interplanetary vehicle can be determined as accurately as is desired.
2. About 21 min.
3. The satellites are in nearly the same orbit and are phased apart by 180°.
4. Use (6.19) to obtain v_1 and v_2 and substitute these values into (6.20) in order to obtain the sign of R^*. To obtain the transmission interval let $R^* = 0$ and solve for τ in (6.20).
5. Periapsis of the orbit is on the central axis of the cylinder defining the shadow envelope.
6. Minimize the function $F = [(v - v_0)/n] + \lambda S$, where λ is an unknown multiplier and S is given by (6.32).
7. Evaluate (6.26) when $\mathbf{r}$ points away from the Sun along the Earth-Moon-Sun line.
8. Use proportional triangles in Figure 6.5. Approximate the Moon's arc by $r_☾ \tan a_l$. Then, by proportion,

$$\frac{a}{R_\odot} = \frac{r_☾ \tan a_l}{R_\odot - r_☾}.$$

9. The angle ζ compensates for the geometric oblateness of the Earth at a given ground station. It is indicative of the maximum possible interference due to geometric oblateness, and, in this respect, must always be included in the calculations as the nominal elevation angle constraint. For the derivation see [1], Chapter 6. The derivation is given as an answer to problem 11, Chapter 8, of the reference.

10. $$t = t_0 + \left[\frac{e \sin E_0}{2n} E_0^2 - \frac{(1 - e\cos E_0)}{n} E_0\right] + \left[\frac{(1 - e\cos E_0)}{n} - \frac{e \sin E_0}{n} E_0\right] E + \left[\frac{e \sin E_0}{2n}\right] E^2.$$

11. The solution is mathematically feasible if $\beta \cos v + \xi \sin v > 0$, but the physics are wrong. The satellite would be intersecting the geometric eclipse cylinder on the side of the planet that is fully exposed to sunlight.
12. See [1], Chapter 6.
13. See [13], Chapter 6.

CHAPTER 7

1. The techniques of special perturbations are limited by round-off and truncation. The only method of improvement is to carry more significant figures in the computations.
2. A hybrid perturbations technique is a combination of special and general perturbations methods. A specific example of these techniques is the method of Encke with a secularly perturbed orbit.
3. Errors due to precession and nutation (noninertial coordinate systems) can be larger than those caused by the perturbations of remote planets and other forces.
4. No rectification of the orbit will be necessary.
5. In the formulas for **P** and **Q**, namely, $\mathbf{P}(i, \Omega, \omega)$, $\mathbf{Q}(i, \Omega, \omega)$, substitute the secular expressions for the elements, that is,

$$\mathbf{P}(i_0 + (di/dt)\,\Delta t, \Omega_0 + (d\Omega/dt)\,\Delta t, \omega_0 + (d\omega/dt)\,\Delta t), \text{etc.},$$

   and expand.
6. Utilize the results of Problem 5 and also substitute for a and e (in all the appropriate expressions) $a = a_0 + (da/dt)\,\Delta t$, $e = e_0 + (de/dt)\,\Delta t$.
7. Write Kepler's equation as follows:

$$M_0 + n(t - t_0) = E - e \sin E,$$

   and differentiate

$$n\,dt = dE(1 - e\cos E) = dE\frac{r}{a}.$$

   Note that for the case in which the anomalistic mean motion $\bar{n}$ is used

$$M_0 + \bar{n}(t - t_0) = \bar{E} - e \sin \bar{E}$$

   and that by differentiation

$$\bar{n}\,dt = d\bar{E}(1 - e\cos E) = d\bar{E}\frac{\bar{r}}{a}.$$

   Finally, divide the two previously used forms of Kepler's equation to yield

$$\frac{d\bar{E}}{dE} = \frac{\bar{n}r}{n\bar{r}}.$$

8. Orbital rectification is the process of abandoning the old reference orbit and determining from the osculating state vector a new reference orbit. It should be performed when the difference in accelerations between the true and reference orbit becomes large.
9. No. The influence of a small error at the lunar vicinity will propagate forward considerably because of the large mass of the Earth. The accuracy will be much better, however, than the ordinary patched conic method.
10. No.

CHAPTER 8

1. The average secular motion of the equinox, due to the combined motion of the ecliptic and celestial equator.
2. The sum of the major oscillatory precessional effects caused by the Moon and the minor oscillatory effects due to the Sun.

3. Lunisolar precession is the part of the total precession that is due to the motion of the celestial equator. Planetary precession is the precession caused by all other bodies in the solar system. General precession is the combined effects of lunisolar and planetary precession.
4. If you have the patience, you can expand these expressions (as was done) to obtain (8.14).
5. For the departure point a geocentric latitude-longitude system will be used (true equator). From this point on the mean celestial equator of epoch C_7 should be used to perform the integration of the equations of motion. To locate positions on the Moon accurately the true selenographic system would be employed.
6. The angle θ is always very small.
7. Only when extreme speed is desired and computing storage space is not at a premium.
8. There is an optical libration caused by the fact that the Moon is in an elliptic orbit around the Earth but has a uniform rotational spin rate and there are dynamic librations caused by external driving forces acting on the Moon.

Index